WEST VALLEY COLLEGE LIBRARY

3 1216 00171 1584

AF173168

WITHDRAWN

Politics in the Laboratory

WEST VALLEY COLLEGE LIBRARY

Politics in the Laboratory

The Constitution of Human Genomics

Ira H. Carmen

The University of Wisconsin Press

Publication of this book has been made possible in part by the generous support of
The Evjue Foundation, Inc., the charitable arm of *The Capital Times.*

The University of Wisconsin Press
1930 Monroe Street
Madison, Wisconsin 53711

www.wisc.edu/wisconsinpress/

3 Henrietta Street
London WC2E 8LU, England

Copyright © 2004
The Board of Regents of the University of Wisconsin System
All rights reserved

1   3   5   4   2

Printed in the United States of America

Library of Congress Cataloging-in-Publication Data
Carmen, Ira H.
Politics in the laboratory : the constitution of human genomics / Ira Carmen.
p.   cm.
Includes bibliographical references and index.
ISBN 0-299-20210-0 (hardcover: alk. paper)
1. Genomics—Social aspects.   2. Genomics—Political aspects.
[DNLM: 1. Genomics—ethics.   2. Ethics, Research.
3. Genetic Techniques—ethics.   4. Politics. QU 58.5 C287p 2004]
I. Title.
√ QH438.7.C376   2004
306.4′5—dc22   2004012818

To Toni,

My inspiration

"It's the stuff dreams are made of."
Sam Spade (Humphrey Bogart) contemplating the Maltese falcon

Contents

    E. Political Genes?   196
6   Consilience   208
    A. Sociobiology and Beyond   209
    B. Biopolitics   225
    C. A Teaching Constitution for Human Genomics   243

    Notes   251
    Bibliography   285
    Index   309

Preface

The race is over. After five years of fierce competitive struggle as compelling as any sporting event, the two mighty antagonists have provided the world with essentially finished descriptions of the human genome. We now know—more or less—the identity of each of the 3 billion letters that make up our DNA. The least we can say about these discoveries is that a new era of research, explanation, and prediction has commenced, investigations that will yield startling insights into the nature of life, into the nature of human life.

As avid readers prowl the shelves of the leading bookstores, they will encounter a fair number of genomics-related titles, several dealing with the race itself. Some of these have been written by sociologists and historians of science; others have been authored by bioethicists or biologists themselves; most have been written by journalists. Taken together, it is an imposing collection. This monograph is hopefully one of many useful additions that will be made to that whole, and yet, by its nature, the story I tell puts a qualitatively different face on the subject. The fact is, my essay is not about the race; at root, what follows is not even a study of human genomics. The book is separate and distinct because I am a political scientist, and I bring to bear the theory and analytic tools of my discipline. Political scientists address power relationships, the implications of their functional attributes. I desire to specify those implications within the context of contemporary laboratory science, to provide an order and coherence to them, to elucidate the extent to which they constitute the rules and norms of a new political game, a new political science paradigm, a constitution of and for human genomics.

I go even further: as I must allude at times to the race itself in order to provide background and context for my discussion, so I must also characterize the power relationships in my own field in order to assess candidly why political science and biopolitics have been for so long at loggerheads, and why the vast majority of political scientists are oblivious in their teaching and research to the scientific breakthroughs at hand. I will hardly be the first person to dissect the current intellectual malaise that afflicts contemporary political science, but I will be the first to do so from the vantage point of human genomics.

The study may be hard going for some political scientists. It will also probably be hard going for genomicists. They are as ignorant of political science as political scientists are ignorant of them. All of this I discovered when I published *Cloning and the Constitution* more than 15 years ago, and it was hammered home to me in 1990 when I assumed my duties as a member of the Recombinant DNA Advisory Committee at the National Institutes of Health (NIH). Geneticists on the committee (there were practically no genomicists in 1990) thought I was either a lawyer or an ethicist. They still do. But that's OK. Whatever they think, it's preferable to what most political scientists think of political scientists studying genomics. Inevitably, this account will contain plenty of law and plenty of ethics. That's because you can't talk about human genomics as a social activity with social consequences unless you appreciate the play of these influences. But to repeat: the pink ribbon securing the welter of facts, events, and intentions is political science.

Yet, my remarks are not intended solely for the academic community, be that community of social science or natural science bent. I desire also to forge a dialogue with the alert, educated, concerned citizenry, people who care about the scope and the pace of contemporary biological research in its broadest dimensions and implications. More even than my political science and genomics colleagues, they will decide if I have done my job as author properly.

Mercifully, I don't have to take deep breaths thanking those who have assisted me. There is no need to doff my hat to any of the various funding agencies and foundations; I have yet to find one willing to subsidize projects such as this. Nor am I obliged to acknowledge my literary agent for a job well done, because I do not employ a literary agent. Finally, I can ignore my army of research assistants; truth to tell, I don't have a solitary research lieutenant to my name.

I tell readers these things to make two points. One is that I am not the chief executive officer of Carmen Research, Inc. I am a professor of the old school, who does his own reading and library foraging, who writes everything out in pencil using yellow legal sheets (my colleagues accuse me of putting quill to parchment), and who is fortunate enough to have a typist,

Delinda Swanson, who can read my handwriting and spot my spelling errors and footnote misnumberings. The second is that I regard myself as an academic outsider, not a member of some "in" disciplinary fashion, a sort of fossil left over from the days when the time-honored liberal arts and sciences held sway before they were corrupted by some things modern and by all things postmodern, and yet, as I hope these pages reveal, still sufficiently curious to let the light shine through.

The University of Wisconsin Press deserves my unbounded thank-yous for planting the seeds of this project in my brain. Their initial suggestion was to revise and expand *Cloning and the Constitution* in the light of Dolly and all that has followed from her birth. I declined that invitation because I didn't want to fall into *Rocky I–Rocky II* sequel wheel spinning. What I did want to do was write the much broader, much deeper account readers will discover here. At that point, Steve Salemson, associate director of the Press, took my proposal by the horns and sent a sample chapter along with other supporting materials to a University of Wisconsin geneticist, James F. Crow, who strongly endorsed the project. Later, Professor Crow would read the entire manuscript and provide invaluable assistance. I cannot thank him enough. I also cannot thank enough Professor Gary Johnson of Lake Superior State University, selected as well by Salemson to read this essay. His reflections were similarly of great importance in making my plans a reality. For his oversight and efficiency in shepherding the book to fruition, my appreciation to Steve Salemson could never be overstated. Finally, by way of accolades, I owe a considerable debt to my University of Illinois colleague Professor Gene Robinson, who has shared with me his extraordinary scientific expertise and with whom I was able to develop an interdisciplinary genomics presence in my department and on our campus.

I hope readers will find the following chronological chapter summary a useful guide. Chapter 1 traces the history of the political science discipline from my particular (peculiar?) perspective, introduces competing notions of "constitution," and captures the evolution in research among human geneticists from Mendelian investigations through DNA splicing to today's genomic frontiers. Chapter 2 develops the constitutional parameters of genomics political oversight, describing the Human Genome Project at NIH and the "race" between HGP and Celera Genomics to publish the genome sequence. The emphasis here is on the dynamic personalities who shaped the context of decision making: James D. Watson, Francis Collins, and J. Craig Venter. Chapter 3 examines the evolution of human genetic manipulations and their constitutional politics dimensions from artificial insemination, sterilization, and abortion to the preimplantation diagnostic tools attendant in IVF and preembryonic stem cell research to the world of therapeutic and reproductive cloning, which, of course, has to do with creating or not creating new genomes. Chapter 4 investigates the world of human somatic cell gene therapy and the possible new world of human germline manipulations. Changes in our germlines would lead to intergenerational genomic reconstitution (note another meaning of "constitution"; in this context, it refers to the basic nature of the human species). As a member of the Recombinant DNA Advisory Committee at NIH from 1990 to 1994, I participated in many dialogues and much decision making on this general subject. Chapter 5 addresses the paradigmatic shift from behavioral genetics to sociogenomics. No longer is the emphasis on tracking down discrete genes; now, the task is to define the place of the genome in the broader context of human evolutionary theory and to locate the many thousands of single nucleotide polymorphisms (SNPs) that make us all different and characterize their functions. Global sociogenomic delineation is currently the preserve of experiments featuring lower-order species such as the honeybee. I am a member of a teaching and research team studying sociogenomics in broad compass with special emphasis on the honeybee, and NIH researchers have now provided a complete sequence characterization of this model organism's genome. The honeybee engages in patterns of complex social behavior; the genes expressed during these routines can be specified and then compared with human genes already in the public domain. Chapter 6 calls for another kind of reconstitution, namely, the synthesis of ethical, legal, and social genomic parameters under the banner of Edward O. Wilson's grand consilience worldview. A revitalized political science as Aristotle's "master discipline" would lead the charge.

Abbreviations

AAAS	American Association for the Advancement of Science
ABR	Anderson-Blaese-Rosenberg
ACT	Advanced Cell Technology, Inc.
ADA	ade nosine deaminase
APLS	Association for Politics and the Life Sciences
APSA	American Political Science Association
BA	Blaese-Anderson
BAC	bacterial artificial chromosome
BIO	Biotechnology Industry Organization
cDNA	complementary (copy) DNA
CEPH	Centre d'Etude du Polymorphism Humain
CF	cystic fibrosis
DOE	Department of Energy
EAB	ethics advisory board
ELSI	ethical, legal, and social implications (of genomics)
ELSPI	ethical, legal, social, and political implications (of genomics)
ES	cells embryonic stem cells
EST	expressed sequence tag
FDA	Food and Drug Administration
FGLGT	fetal germline gene therapy
FGT	fetal gene therapy
FISH	fluorescent detection of in situ hybridization
GTAC	Gene Therapy Advisory Committee (UK)
HD	Huntington's disease
HERP	Human Embryo Research Panel
HGE	human genetic engineering
HGLGE	human germline gene enhancement

HGLGT   human germline gene therapy
HGP   Human Genome Project
HGS   Human Genome Sciences
HGTS   Human Gene Therapy Subcommittee (of the RAC)
HHS   Department of Health and Human Services
HLA   human leukocyte antigen
HSCGE   human somatic cell gene enhancement
HUGO   Human Genome Organization
IBC   institutional biosafety committee
IRB   institutional review board
IVF   in vitro fertilization
MRC   Medical Research Council
MRI   magnetic resonance imaging
mRNA   messenger RNA
MRS   midbrain raphe serotonin
NAS   National Academy of Sciences
NBAC   National Bioethics Advisory Committee
NCGHR   National Center for Human Genome Research
NeoR   neomycin resistance
NHGRI   National Human Genome Research Institute
NIH   National Institutes of Health (Bethesda)
NRC   National Research Council
NSF   National Science Foundation
OTC   ornithine transcarbamylase
PCB   President's Council on Bioethics
PCNT   preembryonic cell nuclear transfer
PCR   polymerase chain reaction
PD   Prisoner's Dilemma
PEG-ADA   polyethylene glycol-modified adenosine deaminase
PND   prenatal diagnosis
QTL   quantitative trait loci
RAC   Recombinant DNA Advisory Committee
RFE&G   Racial, Feminist, Ethnic, and Gay Studies
RFLP   restriction-fragment-length polymorphism
SCID   severe combined immunodeficiency
SCNT   somatic cell nuclear transfer
SES   socioeconomic status
SNP   single nucleotide polymorphism
SRC   Survey Research Center (Ann Arbor)
STS   sequence tagged site

TIGR	The Institute for Genomic Research
TIL	tumor-infiltrating lymphocytes
TNF	tumor necrosis factor
TSC	The SNP Consortium
USDA	United States Department of Agriculture
WBS	whole blood serotonin
YAC	yeast artificial chromosome

Politics in the Laboratory

1

Political Science, Constitutional Politics, Genomics

A. The Strange World of Political Science

On December 30, 1903, an intellectually hardy and courageous cluster of social scientists convened at New Orleans and voted to establish the American Political Science Association.[1] The dawn of a new era in academic organization and professional self-identification had arisen; more fundamentally, these men had deliberately broken loose from the established intellectual order. Gone were the moorings that bound them to history, economics, and sociology, a common sense of purpose to this day essentially a dim memory.

The APSA did not spring full blown from some Zeus's brow. It was 20 years in the making, the combined contribution of distinguished scholars, thinkers, and leaders—some well remembered, some (for no good reason) not even a footnote in today's political science seminar discourses—whose writings, taken together, constituted a new knowledge paradigm. Who were they, and what were the lineaments of this paradigm? Moreover, what does that paradigm of old have to do with genomics, human or otherwise?

The leading light, the founding father of political science as we today recognize it, was John W. Burgess. His essential argument was that politics was a science, that students of politics should study political phenomena in the same fashion as a biologist studied life forms. In his view, the term "political phenomena" included "political science proper" (i.e., the stuff of community), "constitutional law" (i.e., "the rules of the game"), and "public law" (i.e., public policy). His graduate program at Columbia emphasized political science as *primus inter pares:* only through multidisciplinary conversation could a true science of politics take root. Also to be cultivated was a "theoretical and philosophical treatment of public law and jurisprudence." Through the prism of analysis comparing "living people, real events, and functioning governments," a proper application of social science theory would disclose the fundamental laws underlying the workings of communities, the rules of political games, and the mainsprings of efficacious public policies.[2] All this he presented in his treatise, *Political Science and Comparative Constitutional Law.*

Three other pathbreaking figures were Lord Bryce, A. Lawrence Lowell, and Woodrow Wilson. For Bryce, "every political *organism,* every political force, must be studied in and cannot be understood apart from the environment out of which it has grown and in which it plays."[3] Lowell's APSA presidential address bore the remarkable (certainly remarkable by 2004 standards) title, "The Physiology of Politics," in which the term "physiology" was denoted to mean "the science concerned with the functioning of [political] organs." To these insights, Wilson added the notions that "the proper concern of political science was the development of public policy" and "constitutions are vehicles of life, but not sources of it."[4]

From these interrelated themes and pronouncements, I posit the following generalizations regarding, and draw the following inferences from, the APSA founders' paradigm: 1) *Homo sapiens* is but one of a multitude of creatures on this planet; to know how all function in their sundry behavioral repertoires would provide fundamental knowledge about our own political natures and activities; 2) one of these basic repertoires, perhaps the most basic, is the functioning constitution, by which would be meant the rules, norms, expectations, and belief systems of political cohorts; these governing patterns would be dynamic, evolving systems hewing in some fashion to the Darwinian model; 3) public policy must be judged by its adaptive consequences, that is, it must first and foremost enhance the survival capacity of our species in specified environments; moreover, public policy, over time, could conceivably alter the parameters of these environments and therefore work a change in what exactly it means for constitutions to be adaptive instruments; 4) to the extent our species differs from all other species, it is largely because we have created, under certain environmental conditions, a variegated set of cultural trappings, the full measure of which can only be grasped by aggregating knowledge of, and building theory from, all relevant disciplines, the lynchpin discipline for such investigations being a full understanding of *Homo politicus.*

And what would the founders—Burgess, Bryce, Lowell, and Wilson— have said had they been told, as we could tell them today, that we now know that human evolution, like all species evolution, is a function of *genetic* adaptation; that scientists have essentially sequenced the human genome and hence stand ready to characterize in the coming decades each and every human gene complete with a statement of its function; that genes and culture combine in various ways to produce the baselines of social behavior and therefore of political behavior; that humans have the power to create new life forms, indeed to alter their very own genome and that of their progeny; that humans could conceivably reproduce asexually, therefore keeping their genome intact in the face of Darwinian wisdom; and that, for better or worse, the state, the American people, have vested in themselves through their various institutional arrangements the authority to stop this science, to stop this experimentation, perhaps even to stop the dissemination of all this new, this frightening knowledge. What would they say?

I make bold to announce what they would have said: "Everything we perceived, and wrote, and taught about the nature and behavior of political man and how to study that human nature/behavioral matrix has proven to be true. And it is knowledge of those laws of human propensity that will provide the wisdom for our salvation as a viable species."

This dialogue I have contrived is more than pure fantasy; it is tragedy. I do not believe much in golden ages, and although I am prepared to argue that the founders were indeed "giants in the earth," it cannot be that we political scientists of this era have not produced our own giants. But we have also managed to produce our own intellectual mischief. In this we have no monopoly, but for us the stakes are much, much higher. Science did not stop for Canute, and it will not stop for us. We will either learn not merely to live with it but to use it and use it well, or scholars will walk out on the APSA as the founders 100 years ago walked out on the historians and their ilk.

It is beyond necessity to trace in detail the intellectual mischiefs of political science, though in all honesty I should really have to call it American political science. But the phrase "strange world" in the title to this subsection requires some amplification.[5] The initial mischief occurred even as the founders spoke, a consensus among their peers opting for the investigation of institutional structures and the explicitly declared practices of the formal units of government. These scholars didn't know any biology, so how could they do research like biologists? They didn't even know much sociology, as Arthur F. Bentley would soon declare. Rather than learn new tricks, they took the easy, well-traveled road. It is an old story but a saga to be repeated many times over in American political science.

For political science readers, the rest is standard convention cocktail-hour fare. The behavioralists put the institutionalists to rout after the Second World War, marching under the banner of empiricism, sociology, a little social psychology, a lot of public opinion polling data, particularly fleshing out who votes for whom and why, and a whole bunch of numbers.

Fifty years earlier, Bentley had chided political scientists for spending their time describing institutions. What they should be doing, he said, is forsaking institutions and even forsaking the study of "great men," that is, leadership. The locus of analysis, he insisted, should be the group, with heavy emphasis on the *process* through which groups attempt to manipulate the power structure. Bentley talked a good natural science game, but his vision suffered from all the weaknesses attendant in unidimensionality. Like the institutionalists, he didn't know any biology, so how could he swap ideas with biologists? He also didn't care about constitutions, public policy, or multidisciplinary research. In fact, he was a group determinist, and the first decade of behavioral preeminence, sparked in large measure by David B. Truman's tour de force in repackaging Bentley and selling him to a new generation of antiformalists, was steeped in process orientation and group determinism.[6] During this period Watson and Crick discovered the DNA double helix structure, but there is no indication that any American political scientist paid one whit of attention to that event.

Gradually tiring of groups, process, and sociology, the behavioral gurus turned for succor to social psychology, focusing on a concept already slated for that discipline's backburner: attitudes. By the time I reached Ann Arbor to study political science in the Michigan graduate program (1958), the Survey Research Center (SRC) already held sway. Landmark analyses purporting to track down the sociopsychological influences—attitudes—of a national voting sample were very hot stuff. People studying constitutional law, however defined, and public policy were considered curiosities. The behavioral paradigm brought rigor and generalization to political science but denuded the field of substance and content that were not considered useful repositories of data. During this later period of behavioralist hegemony, geneticists learned how to splice the DNA double helix and insert the genes of one organism into the nucleotide sequences of another organism. Leading practitioners in the new field of recombinant DNA experimentation were so fearful of their newfound knowledge and power that they formulated a concordat with the federal government, essentially persuading the National Institutes of Health to establish a censorship board, the Recombinant DNA Advisory Committee, to oversee their work by approving or denying financial support and refusing, outright, to permit especially provocative experiments. This and related events made national headlines; however, they made no impression whatever on the American political science profession.[7]

Today, one can search long and hard for a political scientist who would not call himself or herself a behavioralist. To say "I am not a behavioralist" is to say "I read the speeches of congressmen to assess the logic of their value judgments," or "I don't use any statistics," or "I am subjecting Aristotle to scrutiny worthy of Crane Brinton." There still are political scientists who do (or don't do) these sorts of things, but they are not invited to the right cocktail parties. And yet to say "We are all behavioralists" is to say nothing. Nobody reads Bentley and Truman much anymore—though they should—and nobody reads such SRC classics as *The Voter Decides* and *The American Voter* much anymore—though they should. (You can't improve on the masters without first learning from the masters.) Reacting against what has become the Behavioralist Wastebasket Paradigm, various researchers have developed what I call spin-off versions. One of these is labeled cognitive psychology. Another is labeled new institutionalism. Cognitive psychologists devise questionnaire rosters designed to plumb the depths of attitudinal gestalts. What cognitive-psychologist political scientists do not do is learn anything about what Lowell might have called the physiology of decision making, thus paving the way for MRI studies (or the like) of the brain in action during the power allocation process. New institutionalists argue that the behavioral process cannot be ripped from the organizational context, that the roles and the constraints of the institution are critical independent variables in the decisional or policy process. This is a step forward but not exactly a paradigm buster.

Where, then, is the political science "action"? (Name me a discipline unfeatured by loci of "action," and we are far removed from the groves of academe.) I shall cite what I call the "big two," though I want to confess right at the outset that the term is a blatant misnomer. I call them the "big two" not because they are competing paradigmatic or even programmatic visions (they are not) but simply because they are the current disciplinary "hot spots." They are where the jobs are; they are the favored publishable themes. I shall also argue, bearing further witness to the strangeness of the political science world, that these two intellectual pastimes with nothing in common and, not merely that, living in two antithetical intellectual domains have concluded an unspoken Treaty of Brest Litovsk with one another.

The first of these scholarly agendas is called Rational Choice Theory. "Rat. choice" (as some pundits label it) should be seen as a flat-out, no-holds-barred challenge to behavioralism: it is deductive, not inductive or empirical; it is highly mathematical, formulating elegant equations and proofs based on generalized assumptions about political behavior, not statistical and inferential; it looks to economics and "economic man" theory for guidance in the formulation of general principles, not to sociology and psychology for guidance in anything; it studies humans who are presumed to act rationally on their preferences, that is, humans who in all likelihood may not or do not exist, not the humans of "bounded rationality" theory, who are oftentimes a bundle of contradictory emotions and logic chains but who do exist; and it constructs models, usually quantitative, against which to measure actual patterns of behavior, not standards of assessment that could be challenged as value-loaded. Rational choice theory is the new paradigm in political science, though precious few members of the discipline can manage its formidable methodological turnstiles. It is the easiest road to Harvard and Stanford.

Where did rational choice theory come from? In political science, its major advocate was William Riker; in economics, its most cogent defender might be Gary Becker; but it is in philosophy—always philosophy—where we find its principal exposition in the writings of John Rawls. Rat. choice has many enemies, of course, but the core criticism does not come from philosophy itself, strangely enough; it comes from evolutionary biology. Here is Edward O. Wilson on the subject:

> [Utility maximization] is not an adequate picture of how people think. The human brain is not a very swift calculator, and most decisions have to be made rather quickly, in complex settings with incomplete information. . . . [Rational choice] pays too little attention to the properties of the real brain, which is a stone-age organ evolved over hundreds of millennia and only recently thrust into the alien environment of industrialized society. . . . [In short] the correct answer may lie in genetic evolution."[8]

But rational choice theorists, whether philosophers, economists, or political scientists, care little about genetics and evolution, not to mention constitutions and public policy as substantive, content-loaded phenomena.

The second of the scholarly agendas might well be identified by two names: the more specific title would be Racial, Feminist, Ethnic, and Gay Studies; the more general title would be Deconstructing the Politics of Western Civilization. No matter which of the 22-odd Kuhnian definitions of "paradigm" one cares to defend,[9] it would stretch the concept "paradigm" beyond the bounds of credulity to call either of these intellectual visions a paradigm. At its best, the new field is a modest extension of Truman's "group" theory. At its worst, the agenda is an ideological fulmination. Naturally, any competent political scientist would include leading civil rights cases in a course on the United States Supreme Court. Any competent political scientist would include the voting patterns of Latinos in a course on American electoral politics. And so on. This new agenda, however, shows little interest in integrating an investigation of black politics or Latino politics into a larger study of American group politics. That would smack too much of theory building. Rather, this field—its courses, its faculty, and its literature—is premised on the conviction that these groups have been victimized and, hence, require special, separate, and discreet treatment. Moreover, members of these groups by and large are recruited to teach these courses, because only they can display the requisite sensitivity toward the unique subject-matter phenotypes. But if these groups have been indeed victimized, who are the victimizers? That is where the process of deconstruction comes in. The "exploiters" are the white, male, European, homophobic, colonialists (sometimes Zionists are also thrown in) and their suffocating ethnocentric culture. To translate theory into "action pedagogy," young scholars, well versed in Derrida and Foucault, have also emerged as cutting-edge commentators; for in this European underground of scholarship lurks the truth about the Enlightenment and its curses. As Ivy League institutions and their peer academies were founded by male Anglo-Saxons, they must make sure these victimized groups receive their "fair share" of faculty appointments or risk a spanking in the *New York Times*. Membership in this club is the second easiest road to Harvard and Stanford.

What mutuality of interest could these subdisciplines possibly share? The rational choicers along with other super-methodologists in the behavioral camp run the APSA, the major journals, and the prestige departments. They have the status. The RFE&G folks have the political connections. They must be represented at all levels of policymaking in accordance with some heuristic "fair share" algorithm, whether it be on the APSA Council, journal editorial boards, or promotion and tenure committees. The one does not challenge the other because that would be not merely "politically incorrect" but also political suicide. The other does not challenge the one because the one poses no threat, and the other does not care to understand what the one writes and teaches about anyway. Together they can blockade, or even root out, the main-line liberal qualitative "softies" who are contemptuous of "quantoids" and who were taught by Hubert Humphrey, Martin Luther King Jr., and their political progeny that civil rights do not mean racial and ethnic preferences. (For the uninitiated, main-line conservatives are generally well closeted if they exist at all in political science departments of choice.)

All of which brings us back to the APSA founders and their commitment to science. Consider the broad themes of their calling: classification and comparison in the fashion of biological sciences inquiry, rules of real political games, prudent public policy forged in the crucible of experience, and multidisciplinary interchange. Rarely will you see these themes emphasized even in the prestige departments or written about even in the elite journals. But particularly alien, particularly threatening, is the word "genetics." The rat. choicers and the heavy-duty behavioralists fear genetics because their models, perspectives, and subparadigms deliberately eschew the possibility of genetic influence on political behavior. The RFE&G-ers fear genetics because of the embarrassing questions it raises, forget how it might answer those questions.[10] Politically correct faculty do not want to discuss Arthur Jensen or Charles Murray; it is enough that they have to put up with James Q. Wilson. They do not want IQ scores even to be part of the political science conversational agenda, nor do they want to read about the differential brain structures of men and women. For them, one could not contemplate a greater paradigmatic enemy than sociobiology (now known as evolutionary psychology in order to defuse opprobrium).

Amid this academic melodrama and byplay, President Clinton and Prime Minister Blair convened a joint news conference at which they declared the international public consortium known as the Human Genome Project and the privately funded Celera Genomics Company cowinners in the race to sequence the 3 billion base pairs of the human genome. From this monumental development, the political science discipline stood predictably aloof.

Into this storm center of indifference and nascent hostility would stride anyone committed to the reciprocal relationship between human genomics and a truly human political science.

B. Bioconstitutional Politics

Constitutional law is "hot stuff" in the law schools these days, as professors discover (invent?) new fundamental rights and airmail their epiphanies to the American Civil Liberties Union.[11] In prestige political science departments, unfortunately, constitutional law is virtually a dead letter. When was the last time the APSA Personnel Newsletter carried a job description from a Big Ten university reading "Wanted: A political scientist who studies the implications of constitutions; any constitution will do"?

Such was not the case 40 and more years ago. Edward S. Corwin and his successors, Mason and Murphy, taught a graduate seminar entitled Constitutional Interpretation at Princeton that not only was a flagship departmental course requiring the most intense dedication by budding Ph.D.s but was looked upon with envy by public law scholars around the country. At Harvard, Robert McCloskey's lectures and writings on the Constitution were praised for their perspective and balance, while at Chicago, Herman Pritchett excelled both in quantitative studies of the Supreme Court and in qualitative studies of constitutional values. Today, the most we can expect from a political science constitutional law expert by way of publication is the standard casebook, hopefully to be adopted for use in the large classes devoted to educating a never-ending stream of prelaw undergraduates. Inevitably, these tomes treat the American Constitution as a scrap of paper that the Supreme Court is charged with interpreting, and the grist for the mill are the cases the Court processes and the opinions the justices generate.

Every now and again, a noteworthy discussion of the topic does surface, and a decent respect for the importance of constitutionalism to my aims and goals requires that I touch on some of these. Carrying forward Princeton's commitment to constitutional scholarship, Walter Murphy and colleagues have provided us with a new and different sort of casebook.[12] In the usual collections, the editors categorize cases under conventional doctrinal topics, for example, free speech decisions, due process decisions, commerce clause decisions, and so forth. Murphy et al. begin by asking *what* is the Constitution and then go on to inquire *who*, under our system, has the authority to tell us what it means. The important point I would emphasize is that we do not know the "who" unless we can identify and agree upon the "what." These questions, and the answers provided in the book by various commentators, help break down the notion that there is only one American Constitution (the text of the document) and only one authoritative interpreter (the justices on the Court). Yet another member of the Princeton research cluster, William Harris, takes a step further by distinguishing between the Constitution as a "verbal order" and the Constitution as a "political order."[13] What he calls the "constitutional enterprise" is a synthesis of these two orders. Although Harris devotes his attention to comparing and contrasting "styles of interpretation," I think more needs to be said about his essential argument, namely, that there are two separate and distinct American constitutions. The first is what somebody thinks the Constitution is in theory: its rationalizing, transcendent principles; its core value structure; if you will, the logic chains informing its salient parameters. The second is how somebody thinks the American Constitution, in fact, operates: its roles as power allocator among the three branches of governance; its influence on the struggle between contending groups; its actual effectiveness in guaranteeing individual rights. The way in which the constitutional polity "muddles through," he suggests, is a function of competing models of synthesization, but the unanswered questions revolve around the identities of the "somebodies." In general, the Princeton school favors notions of constitutions as reifications or abstract systems to be made sense of by noted authorities within these postulated systems. The "somebodies" don't really come alive, because there is missing a salient linkage with modern political science, that is, an empirically verifiable universe of real people, living out their competing visions of constitutional propriety in the context of power relationships amenable to study by scientific analysis. Only in that light does the study of constitutions truly become a study of constitutional politics.

I now turn from the world of unconventional legal theory (unconventional because it is informed by serious considerations of things political) to the world of conventional political science. In one study, formal theorists develop a set of lemmas, theorems, and games to test whether various checks and balances make for stability in the Constitution as devoutly wished for by Madison and Hamilton.[14] The authors conclude that the founders' scheme for constitutional stability proves valid. But does it? For one thing, note the step backward in theoretical conceptualization: the American Constitution is conceived as the American Constitution, the document. For another thing, note the step backward in political implication: the decisions real people have made over the years—decisions now reflected in rules and policies—are irrelevant to the investigation. Indeed, if we don't want to study what people do as constitutional phenomena, then why shouldn't we narrow the discussion to the logic of documentary argumentation? In another study, two leading new institutionalist advocates press their case for the proposition that political behavior cannot be well understood unless the trappings of organized environments and their influences on behavior are accounted for.[15] Among these trappings are constitutions and rules. So far so good. What, however, do they mean by constitution? They comment:

> By constitutional principles we mean those aspects of a polity that are generally viewed as not subject to routine political determination through various forms of majority rule.

For them, the quintessential political behavior not subject to majoritarian disposition is "the comprehensive reform of political institutions," such as would occur when duly constituted authority decides whether the model for governance be that of the "corporate-bargaining state," the "sovereign state," the "institutional state," or the "supermarket state."[16] I think this tack both broadens and narrows the constitutional politics playing field. It is broadened appropriately by moving from documents to actual power structures; it is narrowed inappropriately by relying upon application of the majority rule mechanism. The question whether particular decisions should be made by majority rule is itself a function of deeply embedded constitutional understandings, and in point of fact, the American political system provides for many loci of nonmajoritarian determination, most of which were undreamed of by the principal constitutional architects. As I shall elaborate in time, the majority rule mechanism does not well fit the manner in which the federal government has regulated the "new genetics"; besides, there is always the nagging question: majority of whom? The authors themselves point out the importance of these "deeply imbedded understandings" when they set out the importance of rules:

> By "rules" we mean the routines, procedures, conventions, roles, strategies, organizational forms and technologies around which political activity is constructed. We also mean the beliefs, paradigms, codes, cultures and knowledge that surround, support, elaborate, and contradict those roles and routines. . . . Rules are learned as catechisms of expectations.[17]

Rules are for them what constitutions mostly mean to me. As big as is the decision to formulate and administer a certain species of state, the scope of the genus constitution is bigger. In fact, it is wrong to speak of an American constitution. It is far more accurate to speak of American constitutions, each expressing the rules of the particular political game being played. Naturally, there will be disagreement as to what those games are, the identities of the players, the rule structures and functions relevant to those games, and the like. To ferret these out and provide theoretical grounding for the data is what empirical political science is all about. Citing my previous work, I would add three dimensions to the term "rule" so as to highlight precisely which rules rise to the level of constitutionally relevant phenomena.[18] A rule must be fundamental, of first-order importance. A rule must be regularized or patterned in its application, not selectively or randomly in play. A rule must be legitimate, must possess the quality of unquestioned rightness. And who are the "somebodies" that will bring political meaning to these constitutional rules? It is the players of the respective games themselves, but it is the unique craft of the political scientist to characterize rigorously these idea and action sets. And it is precisely here where our authors fail in their mission. Nowhere do they analyze a real political universe for the purpose of ascertaining what their terms "constitution" and "rule" actually mean in the field. Without the empirical dimension the propensity for error in theory construction—which includes the definition of terms—becomes unduly magnified.

I propose to provide a real political universe and then some for study. Were I faithful to the prevailing political science paradigms, my context of analysis would be some political process, some empirically verifiable way of doing things. That is because Bentley and Truman live on in altered, perverse modern dress, disguised by more elegant models of psychological mindset or more biased accounts of interest group hegemony. And, of course, process is important: the formal constitutional document is mostly a set of procedures, ways of achieving consensus or maintaining stability, as the case may be. Harking back to the disciplinary founders, however, I argue the importance of policy, by which I mean the subject matter, the content of decision making. It is the content of policy that arrests the attention of the politically minded citizenry, not the study of rules divorced from subject matter, which is the domain, or should be the domain, of constitutional lawyers. I submit that each important debate over a subject matter class constitutes its own game with its own set of winners and losers, and the play of these games can be so important for the health, for the *adaptability* of the culture, that they require the application of a broad array of multidisciplinary foci. As the keys to "successful play" will ultimately demand the proper deployment of "power algorithms" and will ultimately lead to power allocations and reallocations, the thrust of my inquiries will be inherently political. Only by bringing content back into political science will we bring the Burgesses, the Lowells, and the Wilsons up to date, providing the substantive tissue that binds the investigation of power, the investigation of rules, and the investigation of the policy process, at its best guided by the wisdoms of the natural sciences. What better way to test the efficacy of this new synthesis than to use the natural sciences as the subject matter laboratory of research and theory building?

I herein place on the table the subject matter of biology, and I shall treat biology as both an independent and a dependent variable. That is, I will operationalize biology as a cause of human activity and as an effect of human activity. All biology? No. And do I intend to discuss all political activity relating to biology in this country and elsewhere? No. The biology of central concern to me is genetics. Even a casual examination of today's intellectual docket makes manifest the mix between genetics and politics. The important questions on the conversational agenda are these: 1) To what extent does DNA orchestrate how we think and how we behave? 2) To what extent should the state encourage and to what extent should the state constrain research involving human DNA? 3) To what extent should the state utilize the fruits of human DNA knowledge to achieve policy goals? The first question is of obvious relevance to political science because if we are going to talk about the role of sociology, the role of psychology, and the role of economics in decision theory and inevitably policy formation, we had better address the possible role of genetics in these dynamics. The second and third questions are deliberately phrased in normative terms because the informed citizenry views politics normatively. Thus, people want to know what government should do about such and such. For political scientists, the questions should be rewritten analytically. Thus: How has the state encouraged and at the same time constrained human DNA research? How has the state harnessed human DNA to suit its sundry purposes?

Still, the topic is too broad, too hard to manage, lacking in political science perspective as I have tried to weigh competing political science perspectives. My delimiting concept or theoretical organizing principle is constitutional politics. I desire to address the constitutional politics dimensions of the above questions, for that is my way of getting to the guts of the matter. I shall attempt to elucidate the political struggle over how to define the rules of the game for human DNA research, for human DNA understanding, for human DNA manipulations and applications. What fundamental standards obtain? What patterns of practice have ensued and successfully adapted? What notions of legitimacy, of right conduct, bear upon the play of the human genetics game? What are the genetic attributes of thought and action central to the concept *Homo politicus,* and how has the world of politics sought to confine debate over the influence, even the existence, of such phenomena? I call this general field of research and teaching "bioconstitutional politics."

Unbeknownst to political science, the term "constitution" has been making a comeback of late, chiefly in the world of philosophy. Thus, Baker uses the notion as follows: "When certain things with certain properties are in certain circumstances, new things with new properties come into existence. For example, when a combination of chemicals occurs in a certain environment, a new thing comes into existence: an organism."[19] If we talk about "constitution" in one context and then talk about it in a different context, the whole notion of constitution—its parameters and the reciprocal relationships among those parameters—changes dramatically. As well, the constitutional politics undergirding and informing human DNA as an amorphous entity whose structure is endlessly searched for will not be the same once human DNA is captured and processed according to the rationalizing principles of the human mind and the cultural values that bring life and meaning to the objects we can perceive. In somewhat the same spirit, Schechtman describes how persons develop self-constitutions, that is, self-narratives that make coherent their essential identities as complexes of traits, actions, and experiences. Such "constitutions," however, must be objectively verifiable; they cannot be steeped in self-delusion.[20] I believe human DNA as a political science term of reference in the year 2004 creates in our minds and in objective reality its own unique constitution. This constitution can be treated as a form of narrative through which what makes it a constitution, what makes it of unique political relevance, becomes a well-lighted place.

C. From Genetics to Genomics

Political scientists ought to crow about the fact that Aristotle—loosely speaking, a disciplinary founding father extraordinaire—was in a very real sense the discoverer of the DNA principle. He observed, in the highest spirit of empirical insight, that chickens came from eggs and that oak trees came from acorns; there was some plan or process that inevitably caused A to become B.[21] The master was so busy constructing other paradigms that he may be excused for not making of his casual remark the first law of genetics: Biology, at root, is the study of information and its transmission.

If Aristotle was DNA's first metaphysician, Mendel was its first field practitioner, noting in his pea plant experiments the role of dominant and recessive traits. He published his findings in the 1860s, but they remained "undiscovered" until 1900.[22] While Mendel was working in his cloistered garden, Charles Darwin was exploring the Pacific islands, and upon these investigations would be constructed his notion of species evolution including human evolution.[23] I share Thomas Kuhn's enthusiasm for the term "paradigm," and if I am not careful, I will permit his notion of paradigm to swallow my notion of constitution or vice versa. But Darwin's sweeping conception embraces far more than either expression was ever meant to capture. The theory of natural selection is what I call a "worldview." As we shall see, the paradigms of DNA have come and gone, but DNA itself remains. So also does the evolution of life forms. They are akin to the axioms of plain geometry.

Mendel, to our knowledge, probably had only a passing familiarity with Darwin, and Darwin certainly knew precious little about information dynamics. But over the years, the two worlds they founded would conjoin: William Bateson coined the term "genetics"; genes were tracked to chromosomes; a fertilized egg was found to contain a full array of chromosomes, 50% from each progenitor; scientists linked Mendel's dominant-recessive model to eye color; color blindness proved to be a sex-linked characteristic; most fundamentally, the answer to the question "Survival of the fittest *what*?" is now considered to be the individual and that particular individual's genetic *constitution,* that is, the individual's haploid set of chromosomes or genome.[24] A genome adapts, or does not adapt, to environmental strains and stresses. If it adapts, then its salient parameters proliferate; if it fails to adapt, it becomes extinct. Thus do genes, thus do individuals, thus do societies, thus do species, evolve. Darwin's theory of evolution is a theory of genetics; the study of genomics today and in the future is inevitably a fleshing out of Darwin.

The first step toward formulation of a paradigm for today's biology occurred in the 1940s when Avery showed that genes were not made of proteins, as most had speculated; they were made of DNA.[25] Now, at least, scientists knew where to look to find the secrets of heredity. But how was DNA organized? Here, Watson and Crick came to the fore with their discovery of the double helix. The DNA of all organisms consisted of base pairs or nucleotide sequences, in which A (adenine) always bound to T (thymine) and C (cytosine) always bound to G (guanine).[26] With this key to the puzzle in hand, researchers had something to count, to measure, to manipulate. If by paradigm one means "universally recognized scientific achievements that for a time provide model problems and solutions to a community of practitioners,"[27] then the first paradigm of heredity was "classical genetics" in the grand Mendelian manner. The "winner and new champion" paradigm of heredity ought then to be Watson and Crick's "molecular genetics," which emphasized the flow of information from gene to protein to phenotypic structure and function.[28]

The "normal science" of the subsequent two decades brought major breakthroughs, some at an agonizingly deliberate pace. Researchers discovered that DNA instructions were converted to messenger RNA instructions that in turn were converted to amino acid orderings, ultimately forming protein structures. They also discovered that the DNA code must be read in triplets, that is, a triple codon (say CCG) triggers a particular amino acid called proline to fall into place on a protein chain. Of greater relevance to us is the way human genes were discovered. By 1968 approximately 68 of them had been identified, all on the X chromosome and all mapped through the time-honored methodology of the trade—family pedigree charts. In that year, an old University of New Hampshire classmate and colleague of mine, Roger Donahue, succeeded in tracking a gene to an autosome. By studying his family members' chromosome 1s and their blood types, he determined that the gene for blood group Duffy-a was located on that chromosome.[29] In 1979 Nancy Wexler visited Venezuela for the first time and began collecting blood samples from natives at high risk for the fatal adult-onset disease Huntington's chorea. Four years later, in a stroke of sheer fortuity, laboratory investigators mapped the Huntington mutation to chromosome 4. Some thought the process might take 50 years; it took one. But mapping a gene to a chromosome is one thing. Isolating the exact location of the gene, that is, pinpointing the precise constellation of base pairs making up the gene, is quite another thing. It took a decade of "sequence creeping" to find the toxic devil.[30] Meanwhile, in the 1980s a young scientist named J. Craig Venter set up shop at NIH in Bethesda and began a search for the "fight-or-flight" gene. After several years of purifying protein samples, he succeeded.[31] As of 1990, about 3,000 genes of the then-estimated 100,000 had been cloned and rigorously characterized (copied and sequenced).

In 1973, the recombinant DNA revolution began. Some would call this new initiative a branch of "molecular genetics" and hence not of paradigmatic dimension.[32] They can't be right. First of all, synthesizing the DNA of disparate organisms, of disparate species, and creating genetic chimeras is obviously a major departure from anything intrinsic to the Watson-Crick formulation. Second, and much more to the point, the term "paradigm," like the term "constitution," is not a scientific term at all. It is a social, a *political,* construct. Scientific paradigms are classifications of ideas, of experiments, of ways of looking at reality, and these classifications are based on larger conceptions of reality such as political reality. When Watson and Crick found the double helix, nobody argued for a moratorium on the research that would inevitably follow, and nobody argued for creation of a censorship board to monitor the state of the research once a moratorium had been lifted. But that is exactly what happened with recombinant DNA. The geneticists themselves voted for a moratorium, and through the good offices of the NIH, they organized such a censorship committee, manned largely by themselves, to supervise proposed experimentation. At that time, as now, virtually all serious work of this nature in the United States was funded by NIH. Under the terms and conditions put in place in the middle 1970s, all investigators whose institutions received one penny of federal government assistance were obliged to submit their gene-splicing protocols to the Recombinant DNA Advisory Committee for approval, disapproval, or amendment. The RAC's rulings were tantamount to law and practically never overruled.[33] A new paradigm had come to town.

In a research report that March and Olsen might have gleefully embraced,[34] I drew upon my experience as a RAC member to formulate several competing constitutional politics models and apply these to RAC decision making. The idea was to search for closeness of fit, to determine which model best explained how the RAC came to approve in 1989 the first-ever human gene transfer protocol.[35] It is a worthwhile exercise to review my commentary, because, as shall become very clear later, the RAC model, even the RAC itself, remains available as a forum for solving the policy conundrums arising from human genomics.

Model 1 is the "establishmentarian model," which is a rough equivalent to Lowi's "interest-group, bureaucratic model."[36] Here, government decentralizes power and places it in the hands of technical experts who represent the interest groups to be constrained. Model 1, I found, explains much of the RAC's policymaking process. But while the RAC carries the force of legitimacy for genetic researchers, it lacks a significant kernel of legitimacy for the body politic generally because the powers-that-be permit it to reject or accept protocols largely as it sees fit. Model 2 is the "accountability model," which bears a resemblance to Lowi's "juridical model." Here, these powers-that-be essentially formulate standards, and the RAC would then faithfully implement them. Clearly, this model provides little insight into RAC behavior, and that is not all bad because, as I shall emphasize later, scientific research—even scientific experimentation—is a free-expression activity entitled to special solicitude under the American Constitution. Model 3 is the "consensus-building model." Here, public policy takes whatever form a consensus will tolerate, and indeed the RAC deliberately measures consensus by holding open meetings and permitting a cross-section of opinion to surface. The RAC also reads the newspapers, listens to the voices of elected officials, and rummages through the recommendations of blue-ribbon commissions. While this model also has legitimacy because it owes an intellectual debt to the time-honored marketplace of ideas motif, and although it certainly explains much of how the RAC conducted its affairs, it contains a nagging flaw, namely, no guarantee of closure. This lack of an end-game rule limited the RAC's capacity to develop a broader sense of mission beyond merely reviewing specific research designs. Model 4 is the "new style liberal model." Here, the RAC's job would be to protect the moral order in the face of geneticists who have allegedly sold their souls to either the power structure or the theology of scientific discovery. Of course, the RAC displayed little sympathy for this model and spent a good deal of its time listening to, but inevitably fending off, the Jeremy Rifkins of this country. Model 5 is the "Federalist model." Here, the RAC would be guided by the Alexander Hamilton, Ronald Reagan worldview by encouraging science in the name of technology, business enterprise, market success, and American competitiveness. There is no doubt that RAC discussions featured plenty of cost-benefit, risk-assessment weighing, in a fashion consistent with this approach; however, most RAC members are basic researchers who realize that "science" and "profits" are very different motifs. Model 6 is the "scientific estate model." Here, we swap the rationality of John Rawls for the rationality of Robert Merton. Under this constitutional umbrella, the RAC would permit geneticists to probe dispassionately—guided only by the rigorous standards of the profession—in the search for general truths.[37] That is, *Homo scientificus* should monitor *Homo scientificus*. While the RAC did and does operate at times as though it were a study section, the model is fatefully inadequate in this context because it is fundamentally apolitical. For the RAC to discount the politics attendant in its mission would be to court disaster. Model 7 is the "civil liberties model" or, more accurately, the "correlative rights and duties model." Here, RAC members play the role of judges, balancing genetic engineering research as a form of protected expression against society's interests in regulating this expression. As the aims and methodologies of the research vary, so also do the competing interests vary. Only upon a careful examination of the totality of facts presented in each protocol—a totality to be weighed in the manner championed by Justices Felix Frankfurter and John M. Harlan II—can a proper scientific, ethical, and political decision be made. On this lofty and preferred count, the RAC deliberations I addressed in my report tended to come up short. Although the NIH has stripped the RAC of its virtual autonomy to veto human gene transfer and therapy initiatives in a most controversial order issued in the mid-1990s by then-director Harold Varmus, the committee continues to receive all truly innovative research designs arising from NIH-funded sources and to provide advice on whether these should be permitted, not permitted, or amended in specific ways.

As the recombinant DNA paradigm was slowly becoming the gene therapy paradigm, a change in name and in emphasis but not any sort of qualitative displacement, much less research-oriented insurrection, the third bona fide genetics paradigm, genomics, was commencing to command attention. Genomics has no more obliterated recombinant DNA than Watson and Crick obliterated Mendel and Thomas H. Morgan, perhaps his most famous disciple. And certainly much is still expected of gene therapy, though it has thus far saved precious few lives. It is a question of center stage or, to use the old term, "action." It is also a question of newness, of unbounded potential, of inquiries never before posed simply because no researcher had ever before been able to pose them. Yes, over the years they had been posed but without benefit of a proper information base and the proper tools of inquiry—matters for philosophers, not scientists. I now provide a brief history of the rise of genomics, focusing on the great debate between the yea-sayers and naysayers, with special attention accorded to the constitutional politics of that debate.

What exactly is the human genome, and what do I mean when I talk of human genomics? By "human genome," I refer to the 3 billion base pair sequences that comprise the DNA in the nucleus of each and every cell in the body (with the exception of mature red blood cells) of each and every human, and I also mean all of the genes sitting, somewhere, among this vast array of sequences. (Each somatic cell actually carries two double helixes or two genomes, so the total number of nucleotide sequences in these cells is 6 billion.) By "human genomics," I have in mind the scientific explorations designed to flesh out the characteristics of the genome and the scientific implications arising from knowledge of the genome. But of equal importance to my concerns, I also have in mind the ethical, legal, social, and political issues and consequences arising from the scientific enterprise itself.

As best as I can tell, the first geneticist to have a meaningful grasp of genomics amenable to operationalization in the laboratory was David Botstein.[38] It was well established that all of us share 99.9% of our DNA sequence. Polymorphisms are stretches of DNA we do not share, thus ensuring that we are not all clones. They might take the form of alleles, let us say two, though the number could be much higher. I might have allele A of a certain gene, and my brother might have allele B. Then again, some of these polymorphisms do not code for any proteins. Botstein understood, as did several others in the late 1970s, that these noncoding polymorphisms very often were proximately located or tightly linked to certain genes whose chromosomal sites could be ascertained via pedigree studies. These polymorphisms could even be removed from a genome by employing restriction enzymes, thus facilitating length-wise comparisons to see who carried which configuration. In brief, the "cutting and pasting" that had become the staple of recombinant DNA analysis led naturally and inevitably to small-scale sequencing expertise followed by large-scale sequencing visions. Botstein's insight was that these RFLPs (restriction-fragment-length polymorphisms) might well serve as genetic markers along the full length of the genome and could therefore be likened to road signs for tracking down any gene anywhere. He published a paper arguing the point in 1980.

In the mid-1980s, momentum for a human genome initiative began to build from a variety of political constituencies. In 1985, Robert Sinsheimer, the chancellor of the University of California at Santa Cruz, hosted a brainstorming session for the purpose of drumming up interest in a major sequencing center to be housed on his campus. Sinsheimer's aims appear to have been grounded in knowledge for knowledge's sake with a eugenics twist.[39] Present at the meeting was Nobelist Walter Gilbert, who shared Sinsheimer's enthusiasm but who had a far different vision of the project's scope and magnitude. At the Santa Cruz conference, Gilbert threw down the gauntlet, declaiming that "the total human sequence is the grail of human genetics . . . an incomparable tool for the investigation of every aspect of human function."[40] At a later meeting convened by James Watson at his Cold Spring Harbor digs, Gilbert provided refinements: He would found a company, employ a bevy of coworkers (technicians?), sequence the genome at a total cost of 3 billion dollars, and sell or license the information to the highest bidders.[41] On a third front, Charles DeLisi saw a human genome initiative as the perfect fit for the organization he represented, the Department of Energy. The DOE had been working for some years on studying the effects of radiation on mutation rates among Japanese atomic bomb survivors. Also, with the threat of nuclear war and the energy crisis of the 1970s having abated considerably, the agency was, quite simply, looking to reorient its mission. DeLisi hosted his own workshop at Santa Fe where the ubiquitous Gilbert provided an entranced assembly with another of his "grail" allusions.[42] During precisely that week, a fourth front appeared. Its advocate was Nobelist Renato Dulbecco, who, in an editorial in *Science* magazine, argued that President Nixon's "war on cancer" could best be fought by sequencing the genome and then tracking down oncogenes. He likened the project to the space program, and his essay galvanized the biomedical community.[43] So if we assumed we were to have a full-blown human genomics research institute, four very different political modes of governance had been put forward to orchestrate the science: the university model, the corporate model, the DOE model, and the medical sciences model. Two of these quickly fell by the wayside: no campus—no consortia of campuses—could finance and run such a large program, and at that time, with the economy sputtering along, it was simply not possible to harness the private capital necessary to make the project a viable business venture.

I want to pause for a moment and return to Gilbert's oft-cited quotation. When a Harvard Nobel Prize winner labels some far distant goal the "holy grail" of his highly visible and important research field and then says that the grail can be reached with a little time (a decade) and effort (a few billion), and when his colleagues begin to rethink what they are doing in the light of this preachment, a new paradigm is at hand, to say the least. But a paradigm requires not merely some invocation to mythology. It also requires more than scientific grounding. Because paradigms are social constructs, they beg for analogies with other known social constructs and particularly those of parallel scientific dimension. In this spirit, the human genome attracted the best efforts of other analogists besides Gilbert. Thus, it was called, and still is variously called, the "code of codes," the "human blueprint," and biology's "periodic table." All bear witness to the fact that human genomics is unique among most other biological paradigms, but all are inadequate in appreciating the parameters of that uniqueness. The human genome is not some cryptographic message, because the Morse code, for example, does not evolve. The human genome is also not some blueprint, because blueprints are two-dimensional charts, not one-dimensional arrays of only four digits.[44] Finally, the human genome is qualitatively distinct from chemistry's universal table of elements, not simply because life on some distant planet might bear little resemblance to life on Earth[45] but because humans now possesses the capacity to alter their own genome and the genomes of species around them. Not to see the genome in this awesome light, not to see humans in this awesome light, is to miss the whole point of where human genomics sits today in our cultural vocabulary.

It hardly took time for battle lines to be drawn between the medical geneticists and the DOE. The former gained an early advantage when those "pure science" biologists who favored the project as a boon to understanding human behavior sided with the medical group simply out of fear of the "bomb makers" getting their hands on such precious material. The former also had at their disposal the influence of the NIH, their bureaucratic spokesperson and generally a very competitive spokesperson in the contest for Washington dollars on account of its disease-fighting mission. The NIH had long since trumped the NSF as underwriter of basic science biological research precisely because of its utilitarian focus, so in a sense the "non-medical, non-DOE-ers" had no place else to go. All this was bandied about at the Cold Spring Harbor conference, with DOE personnel scoring the important counterpoint that if NIH got the project, it would cost nongenomicists dearly as they jousted for NIH grant funding.[46] To this argument came the response that the NIH's peer review system would provide adequate quality control for the new research agenda, whereas DOE work was generally done in-house.[47] All parties seemed to think the program would have to be international, as the genome belonged to everyone.

In 1987 the National Research Council, an arm of the National Academy of Sciences, appointed a committee to address the pros and cons of a human genome initiative. The committee minced no words on the questions of feasibility and justification: a full-scale mapping and sequencing effort should most assuredly be undertaken. But the panel minced not merely words but also politics when it declined to address the question of governance: "the committee admits that it has little expertise in this area." Still, it was prepared to say that a single agency should be put in charge, informed by an advisory board of distinguished scientists who would be guaranteed a substantial policy role. However, the committee deliberately and not very artfully declined to cast its support for the NIH, the DOE, or a new agency.[48] The panel's decision not to bite the political bullet might have sad-dened political scientists had they even known of this exceedingly impor-tant moment in the history of science. But the biologists themselves were partly to blame. Did the NRC put any political scientists on the committee for the precise purpose of providing expertise on a fundamental issue all knew would arise? This issue, having to do with federal institutional ar-rangement and administration, would have been up the alleys of several APSA founders and their successors, though their lack of knowledge about the intimacies of biological science work ways would have been a handicap. So there was interdisciplinary ignorance aplenty to go around. One thing was certain, however: from this point on in the dialogue, those opposed to a separate, distinct, and heavy-duty project would be on the defensive.

Already the DOE was pushing ahead: DeLisi received a congressional budgetary recommendation in the millions; the secretary of energy moved to establish human genome centers at three national laboratories; Colum-bia's Charles Cantor was named to head the new center at the Lawrence Berkeley site; various chromosomes were assigned to these centers, where research teams would chop them up into fragments in hopes of recon-structing them into chromosomal physical maps, a precursor to sequenc-ing. All of this horrified the gene mappers, those who thought the essence of the game was to track down disease genes and locate their chromosomal homes rather than identifying DNA sequences, 97% of which appeared to be "junk," that is, coded for no proteins.[49] (The sublime aesthetic dimen-sions of DNA structure and function should have counseled us not to jump to such hasty conclusions. Recent research demonstrates that "junk" se-quences may very well integrate coding exons, hence magnifying their adaptive capacities.)[50]

NIH would more than prove equal to the challenge. Under much pressure from his research constituency, Director James Wyngaarden—earlier fearful that a genome shop located at his front door might devour his agency's re-sources—committed himself to establishing an NIH Office for Human Ge-nome Research. Meanwhile, NIH senatorial sponsor Edward M. Kennedy, who chaired a key committee, outmaneuvered DOE's principal advocate, Pete Domenici, a Republican senator without a committee chairmanship, the bottom line being that the NIH appropriation for genomics in fiscal 1988 was almost 50% superior to the money given DOE. The big question remained: Whom would Wyngaarden name to head up the new genomics office? In May 1988 the news broke that he had tapped Watson, the "impre-sario of molecular biology."[51] In October Watson accepted. The die was cast. DOE had lost the energetic DeLisi to the medical school community, while NIH was getting Watson. The Office of Technology Assessment, a congressional research unit, had come down in favor of an interagency task force to coordinate and mediate NIH and DOE genomics efforts, thus finessing in its own way the very conundrum in Washington power politics the NRC committee had previously tried to finesse.[52] In fairly short order the two rival agencies signed a memorandum of understanding essentially giving NIH responsibility for mapping and DOE responsibility for sequencing.[53] But all this was for public relations and the keepers of organization charts. On the prestige pecking order and therefore on this particular influence pecking order, the NIH would henceforward call the key genomics science shots.[54]

I have spent a little time discussing NIH-DOE competition in part so that we all understand the germination of human genomics as a political phenomenon in the United States and in part because we can see firsthand how the constitution of American biological science operates. That NIH "defeated" DOE was a function of how the leading genomicists perceived their activity. They were "basic scientists," not "applied scientists." Of course, many were M.D. disease fighters, but curing a disease, to them, was a higher calling than making America stronger and safer, as if the two were inevitably at tension. Such was the mythology of the moment—probably even of this moment—and constitutionalism is suffused with sentiments mythological. In the next chapter, we will see how this logic of NIH primacy in human genomics investigation begins to unravel in the face of bold new conceptions of "basic research."

I have thus far paid the "other side" scant attention. Many were the geneticists and biological science commentators who opposed the human genomics initiative, no matter which agency would sponsor and nurture it. The critics fall into two broad clusters: the instrumentalists and the ideologues. I conclude this chapter by taking each in turn, offering as well my criticisms of both.

The instrumental arguments seem to reduce to these: 1) the NIH human genome program (by 1990, sophisticated observers were no longer even bothering to discuss the DOE side of the ledger) had already begun to siphon off money badly needed by investigator-initiated programs, be they new projects or renewals; 2) the program was "big science," and the biological sciences have never needed and do not need now the centralized administration that is the hallmark of "big science"; 3) sequencing 3 billion base pairs—counting every A, every C, every G, and every T—will mean spending a lot of money to learn precious little about human genetics, given that 97% of DNA is trivial.[55]

I find, after reviewing the literature of that period, that the counterarguments successfully rebutted the allegations. On the question of funding priorities, human genomics accounted for a paltry 1% of the NIH budget in 1991, while the AIDS program attracted almost 10 times that level of NIH funding.[56] On the matter of "big science," the romance of "small science" conjured up by reflections of Watson and Crick "tinkertoying" in the laboratory had given way to a genomics agenda requiring substantial capital and operating costs, graduate students, postdocs, technicians, equipment, and sheer space. More fundamentally, the NIH genomics initiative was not "big science" at all when compared with large-scale research enterprises in other disciplines. A truly centralized form was exemplified by the Manhattan Project or the Apollo program. The human genomics effort would see NIH farm out assignments to various laboratory centers around the country, all academic in structure and function, and all subject to the peer review process. Even laboratories abroad—in the United Kingdom, France, and Japan—would be allocated responsibilities presumably by negotiation. Each principal investigator would orchestrate how best to go about particular mapping and sequencing tasks.[57] On the issue of sequencing "junk," Watson himself provided the answer: "the project will be over when we can identify the genes . . . [but while it] would be nice if the whole program could be done by copy DNA . . . we will never know whether we have all the cDNAs [therefore] I think that we have got to sequence the whole thing."[58]

One must approach a discussion of ideological opposition to the initiative with caution. I do not count among the ideologues those who saw, in my view absolutely correctly, that human genomics information could be used for deleterious purposes and that, therefore, appropriate prophylactic measures were at once required even before such information saw the light of day. Consider the warning posted by Nobelist Salvador Luria: "Will the Nazi program to eradicate . . . inferior genes . . . be transformed here into a . . . program to 'perfect' human individuals by 'correcting' their genomes?"[59] Hyperbole? Then consider also the following, representing the editorial judgment of perhaps the nation's most prestigious scientific journal:

> Illnesses such as manic depression, Alzheimer's, schizophrenia . . . are at the root of many current societal problems. The costs of mental illness . . . all cry out for an early solution that involves prevention. . . . To continue the current warehousing or neglect of these people, many of whom are in the ranks of the homeless, is the equivalent of providing iron lungs to polio victims at the expense of working on a vaccine. . . . [We must beware] the immorality of omission—the failure to apply a great new technology to aid the poor, the infirm, and the underprivileged.[60]

It is amazing and disturbing the extent to which scientists can become totally nonscientific when discussing social issues. Nobody then and nobody now knows whether the mentally "ill" (a potentially value-loaded term not to be confused with the mentally retarded), the homeless, the poor, and the underprivileged suffer from dysfunctions genetic in nature to a greater degree than anyone else. How then can one argue that the human genome effort would be of special benefit to them? I could label this a contemporary example of eugenics thinking. I will content myself with labeling it genetic determinism of the social right wing. When translated into government policymaking, the theory becomes straightforward political ideology. To challenge the credibility of what I call a species of right-wing ideology by pointing out the pseudoscience in what masquerades as scientific commentary is not itself ideology. It is peer review.

The architects of the project had the vision to provide a mechanism for checking and balancing the extremist views of their own supporters. In that spirit, Watson proposed that at least 3% of NIH's genome budget should be earmarked to address what he called "ethical and social implications." Further, "We must, . . . if necessary, pass laws . . . to prevent discrimination on genetic grounds."[61] Watson's successful attempt to plug in at the ground floor of paradigmatic application a parallel research track directed toward the social science and humanistic dimensions of the paradigm itself was unprecedented. It was a unique policy suitable for a unique initiative.

If Watson had headed off right-wing sentiment, he would be more severely tested by left-wing sentiment. From the outset, this cadre let its antipathy be known. Here is a comment from Sheldon Krimsky, representing the Committee for Responsible Genetics: "[We are concerned that researchers will] find statistical correlations between genes and social behavior and these will be used for social policy purposes."[62] Well, suppose scientists *do* find these correlations—why shouldn't such information be part of the data mix of policy formation? As a matter of fact, armed with reliable and valid correlations, we might be able to provide reliable and valid tests for causation. The opposition was ready to challenge the quest before it had even begun. To quote Jon Beckwith: "The argument that a high proportion of human disease is genetic is problematic. . . . The claim that [the project] will be a source for understanding human behavior is even more questionable."[63] As I shall eventually show, this statement flies in the face of substantial evidence to the contrary even as it had accumulated to that time. What these various arguments add up to is that attempts to show correlation and causation are forms of "geneticism" because the "roots are often in failings of the society itself."[64] In fact, the opposition had taken to these barricades on an earlier occasion. Upon the publication of Edward O. Wilson's mammoth tract extolling the virtues of sociobiology, which had suggested that much human social behavior is affected by DNA, Beckwith and several of his colleagues founded a group called Science for the People, its avowed purpose being to hold Wilson up to public obloquy.[65] According to Science for the People, any attempt to frame hypotheses regarding the genetic bases of human social behavior does nothing more than perpetuate the idea that certain people representing a certain race, a certain class, and a certain gender are more fit than other people.[66] I could call this a contemporary example of Marxism-Leninism given the group's fixation on "capitalist-imperialist ruling class oppression."[67] I will content myself with labeling it environmental determinism of the social left wing. When translated into government policy, the theory becomes straightforward political ideology. More peer review.

Unfortunately, this particular brand of left-wing ideology has an even more sinister dimension. One of Beckwith's Science for the People associates was Richard Lewontin. While Beckwith had conceivably mellowed in 15 years' time, Lewontin most assuredly hadn't.[68] In his no-holds-barred denunciation of the concept of human genomics was included the following:

> Molecular biology is now a religion, and molecular biologists are its prophets. Scientists now speak of the "Central Dogma" of molecular biology, and Walter Gilbert's contribution to the collection *The Code of Codes* is entitled a vision of the grail. . . . It is a sure sign of the alienation from revealed religion that a scientific community with a high concentration of Eastern European Jews and atheists has chosen for its central metaphor the most mystery-laden object of medieval Christianity.[69]

What we have here is a classic case of Jewish self-hatred projection. It is too easy to counter Lewontin by pointing out that Watson, Crick, and Venter aren't Jewish. It is also too easy to simply point out that Krimsky and Beckwith are Jewish. The essential point is that Lewontin, the Jewish Marxist who is infinitely more Marxist than he is Jewish in his role as a public intellectual, is accusing Gilbert of selling his Jewish identity to the God of Genomics.

This is important stuff. To understand the full scope and effect of human genomics in our time is going to require that we penetrate to the bedrock of our convictions about how we define science, how we define politics, and how we define ourselves. To all this we must bring a sense of system, a sense of constitution. As I end this chapter at the turn of the 1980s into the 1990s, the quest and the project itself have achieved mainstream consensus. They are deemed to be fundamental to our grand cultural purposes. They are firmly established in our psyches and our pocketbooks. And the power structure has blessed them with high purpose and station. Who would have predicted the process of genomic discovery as it ultimately unfolded?

2

Conducting Genome Research

A. Watson at the Podium

Studying leadership is not de rigueur among today's political scientists. For one thing, it is too much the stuff of biography, and biographies are considered case studies not readily amenable to broader behavioral generalization. For another thing, the decisions of a president, a civil rights leader, or a corporate executive (when was the last time a political scientist wrote a book about a modern-day business tycoon!) are not easily quantified.

I, however, consider studies of leadership fundamental, not because they make for interesting psychological profiling (though they do), but because leaders are critical figures in understanding the concept of constitution. Given proper environmental circumstances, political leaders can take constitutions in new directions, that is, they can mold the rules of the game to their own purposes and forever change the adaptive possibilities for those rules. Conversely, a lack of leadership at crucial moments can result in maladaptive consequences for players of particular games in particular environmental contexts.

Human genomic research agendas may not be "big science" in the abstract, yet they require a qualitative shift in leadership style and responsibility from the halcyon days of single investigators in their solitary surroundings. For example, a characterization of the discovery of the transistor would be incomplete lacking an appreciation of the political relationships (read: power relationships) among Shockley, Bardeen, and Brattain. Even that dynamic was low-key compared to what we are about to encounter. I liken the NIH Human Genome Project (HGP)—its many principal scientists, laboratory centers, advisory boards national and international—to one symphony orchestra, with the HGP director at NIH as its conductor. So the title of this chapter has at least two frames of reference: to conduct genome research, of course, is to *do* the research, but to conduct the research is also to orchestrate it, to infuse it with one's own sense of appropriate research styles, strategies, and performance levels, to govern it, in short, to provide constitutional guidance for its mission.

If we are to assess the constitutional politics profile of the Human Genome Project during its formative years, then we had better meet, up close and personal, the conductor of its symphony number one phase, James D. Watson. Edward O. Wilson has called the J.D.W. of the 1950s and 1960s "the most unpleasant human being I had ever met." "He shunned ordinary courtesy and polite conversation. . . . His bad manners were tolerated because of the discovery he had made."[1] When Watson became director of the Cold Spring Harbor laboratory in 1968, Wilson told friends: "I wouldn't put him in charge of a lemonade stand [but] [h]e proved me wrong." Said Wilson, he could inspire, he could raise money, and he could spot talent.[2] Robert Cook-Deegan read Watson's book *The Double Helix* while a student at Harvard, attended one of his lectures, and "decided he was a jerk." Years later, during Watson's NIH tenure, Cook-Deegan worked closely with him and reported in 1994 that "I was wrong."[3] In 1992 I gave a speech at Cold Spring Harbor on the political implications of human gene therapy. One week later, Watson visited Urbana, and while I didn't get a chance to meet him in his environment, our paths crossed in mine. We chatted for 15 minutes uninterrupted and unmonitored; the subject was my visit to Cold Spring and the current state of gene therapy. The conversation was difficult. Watson was polite, and he is not a jerk. When a person has opened a door of knowledge, that person faces a profound challenge in social relations. If the individual lacks certain social skills, the results can be counterproductive. My colleague John Bardeen, who had two Nobels to Watson's singleton, was very pleasant socially but painfully shy. Maybe he wondered, "Can a person whom everyone considers a genius sound like a genius when he's talking about the weather or politics?" Watson took a different road. He decided to sound like the genius he knew everyone thought he was no matter what the conversational tack. After all, why can't the objectivity of science be the template for an objectivity in social relations? The problem is that not all great scientists have boarding school manners, a salient element of which is noblesse oblige. Without those fall-back systems, they can make trouble for themselves, as the following commentary shows.

Watson got off to a good start, outflanking his potential ideological adversaries with his "3% goes to ELSI" promise (ELSI is insider lingo for the ethical, legal, and social implications of genomics). More on the implementation of that promise later. First, though, consider Cook-Deegan: "[I] have never seen a scientist so quickly adapt to the policy process. . . . I was astonished at . . . the facility with which he began to amble through the halls of power.[4] Watson seems at once to have grasped the need to establish bureaucratic legitimacy for the project. It would not do to run what might appear as a shoestring operation out of the director's office; he pushed for the creation of the National Center for Human Genome Research (NCHGR). The new center could make contracts, authorize its own study sections, and award grants. Human genomics at NIH now had the Washington look of a big time operation.[5] The next ingredient for bureaucratic success inside the Beltway is money. In 1989, Watson's operation received $28 million from Congress; the going was tough, but in 1990 the figure jumped to $59 million. In 1991, the president's budget allocated his shop $110 million.[6] In Washington, D.C., money often follows networks. Watson had no networks when he began, but he was an exception to the rule: he didn't need them, given his "impresario" status. Armed with that persona, Watson sold Congress on the proposition that his project was "our best go at diseases."[7] Long on reality and short on theory, the nation's lawmakers were only too pleased to help the refreshingly candid superstar if he permitted them to glimpse the utilitarian pot of gold.[8]

Watson also excelled in working with top flight "insider" science notables. In the brief pre–J. D. W. genomics period, several investigators were constructing chromosomal physical maps using different strategies for locating landmarks. On Watson's watch, a select four-person panel settled on a uniform definition of "landmark," namely, sequence tagged sites that would yield complementary DNA for the base pairs between the sites when made subject to the new polymerase chain reaction technology. The committee envisioned as a five-year goal constructing a physical map of the genome consisting of one landmark every 100,000 nucleotide sequences, with the hopefully much simplified end in view of isolating genes between landmarks.[9] A year later, the same cast of heavyweights agreed to parcel out the chromosomes among themselves in order to facilitate construction of a parallel genetic map consisting of "highly informative" landmarks for gene detection lying about 10–15 million bases apart. The ends of these signposts would also be sequenced so that the two maps would be synthesized through the medium of common STSs.[10]

These steps brought to the fore more overtly political questions, questions that at this writing are still contentious. One is data sharing. Scientists generally do not enjoy parting with their private stock of information, some of which takes years to accumulate. But it just didn't make sense for a handful of laboratories to be slogging away at the ever-enticing chromosome 21 (enticing because of its known role in generating Down's syndrome and its suspected role in causing some forms of Alzheimer's), each using its own set of tools. So they agreed not only to pool but also to collaborate in creating a complete set of overlapping sequences using the yeast artificial chromosome (YAC) cloning technology. The sequence clones would become the common property of the group. Another issue was data release. Watson enthusiastically supported the chromosome 21 alliance but thought large-scale sequencers deserved more leeway. "The stuff should be released immediately," said some. Others said a three-month window would suffice; still others said six months. Watson thought a hard-and-fast set of rules was counterproductive but ran into the argument that the public was footing the bill so public accountability should be maximized. Somewhat ominously, Craig Venter, who was beginning to establish himself as a big-time sequencer, declared that principal investigators should have one year to examine the fruits of their labors before divulging results.[11] At around this time, Watson took the lead in championing the "big science" gambit of national genomic research laboratories to perform lead and cutting-edge missions, and he instituted a peer review process to narrow candidates. The four winners were the University of California at San Francisco, which would concentrate on chromosome 4; MIT, which would map and sequence the mouse; Washington University, which would work on chromosome 7 and the X chromosome; and the University of Michigan, which would refine mapping methodologies. A howl went up from critics: "Jim Watson is trying to change the social fabric of science." Rejoindered Watson: we can't do our job in 15 years at a cost of $3 billion using the same old, same old research organizational models.[12] In a meeting with Senator Domenici, Watson made just the sort of announcement he knew Congress wanted to hear: Tom Caskey at Baylor had found the gene for the dreaded Fragile X syndrome, the most common inherited form of mental retardation. Taking data from the genetic map of the X chromosome, Caskey had found a marker proximate to the gene, had then cloned a stretch of sequence containing the gene by using the YAC methodology, a technique created especially for that purpose by genomic researchers, and then had pinpointed the exact location of the gene.[13]

Watson's troubles began to surface in his attempts to "go international" with the project. In point of fact, several Europeans had gotten off the mark ahead of U.S. scientists. In 1984, Nobelist Jean Dausset had established CEPH (Centre d'Etude du Polymorphism Humain) to conduct human genomic mapping studies, and Prime Minister Jacques Chirac called the field "a new priority for the nation."[14] The British became visible in the late 1980s when their NIH-equivalent Medical Research Council (MRC) put together a blue-ribbon panel under Sir Walter Bodmer's direction. Even the U.S.S.R., normally not a player in the molecular genetics game, caught the fever; it talked of founding an "institute of man" with the thought of aggregating "all data on man as a biological and social being."[15] This was heady multidisciplinary stuff, going well beyond in theoretical compass the "disease gene" of American preoccupation. Sensing the need for joint action on matters genomic, the British and French governments agreed to assist a collaboration between the Imperial Cancer Research Fund and CEPH which would hopefully spawn dramatic improvements in sequencing technology.[16] This flurry of activity—somewhat disjointed abroad as it was in the United States—led deep-thinking members of the "insider" community to realize that they needed their own council, what one of their number called a "U.N. for the human genome."[17] And so was born the HUGO (Human Genome Organization). The HUGO won foundation support, presumably freeing researchers in their new political capacity from governmental bureaucratic strings, conditions, and priorities.

Surveying this scene, Watson essayed a pitch for uniformity. Why not allocate chromosomes to various countries, he opined. Whoops! Here goes Uncle Sam directing traffic, came back the reply from several of his peers. Said the United Kingdom's highly respected Sydney Brenner: "We do not intend to be assigned a part of a chromosome by some Politburo somewhere."[18] Nor did the HUGO cotton to his idea. It saw the national funding agencies making such allocations, not scientists. Capitol Hillers also were not enthralled. If we are ahead, they asked, why share the spoils?[19] Much more popular was a Watson initiative to fund a collaborative venture between Washington University and the Sanger Centre (Cambridge, England) to sequence *C. elegans*. A prime argument that had brought nonmedical geneticists on board the human genomics train was that lower-order species would also be sequenced. To unravel *Homo sapiens* solely might provide little theoretical knowledge. But if the project were to embrace *E. coli* (the bacterium), *Arabidopsis* (the plant), *S. cerevisiae* (the yeast), and *Drosophila* (the fruit fly), then a new field—comparative genomics—could yield precious findings central to evolutionary dynamics because basic genes are conserved up the phylogenetic tree. *C. elegans* (the roundworm) was Brenner's pet; he was certain it held important clues to *Homo*'s genetic posterity. The collaboration that would ultimately take root was sure to save a lot of money and a lot of time.[20]

Watson's face-off with the Europeans was a sort-of prelude to his confrontation with the Japanese. Back then, Ezra Vogel's "Japan is number one" thesis had raised eyebrows in the United States, and when a prominent geneticist at the University of Tokyo announced that his country planned to outsequence the American sequencers, Bethesda-linked molecular biologists were much concerned. They needn't have worried. The Japanese were well behind in theoretical knowledge, and both their politics and their science were steeped in bureaucracy. All the machines in the world wouldn't give them a mass-production advantage in this field.

While Japan was dithering, the HUGO was starving. As 1990 began, it had only $25,000 in cash reserves. Shortly thereafter, the United Kingdom's Wellcome Trust came through with $350,000, and the Howard Hughes Medical Institute pledged $1 million.[21] But where was Japan? Watson was furious. He accused the Japanese of "free-loading"; they had unfettered access to all human genomics data generally available but were paying none of the freight. He also claimed the Japanese were stonewalling access to their own DNA sequences absent remuneration.[22] In typical "Honest Jim" prose he put his position thusly: "I'm all for peace, but if there is going to be a war, I will fight it. . . . I've found you never get anywhere in the world by being a wimp." He made it quite plain that if the Japanese don't contribute, they will forfeit access to data. The Japanese, of course, were horrified by this outburst. They considered it "Japan-bashing" and wondered if Watson were speaking in his capacity as HUGO member, NCHGR director, or simply as self-appointed "Mr. DNA."[23] Watson's HUGO colleagues were either appalled or exasperated. In a written statement, the elite cluster of genomicists announced that it was the unanimous view of the membership (not counting Watson, evidently) that "all data should be freely accessible to scientists and not used to secure narrow national interests."[24]

Not parenthetically, HUGO continued to struggle after the Japanese flap subsided. For a time, it appeared as though it had found its niche: serving as high-level liaison between the mappers and sequencers, who did, in truth, represent different research cultures.[25] But in January of 1992, HUGO announced it simply could not afford to hire a permanent executive director, adding yet more evidence to the conviction that the organization was "still . . . in disarray."[26]

I now return to Watson's unique and exemplary proposal to set aside what would turn out to be 3–5% of his office appropriation for ELSI studies. One can imagine the snickers and derisive asides uttered by at least some of his colleagues.[27] From my tour of duty on the RAC, I know that ELSI pursuits (I call them ELSPI pursuits out of deference to what I am arguing really counts: *politics*) are a hard sell in the company of self-styled Nobel contenders. ELSI commentary is perceived as "soft," not "hard." That predilection is baseless. This book, for example, does not purport to be a laboratory study, nor could it ever aspire to such elevated status, if elevated it be in this context. More probative is the objection that ELSI commentary is indeed commentary, not research, or at least rarely achieves the status of good social science research. It tends to be normative, philosophical, and anecdotal. Even more probative is the criticism that ELSI practitioners not only don't do science but also don't know much about the theory, even the vocabulary, of science. When I interviewed recombinant DNA specialists for *Cloning and the Constitution* and later on when I worked with gene therapists on the RAC, I had to negotiate a hazing process because, of course, political scientists are presumed (correctly) to know nothing of such things. I make bold to say that I passed those little litmus tests, for I have no complaints whatever to lodge against these cohorts, all of whom I regard as my colleagues.

Watson tapped Nancy Wexler to be chair of the ELSI working group soon after he assumed the directorship. At first glance, she was a splendid choice. Upon closer inspection, optimality of job performance required high marks in three somewhat unrelated skills. Wexler excelled in two of these—very good but leaving something to be desired. To sell ELSI to the widest possible audience at that germinal moment, the chair needed to be reputable in the genetics arena. Wexler had played an important role in mapping the Huntington mutation to chromosome 4, so she qualified on that score. The chair also needed ELSI credentials, of course. This was a Wexler strong point. She had discussed in print the roles of genetic screening and genetic counseling as social and patient policy practices. Many knew that she herself was at risk for Huntington's.[28] Not clear at the moment of her appointment was that a third credential for success was in order: political science acumen. Politics was not her métier.

Consider the modes and purposes of her selection. The fact that "Watson did not fancy himself an authority on ethical issues" only meant that for him the entire appointment process was a trivial matter.[29] There was no search committee, there was no discussion as to the competing credentials of candidates, there was no preliminary dialogue as to the political tasks inherent in the position. Wexler was simply at the right place at the right time and looked to have a broader view than others at that place and time regarding the scope of the program.[30]

What "political tasks" have I in mind? An excellent template for the ELSI working group, officially known as the NIH-DOE Working Group on

Ethical, Legal, and Social Implications of Human Genome Research, was the Recombinant DNA Advisory Committee. The RAC had been in business for fifteen years. It had its own staff, its own budget, and its own clearly circum-scribed policy arena. It formulated and updated what came to be known all over the DNA world as the NIH Guidelines spelling out the do's and don'ts of recombinant experimentation. It screened research protocols and had its own ELSI component. If it wanted to hold a conference on a pertinent issue, it held a conference. The working group would hire a very able full-time di-rector, Eric Juengst, and it reviewed for funding (the RAC has never funded anyone) proposed ELSI research. The salient problem was not that the work-ing group was a study section whereas the RAC was a guidelines interpreter and amender; the salient problem was that the working group had no politi-cal independence whatever. It sat as an ad hoc afterthought in the director's office, beholden to him for not merely political support in the conventional sense but resource support in the full sense of that term.[31]

This posture was particularly precarious when "the boss" who does not "fancy himself an authority" has an opinion on everything and is sure to flaunt that opinion publicly in as clever a way as putative genius permits. Hence, Watson proclaimed, "I don't think *anyone* should have access to anyone else's DNA fingerprints," and he delivered this pronouncement on the heels of providing $50,000 to an NAS study on the subject.[32] Just why the police should not be permitted to secure a search warrant enabling them to peruse the contents of a DNA data repository in order to establish a match with DNA left at the scene of a crime—one of many caveats to the Watsonian dictum on law enforcement—is unclear.

We have seen how a lack of political science expertise on the part of government-sponsored commissions handicapped their vision as to where in the bureaucratic scheme of things the Human Genome Project should sit. History repeated itself in spades with the bureaucratic positioning of the ELSI working group. Recall the memorandum of understanding between NIH and DOE staking out respective genomic pastures. Concerned mem-bers of Congress made it clear that just as NIH had an ELSI component, so DOE had better have an ELSI component. What this meant was that Wexler's shop would serve both masters. But how? Cook-Deegan summed up the organization chart mess as follows:

> [T]he working group was buried several layers down in the NIH and DOE bureaucracies. It was advisory to [the] joint NIH-DOE [interagency] sub-committee, in turn advisory to the main outside advisory bodies to NIH and DOE. The ELSI working group was advisory to an advisory group to an advisory group to two different government departments.[33]

To be sure, this picture understates Watson's role in the process, and mostly he let the ELSI working group muddle along. It drafted several mission statements, a fine-tuned version setting out four broad goals: 1) "anticipate and address the implications for individuals and society" of human genomics; 2) "examine the ethical, legal, and social consequences" of human genomics; 3) "stimulate public discussion" of human genomics; 4) "develop policy options to assure that [human genomics] is used for the benefit of the individual and society."[34] And it did some good work, hosting meetings, dispensing grants to ELSI researchers, and all too infrequently advancing policy positions. One success story involved screening for cystic fibrosis mutations, with Wexler's group persuading Watson and NIH to foster a grant-application process for pilot programs before an ill-considered national testing strategy was put into place. Less successful was its foray into genetic privacy issues, the working group ultimately taking no position on proposed congressional legislation. An outright failure—though only after the panel produced an "old college try"—resulted when the Equal Employment Opportunity Commission rejected the working group's recommendation, concurred in by the joint NIH-DOE subcommittee, that the Americans with Disabilities Act be construed to cover employer discrimination against workers who carried harmful genes but were not themselves affected by those genes.[35]

The fact is that Nancy Wexler's committee never had, and today still does not have, the political potency necessary to play the role of policy mover and shaker. If Watson had made ELSI a front-burner "action concern," and had Wexler bargained with him on matters of influence, wherewithal, and implementation, the working group could have stepped forward, as the RAC stepped forward, armed with marching orders and prepared to engage a concrete agenda of its own orchestration. Impossible? Naive? There is ample evidence that key members of Congress wanted the ELSI group to do just that. Of course, there were political risks. The RAC generally understood its political limitations during these years and lived with them. A specific congressional charge would have helped as it would have helped the RAC, but it was not a requisite to success. The argument is made that what we needed then was a National Genomics Commission, not a working group.[36] I disagree and will elaborate later. Suffice it to say here that independent national commissions lack the political, institutional moorings to enhance success in Washington's hurly-burly environment.

James D. Watson is reputed to have said about Nancy Wexler's committee: "I wanted a group that would talk and talk and never get anything done, and if they did do something, I wanted them to get it wrong."[37] More rhetorical overkill, of course, and, as usual, entirely deliberate. Probably Watson did want a lot of talk and not much action. But if the committee had done nothing he would have looked fairly foolish for insisting on its existence. The trick was for the working group to sell him on the message that its interventions would help the project, not perturb much less undermine the project.

I provide an episode from my RAC experience. Shortly before I came on board the committee, but after I had been officially informed of my appointment, Steven Rosenberg, W. French Anderson, and Michael Blaese presented the first-ever gene transfer protocol for RAC consideration. I received all the papers by mail. Rosenberg's informed consent documents seemed incomplete, indeed fatally flawed. The problem was that he didn't explicitly state that he was inserting into his patients' bodies a bacterial gene recombined with a monkey retroviral vector; this artifact would hopefully transduce some of their cellular genomes (double helixes). To me, the first political scientist to sit in a RAC-review capacity, this omission constituted a possible violation of procedural due process: an NIH-employed doctor was not providing what I would call reasonable notice to his patients as to the unique nature of the medical procedure. What should I do? I could hide behind the fact that I was not yet a voting member of the panel and cop out. Or I could confront Rosenberg with my objections (as a matter of courtesy) and then make a little speech before the committee lodging my objections. I took a third route. Spotting him during the first day's proceedings, I introduced myself and then said: "I've gone over your informed consent papers, and I see some problems." His face tightened. He and his coworkers were having enough trouble getting RAC approval, and now, he was surely thinking, along comes some ELSI pest telling him his documents were no good. I continued quickly: "I've taken the trouble to revise your forms so they say what I think needs saying. Read them over and tell me if you approve my changes." He took them. The next day—decision day—he greeted me warmly. "You really helped me," he said. "Thanks a lot." It so happened that I would have voted to approve his protocol, but that wasn't the point because reasonable people might have seen the merits differently, as, indeed, they did. The point was that I had some expertise that he didn't have, and my job was to see to it that if he was going to perform these experiments, then they would be done properly to the extent I felt myself responsible.

To deal with a Watson requires this sort of political know-how and more. At least on the part of those sitting under him.

In July of 1991, Craig Venter, then a scientist at NIH working in a unit well removed from Watson's orbit, announced at a congressional hearing on the HGP that his superiors were planning to file patent applications on the sequences he was churning out at an unprecedented pace. Watson, present on the occasion, knew nothing of any such patenting strategy and exploded when he shortly took the witness chair. Patenting genome sequences was "sheer lunacy," he stated; furthermore, "virtually any monkey" could do what Venter was doing, he announced.[38] Even by Watsonian standards, this melodramatic, theatrical descent into the nuances of patent law seemed hyperbolic if not tasteless. What exactly had Venter been up to, and what patent law "nuances" were involved?

Recall Walter Gilbert's grandiose plan to map and sequence the human genome and sell the contents to willing buyers. A critical element of his strategy, had it ever borne fruit, would have been to obtain a copyright from the federal government on the genome and its essential parameters.[39] Upon reading of Gilbert's aspirations, I published an essay in the form of a letter to *Science* magazine in which I reiterated the argument I had made in *Cloning and the Constitution* a year earlier, namely, that Congress lacked the constitutional power to give researchers any such monopoly or franchise.[40] We may begin by at once clarifying that there is no difference in the constitutional sense between Gilbert's copyright claim and Venter's patent claim, because neither term can be found in the document. The phrase that is used by the framers of the Constitution is "exclusive right."[41] Under federal statute, a scientist or inventor can receive exclusive right for a time certain (now 20 years) to "'anything under the sun that is made by man,' provided it be original enough and practical enough."[42] But according to the Supreme Court, the statutes have never covered "laws of nature, physical phenomena, and abstract ideas"; these were "free to all men and reserved exclusively to none."[43] The critical question, then, is whether Congress *could* decide by law to allocate what we now call patent rights to Einstein for discovering $E=MC^2$ or to Watson and Crick for discovering the double helix. Surely, the answer is no, but why?

Constitutional authorities talk a lot about the "marketplace of ideas" that freedom of speech is designed to protect. They also debate the forms of speech undeserving of constitutional solicitude because they contribute little or nothing to this marketplace. However, they have spent virtually no time discussing the attributes of the marketplace itself, by which I mean the subject matters that speech is meant to address and enhance. What good is speech if the speaker has to pay a fee to gain access to the topic itself? The speech may have little bearing on reality if the reality is off limits because the government has vested someone with a monopoly on that reality. Now if the researcher has created or invented the reality or subject matter, that is one thing. The law says the person likely has a property interest in the item or entity. Discovering a chemical element, a law of nature, or a body part is quite another thing. The government, under the U.S. Constitution thus perceived, can neither pass laws keeping us from unearthing these natural treasures nor allocate to us rights to any of them. They are ours to be discussed freely by all of us subject to no tariff, no fee, and no regulation except the regulations traditionally attached to speech. Surely, I argued at the time and still argue, the human genome is worthy of such First Amendment protection from patent-seeking entrepreneurs or from anybody else for that matter.

Venter's work was with complementary DNA (cDNA), otherwise called copy DNA. When DNA codes for proteins, it initially is converted into messenger RNA (mRNA) during which process the DNA is stripped of all of its "junk." RNA is no more patentable than DNA according to my theory, but Venter (and already many others) took human mRNA and converted it to cDNA by employing the reverse transcriptase procedure. This method of change occurs naturally in certain organisms such as retroviruses, but Venter's procedure of isolation and purification was a strictly human scientific intervention. Complementary DNA does not occur in nature, in other words. But if one "manufactures" a codon the way Venter did that reads TAG, it must be that the naturally occurring arrangement that appears somewhere on someone's double helix is also TAG. If the human genome cannot be patented, then naturally occurring genes cannot be patented, and naturally occurring sequences cannot be patented either. Yet the courts have ruled (though not the Supreme Court) that complementary DNA can be patented, and so, *a fortiori,* it would appear that sequences drawn from cDNA such as our hypothetical TAG could be patented as well, at least in the sense that the Constitution is part of the equation.[44] The logic inherent in all this is somewhat alarming: by patenting cDNA, one is, in effect, patenting sequences and genes as they truly exist in the "nature" of bodies everywhere. The human genome as protein-coding instrument could theoretically be patented piecemeal and by indirection.[45] As an aside to Venter and his projects, I shall simply say now that Congress should amend the patent laws to allow exclusive rights to human DNA only as used in the context of particular experimental protocols that themselves constitute patentable processes or yield patentable products. Patenting DNA through the back door of cDNA may be constitutional, but its public policy implications are dubious.[46]

Back to Venter. Using state-of-the-art high-speed computer technology, he took a random sample of cDNAs and sequenced about 450 base pairs of each clone (he labeled them "expressed sequence tags," or ESTs). He then deposited the ESTs in public data banks. The tags were dead giveaways as to the identity of genes in the human body, some of which were already on deposit in these repositories while others were unknown in the literature (and thus were the most interesting), and he prepared an article on his investigations for publication. In the abstract of his essay, we find the following: "This fast approach [requiring only one run through the machine] to cDNA characterization will facilitate the tagging of most human genes in a few years at a fraction of the cost of complete genomic sequencing."[47] This succinct bit of self-aggrandizement hit Watson where it hurt. It challenged directly the premises undergirding the HGP and posed the same sorts of caveats that eminent mappers such as Sydney Brenner had been voicing for years, except now the reservations took the form of a tangible, overt confrontation, and from a fellow NIHer to boot! Adding insult to injury, these in-house scientists, with no affiliation whatever in Watson's genomics shop, were going to try to get patents on their EST fruits, and they had not even bothered to discuss it with him! Hence, the "lunacy" and "monkey" tirade.

What were the pros and cons of NIH's patenting strategy? The most obvious reason to patent was to gain competitive advantage, to establish proprietary control over vendibles of value before others did. What "others"? Not academic scientists. They were free to use Venter's ESTs for their own basic research investigations, assuming these tags were made available. And they had been and would be made available, for one of the reasons why the Constitution framers gave Congress the power to provide for patenting was to give the inventor an incentive not to hoard but to disseminate. The "others" might ordinarily have included entrepreneurs both here and abroad, people who would employ ESTs for commercial purposes. These biotechers might well seek their own patents, but if NIH were successful in its application, they would have to pay a fee if they opted to use the information for marketing purposes in this country, where the patents would have legal force and effect. This scenario, however, belied reality; no entrepreneur in the foreseeable future was in a viable position to compete with Venter. And so the "others" meant foreign national genomic projects in the United Kingdom, France, and Japan. Keep in mind the politics: President Bush, a Republican, had named Bernadine Healy, a Republican, to be NIH director. Nothing untoward there. What is relevant is that they shared a common frame of reference on these matters. "The Bush Administration was focusing attention on economic competitiveness, and biotechnology was one of its darlings." "Healy was pursuing technology transfer at NIH as a major policy thrust. She had a long-standing interest in commercial applications of biomedical research."[48] One could well argue that the critical policy question was not whether NIH was able to patent ESTs, but, rather, what licensing terms and conditions would it enter into with pharmaceutical firms, biotech companies, and other producers of genetically based therapies? Would NIH grant exclusive licenses, and to whom, and based on what criteria? Or would it only sanction nonexclusive licensing, thus enhancing a competitive marketplace? But emotions were running too high to permit serious debate on such abstruse matters. Venter wanted to publish; once he did publish, all opportunities to patent, to even test the waters as to whether the U.S. Patent Office would smile on an application of this ilk, would be lost. Speed was the order of the day. The patent was filed, and Venter's work appeared in print.

And the cons? Well, obviously, Watson and the rest of the world outside the upper echelons of the NIH had seen the Human Genome Project as an international cooperative venture, their image fueled by the notion that if ever scientists ought to work together, this was the moment, namely, when the subject matter was the quintessential species' posterity. As the process unfolded, Healy's patenting initiative would cause a rift among our putative collaborators. The United Kingdom felt its only recourse was to file for patents on its own cDNA repository; one tariff barrier deserved another, as it were. The European Community (EC) had adopted policies openly hostile to human DNA commodification, and the European Patent Office had rejected Harvard's application for a patent on Philip Leder's transgenic mouse.[49] Taking the conservative route (from the Greens' point of view, the "radical left" position), France, Germany, and Italy declined to follow the UK lead. Clearly, joint cDNA missions among them were in jeopardy.[50] In the United States, both researchers and biotech firms expressed outrage in reaction to the NIH initiative.[51]

A second line of criticism, of equal concern to genome scientists, was that a patent on an EST might well be construed to be a patent on the gene of which, in a sense, it was a part and even perhaps a patent on the protein for which the gene coded. The first NIH patent application on Venter's collection posed just such a clear and present danger: researchers were horrified when they saw that NIH wanted coverage on all genes found to be tagged and all proteins found to be expressed. Under U.S. law, a proposed patent must meet three statutory criteria: novelty, nonobviousness, and utility. Perhaps Venter's ESTs were novel, and perhaps they weren't obvious, but what utility did they possess? The truly challenging task was to elucidate the function of a gene, not sequence the piece of an unidentified clone sitting in a data collection by "a dumb, repetitive" process.[52] Eventually, the American Society of Human Genetics, the HUGO, and even the joint NIH-DOE

Interagency Genomics Subcommittee turned thumbs down on the Healy patenting proposal.[53]

Political science teaches that this level of falsetto disputation among governing elites probably results from not only heart-felt disagreements over policy but also clashes in personality. Such was certainly the case here. Before Venter began his quest for ESTs, he had applied to Watson's genomics center for funding to underwrite a plan to sequence a gene-rich section of the X chromosome. After protracted hemming and hawing by the NCHGR worthy of academic politics, Venter finally withdrew his protocol from consideration. "Unlike the other investigators rejected in the first review, Venter's group never secured genome project funding," though the reason for his in effect conceding defeat may well have been his growing belief that the EST route was the way to go.[54] All this preceded Watson's "monkey" blast, which, of course, Venter took personally, and Venter's retort in the form of "I can do better than you can, Mr. Genius."

As it turned out, the animus between Watson and Venter was a tempest in a teapot compared to the friction between Watson and Healy. In order to provide appropriate background, I relate another snippet from my RAC experience. The RAC reports directly to the NIH director, and when I attended my first meeting as a voting member Bernadine Healy's term of office was well underway. We heard not a word from her for many months; in other words, she let us go our own policy way, a grant of discretion we greatly appreciated. Ultimately, there arose a RAC-related issue in which she became intimately involved, much to our displeasure. Under the guidelines at that time, there was no provision for expedited review of "compassionate relief" cases. Why should there have been? A good case for compassionate relief could be made for virtually every patient enrolled in a proposed gene therapy protocol. If federal authorities were to grant compassionate relief, thus permitting doctors to administer untested, untried genetically engineered potions to patients at great risk, which, in speculative theory, might do them some good, why have the RAC? In the case at hand, Ivor Royston had a terminally ill patient suffering from an advanced glioblastoma, and he wanted to insert into her brain some of her own cancer cells that had been transduced with the interleukin-2 gene, hopefully boosting her immune response. Royston appealed directly to Healy, who initially rejected the unprecedented request, which in effect had short-circuited RAC oversight, but then, much to the RAC's chagrin, she reversed field and granted the petition on her own authority. It turned out that the patient had a political connection: one of her relatives knew Iowa senator Tom Harkin, a key player in formulating the NIH's annual appropriation, and Harkin had written a letter to Healy urging immediate action. I can still hear my RAC colleague Dusty Miller at a subsequent RAC meeting: "[Now it's okay to] put recombinant split pea soup into people [and our committee is helpless to stop it]."[55] I assert that Healy crumbled in the face of unvarnished, unprincipled political pressure. The moral to the tale, for present purposes, is this: in reciting the Healy-Watson contretemps, for my money there are no heroes and no villains. Let the reader judge.

Healy and Watson first crossed public paths in 1985. Healy was working in the White House as deputy director for biomedical research in the Office of Science and Technology Policy. Watson thought her office was issuing unduly stringent requirements. His comment: "They had to put a woman someplace." Watson would belatedly apologize.[56] Five years later, Healy would be Watson's boss.

In point of fact, the prime mover in the patenting campaign was not Craig Venter, not Bernadine Healy. The prime mover was Reid Adler, in charge of NIH's Office of Technology Transfer. He favored patenting for the obvious nonphilosophical reasons, but how he went about the task was certainly a puzzlement. He informed Watson's office, and discussions took place; later he notified Watson's office of his decision to press forward, but he informed neither Watson nor Healy directly. He thought the entire transaction was "business as usual." The NIH genome center raised no objection to the initiative but did not endorse the gambit. Where was Watson? Healy, when once informed, asked the same question. Watson figured Healy must have known all about this patent planning. Healy figured Watson must have known as well, so why was he yelling to the media instead of pressing her ear? Healy then proceeded to appoint Venter to oversee an initiative to establish an intramural genomics research program, clearly bypassing Watson. Cook-Deegan comments: "The miscues were symptomatic of a general breakdown of communication." The fact is the two major players simply stopped talking to one another, as Healy kept boosting the patent application in public while Watson kept criticizing it in public. Eventually, they did have a meeting, and Watson agreed to keep "company business" private. So he simply complained privately and incessantly. Increasingly, he was seen less and less at NIH.[57]

The dispute between Healy and Watson over patenting ESTs was rancorous, but at least the issue was serious public business. An issue that was not serious public business, but seemingly the straw that broke the camel's back, involved Watson's stock portfolio. Evidently, he had substantial holdings in biomedically oriented commercial outfits and had made full public disclosure. Self-serving complaints from a major entrepreneurial player

about possible conflicts of interest arose when Watson intervened successfully in the businessman's attempt to raid the project and woo some of its
key players into private commercial employment. At first, Watson seemed
well in the clear when NIH chief ethics officers found, as they had found before, no irregularities in his financial position. Healy, however, was not satisfied, though she was prepared to give Watson a clean bill of health if her
superiors at the Department of Health and Human Services were prepared
to do so. Not comfortable with Healy's buck-passing, they balked; nor did
the air exactly clear when Healy's spokesperson at NIH told the *Washington
Post,* "Dr. Healy does not have the luxury of ignoring ethical questions, even
for a Nobel Prize winner."[58] Watson felt compromised. "I don't know anyone who doesn't have stocks," he bemoaned, "And I don't know anyone who
would want to live with my boss." On April 10, 1992, he resigned.[59]

Whether Watson surrendered his position as head of the NIH genome
project on account of disagreements over patent policy or disagreements
over stock ownership or both is a trivial question. From the outset, he was
leery of holding "two demanding positions simultaneously."[60] Cook-
Deegan, who was close to the scene, has remarked that "Watson disliked the
pressures of his NIH job from the beginning, and he found them particularly trying . . . as . . . NIH's internal politics intensified. . . . [T]he . . . program was getting sufficiently large to demand the attention of a full-time
director on site." The fact is that the position cried out for a "full-time director" from the very outset, but Watson was not prepared to take a leave of
absence from Cold Spring Harbor. He must be given credit for getting the
HGP off the ground by adding his name and reputation to the initiative,
and he fought the good fight for international cooperation and good science as he saw it. He could never adjust to his "uncharacteristically subordinate position" inferior to the likes of Healy, and it brought out too often his
propensity for preemptive rhetoric. On the matter of stock holdings, his
reliance upon the notion that everybody owns security shares is baloney.
When I went on the RAC, I deliberately made sure I owned no stock that
anyone could say compromised my role. All Watson had to do was buy
Bank of New York and sell his Pfizer, if he had any. Disclosure, for first-line
public officials, is not enough. Watson thought about divesture but then
correctly concluded that it was just forestalling the inevitable.[61] He might
take pleasure in looking back and knowing that the bureaucratic confusion
at NIH during his tenure occurred on Healy's watch and speaks for itself.
He surely would take pleasure in the fact that the U.S. Patent Office slapped
both Healy and Adler in the face by finding that Venter's ESTs flunked all
three statutory criteria: they were not novel, they lacked utility, and they

were all too obvious.[62] On July 13, 1992, Venter resigned his NIH post to take up private employment. His career was just beginning.

B. The Battle of the Orchestras: Collins vs. Venter

The genomics community watched closely as Healy pondered Watson's successor. And Healy knew she was being closely monitored. She had "divorced" an icon in biological science, or at least her behavior had been a proximate cause for his self-executed divorce decree. She needed to bolster her reputation in the field by naming a scholar with impeccable credentials. In April of 1993, Healy selected Francis Collins.

Collins's contributions to the literature were impressive. Least spectacular but perhaps most significant was his role in developing a research methodology called positional cloning. I have earlier made reference to the practice of walking the chromosomal trail in search of a new gene. Collins and colleagues found a way to leapfrog many thousands of sequences by splicing together the ends of such stretches, cloning them, and then deciphering the sequences that comprised the fused section. It was easy enough to spot whether one had reached the gene of interest.[63] A painstaking process of discovery had become relatively manageable. But it was his headline-grabbing achievements employing this new tool that landed him the job. With Lap-Chee Tsui, he located the commonest of the cystic fibrosis mutations, the delta 508 3-letter deletion on chromosome 7; working alone, he found the malfunctioning allele for neurofibromatosis on chromosome 17. He was also a member of the team that at long last had located the dreaded Huntington's dominant, though positional cloning proved of no avail, given the paucity of chromosome 4 sequences then in hand, and he had teamed up with Mary-Claire King to make a concerted effort to track down the BRCA1 breast cancer mutation also on chromosome 17. Of greater political consequence was that Collins had come up a winner in one-on-one competitions. He and his colleagues had vanquished Robert Williamson's London group to win the cystic fibrosis race, and he had bested Ray White in the search for the genetic root of neurofibromatosis, though the "powers that be" arranged to declare the match a draw, much in the fashion of Clinton-Blair a decade in the future. But in the sports vernacular that so often colors media coverage of contemporary events including the genetic and genomic, he had failed to go "four for four" when Mark Skolnick of Myriad Genetics, a commercial Salt Lake City outfit, "defeated" the King-Collins duo in the race to isolate BRCA1.[64] Success in the arena of "scientific sport" had been a badge of honor in molecular biology since Watson and

Crick's legendary triumphs over both Pauling and Wilkins-Franklin. It bespoke both intellectual stature and what Hemingway called grace under pressure, that is, what I call political toughness. Nor did Collins's flair for the dramatic pale compared to that of his predecessor at NIH: "[T]his job . . . is more important than putting a man on the moon or splitting an atom," he declared.[65] At the time, no one could have appreciated how important these attributes would be to the success of the HGP. But would they be enough?

Perhaps the fact that Collins was an in-your-face born-again Christian would be a sufficient hedge against all margins for error. Not so long ago, national political figures pretty much left religion alone in fashioning their public persona. Some went to church, and some didn't. Mostly you were politically safe if you could call yourself a Protestant or, since JFK's time, Catholic. Even the layperson knows this has changed: politicians these days often have to work overtime to "prove" to their electorates that they are in sync with the deities or, more specifically, God. The reasons for this sea change in the politics/religion equation are numerous; however, I have no doubt as to the identity of one key factor, the abortion issue. In the 1930s–1950s, Democrats and Republicans fought over many critical policy questions, but one of them was not abortion. Then came *Roe v. Wade*. The notion that destroying embryos on the whim of a pregnant female was part and parcel of the Constitution of the United States so enraged (and so enrages) many Catholics and Protestant fundamentalists that they have bent every political energy to reversing, or at least blunting, that decision. Hence, Jerry Falwell; hence, Pat Robertson; hence, "the right to life." And hence, religious litmus tests for national political candidates pro and con.[66] From the perspective of genomics policy even narrowly defined, the abortion issue has its own unique place, which I now simply specify without elaboration: aborting preembryos (fertilized eggs not yet implanted in wombs), embryos, or fetuses involves the destruction of unique genomes and in that sense raises profound implications for genetic variability and ultimately genetic adaptation in the Darwinian sense of the term.

Francis Collins had become, upon his appointment, a national political figure. He would be doing business with the Congress on a regular basis. Several key senior members of Congress were "conservatives" from the South and West, and by "conservative" I mean here in good standing with the Christian Coalition version of political correctness. Watson did not need to spout religion to score brownie points. He might have been viewed by such constituencies as "just another secular humanist scientist," but he was Watson. Also, Watson never threatened anyone with rhetoric purporting to sell the HGP as the road to genetic superiority. In fact, being a good self-styled liberal, he took a figurative oath against all things eugenic, all things that smacked of playing God. Collins did him one better, citing scripture to provide theological imprimatur for the project:

> Jesus went through all the towns and villages, teaching in their synagogues, preaching the good news of the kingdom, and healing every disease and sickness. It is part of our mandate as Christians to pursue such medical advances, attempting to emulate Christ in his healing role.[67]

If Richard Nixon, Jimmy Carter, Bill Clinton, and the Bushes could attend prayer breakfasts with members of Congress, so could Collins. And he did.

Naturally, religion can cut both ways. So was the case here. Collins had received a patent on the cystic fibrosis mutation, an entity he had discovered, not created. Some "born-agains" opposed the patenting of human genes on the ground that they were God's unique handiwork, if anything was. To this, Collins retorted:

> As a Christian, it was a black day for me when the church that I care so much about supported [a view] that was considered ludicrous in the scientific community. . . . To the extent that we speak with reason and love, God is glorified. If we do not, he is not.[68]

Whose reason and whose love?

Collins had ambitious, perhaps even grandiose, plans for the Human Genome Project. Unlike Watson, who maintained no research agenda of his own at NIH, Collins moved his University of Michigan laboratory to Bethesda, where his goal was to launch full-scale quests for disease mutations to the tune of $25 million in FY94. He heaped praise on NCHGR's extramural programs and promised not to foster competition between them and the missions of other facilities. That promise would prove about as feasible as FDR's pledge to fight the Depression and simultaneously balance the budget. Few were so ungenerous as to ask him during his honeymoon period whether there was an inherent conflict of interest between running one's own lab and deciding what monies should go to other labs that might or might not prove useful to his intramural investigations.[69] There followed shortly a new five-year blueprint for the project. The "old" NRC 1988 plan embraced by Watson had argued that the total sequence could be deciphered in 15 years (by 2005, that is) for $3 billion, or $200 million per annum. The project in Collins's first year was receiving $165 million. The mapping strategy was going strong (more on mapping presently); however, the sequencing part of the mission was lagging, ostensibly on account of inefficient, laborious technologies. Therefore, Collins envisioned $100 million a year for sequencing alone, which would get the program back on financial track. He noted that the *C. elegans* initiative had already paid dividends by demonstrating the extraordinary genetic fit among diverse species.[70] All this he announced before anyone would have predicted the Clinton-era economic boom.

As I see it, Collins's basic problem was not money. Rather, he came to NIH at a time when brain power, when technological sophistication, was shifting from the academic sphere to the commercial sphere to the point where even "star" professors were forging links with the business world in order to stay abreast of the game. Let us return to mapping. The prime methodological mover during the Watson era was not an HGP person; he was not even an American. My candidate for that distinction was CEPH director Daniel Cohen. The media description for him was "talented and ambitious." That is code language for "pushy" in the modern jargon of scientific sport, or should I also say political science. Cohen's plan was to take all the DNA culled from the many pedigree studies available to him—DNA covering the entire chromosomal gamut—and clone the sequences into the previously referred to YACs (yeast artificial chromosomes). Yeast, he thought correctly, could house much larger sequence arrays than could the standard viral or bacterial vectors. By 1992, Cohen had produced a rough physical map of the entire human genome through the medium of contiguous YACs, and he was dubbed "the Henry Ford of industrial genetics" because of his deployment of assembly-line robotics.[71]

Cohen's approach shocked American HGPers, who were mostly wedded to the chromosome-by-chromosome strategy. One of the most "talented and ambitious" of the Americans was Eric Lander, director of the Whitehead Institute at MIT. He formed a collaboration with Cohen to build an even fancier physical map, not one using Cohen's gene markers but one using the aforementioned sequence-tagged sites (not to be confused with Craig Venter's ESTs) that had proved so readily detectable by PCR.[72] Eventually, Lander and Cohen became partners in Millennium Pharmaceuticals, a biotechnology outfit committed to ferreting out disease genes. Meanwhile, Harvard's Walter Gilbert had forged an alliance with Mark Skolnick at Myriad Genetics, a small, highly specialized firm with support from Eli Lilly. Skolnick had decided that Cohen's YACs lacked stability as a consequence of unpredictable genetic rearrangements among clones, so he turned to a new medium called BACs (bacterial artificial chromosomes). With this new technology in hand, he reached BRCA1 before Collins and King, who were still using YACs. As Collins prepared to take the HGP helm, private capital was anticipated to provide about $80 million to the genomics experimental agenda in one form or another, with academics (armed with NIH money) holding equity positions in various biotech corporate arrangements.[73] So all in all, mapping was in good shape, yielding not only genes of importance but genetic *functions* of importance, such as the role of trinucleotide repeat mutations (CAGCAGCAGCAG) in causing fearful disorders; on the other side of the ledger, one didn't need an advanced degree in ELSPI to perceive inherent conflict-of-interest conundrums.

If Cohen and Lander are "talented and ambitious," what is Craig Venter? Enemies have gone overboard: "asshole," "idiot," "egomaniac," "extremely shallow," "only interested in [himself] and in making money," "morally wrong."[74] They make Watson's enemies almost seem benign. The question is whether the epithets bear any connection to his record of accomplishments and its scientific/ELSPI profile.

I have earlier recounted that genomics movers were not impressed with Venter's EST "breakthrough." Yes, part of the problem was NIH's determination to obtain patents on the fragments, not exactly Venter's fault. A bigger problem was simply that Watson, and later Collins, had accepted as legitimate the strategy of "you have to know everything to get everything of value," whether the "value" be "disease genes" or evolutionary insight. Nor was the world of commerce taken with Venter's strategy of "let's stockpile genes and tackle the rest later." For biotech firms, payoff was in profits. How profitable were ESTs?

In a story that has been recounted many times elsewhere, Venter departed NIH, with dashed hopes for support to develop further a viable EST program. For my purposes, the essential lineaments of the tale are these: Venter saw himself as a basic scientist, and the proof is that he rejected a $70 million job offer from Amgen; his hostility toward NIH was exacerbated when his wife, Claire Fraser, was denied tenure there because inhouse peer reviewers said they could not assess her work on its own terms even though she and her husband were not on joint projects, thus once again demonstrating that to know the seamy side of academic politics is to descend into the shameful; he received solace and vindication from every researcher's dream come true, a personal benefactor (Wallace Steinberg), who thought his program the yellow brick road to medical interventions through genomics mastery, and who gave him $70 million over ten years to do his work with but one string attached; his end of the bargain obliged him to turn the fruits of his labor over to a partner company that would perform the "dirty, commercial development chores" should opportunities present themselves. Enter from stage left William Haseltine, a Harvard Medical School cancer specialist and collector of more negative press (his one-time colleagues denounce him as a "money-grubber") than Venter himself. Steinberg put Venter, Fraser, and other Venter colleagues in charge of TIGR (The Institute for Genomic Research), which would be nonprofit, and he put Haseltine in charge of HGS (Human Genome Sciences) to make millions. Venter would sequence cDNA, and Haseltine would have six months to inspect the data for proprietary value and twelve months for possible drug spin-offs to present themselves. Haseltine would apply for patents and negotiate licensing terms; Venter could publish, and did publish, where Haseltine signed off. Academic scholars generally had access to the data if they paid a fee—as many did—and were free to publish subject to a waiver of all commercial rights to their discoveries, which would continue to reside with HGS.[75] One month after Collins came on board the NCHGR, SmithKline-Beecham, a pharmaceutical giant, shelled out $125 million for 7% of HGS and exclusive licensing entrée to Venter's cDNAs, and Hoffman-LaRoche swiftly followed by investing $70 million in Millennium Pharmaceuticals.

The burgeoning EST data bank quickly proved its worth to the larger scientific community. With full cooperation from Venter and Haseltine, Bert Vogelstein was able to locate the human MLH1 gene, a mutation of which causes some forms of colon cancer. Shortly thereafter, scientists using Venter's ESTs were able to track an Alzheimer's mutation to chromosome 1. The more ESTs were linked to particular chromosomes, the easier positional cloning became. Collins, however, flat-out rejected all suggestions to fold Venter's fruits into HGP missions. Instead, he joined with Merck—one of SmithKline's chief rivals—to create their own EST trove, with all sequence deposits to be made available without conditions to anyone. That meant that Venter could look at the Merck repository, but Merck scientists couldn't look at HGS-SmithKline's. This would not be the last time Venter would be accused of accumulating unfair advantages.[76]

One of Venter's strong points is the ability to gather loyalists around him. A second strong point is the knack of selecting potential loyalists with exceedingly innovative ideas. A third strong point is the capacity to formulate research strategies grounded in the melding of these innovative ideas. In other words, Venter is a great organizer armed with a keen vision for where things genomic ought to go.

At NIH, Venter worked closely with Mark Adams on making ESTs a functional reality, and when Venter went to TIGR, Adams went with him. In 1993, Venter made contact with Hamilton Smith, a Nobel Prize winner, whose work in harnessing the properties of restriction enzymes had helped usher in the recombinant DNA paradigm shift. Smith had the epiphany that through a process that he called whole-shotgun genome sequencing it would be possible for the first time to sequence a nonviral organism. He suggested for a model specimen the bacterium *H. influenzae*. His colleagues at Johns Hopkins, steeped in orthodoxy, turned a cold shoulder to his idea. Venter, however, was ready to do something the HGP had yet to do—provide a full and complete A–Z species DNA profile—and he jumped on board.

Smith's procedure worked essentially as follows: first, break up the genome into random overlapping pieces; second, clone them into BACs for multiplying and purifying; third, sequence them; fourth, reassemble them in a high-speed computer that would hopefully compare the thousands of chunks each to each and then match them via their overlaps. In a gesture bordering on chutzpah, Venter and Smith applied to NIH for subsidy while proceeding to the task on TIGR money. With 90% of the project completed, they received a definitive rejection. Why? The protocol wouldn't work! Wise heads might have conjectured that academic politics had struck again: that the NIH should give money to fund TIGR when TIGR was going to turn the fruits over to HGS was more than Collins's helpmates could stomach. In a year, the project was finished and the results published. *H. influenzae* contained 1743 genes, and practically no "junk," that is, noncoding DNA. More than 40% of these genes had no counterparts among all the genes to that point found in other organisms. If one wanted to observe the function of an unknown gene, then the obvious tactic was to knock it out of a clone and see what happened. Was it conceivable that the human genome—1,500 times larger—could be sequenced in more or less the same way?[77]

Gathering momentum, Venter's TIGR forces proceeded to sequence *M. genitalium*, an organism featuring the smallest nonviral genome known to exist. The project took a mere two months at a paltry cost of $200,000. The bacterium contained 517 genes, one-third the size of *H. influenzae*. Almost at once, Venter et al. realized a research agenda of staggering consequences presented itself: what was the minimum number of genes required for a life form's subsistence, indeed, for a life form's existence? After a series of gene perturbation maneuvers, the team concluded that such an organism required 250–350 genes for survival and that the functions of 100 or so of these could not be deduced. Nor did it escape the team's notice that a bottom-up methodology might be more efficacious than a top-down strategy: one might build an organism the way one built a house, gene by gene rather than brick by brick, the minimum genome being operationally defined to include only the bare necessities for replication and translation.[78] With a manageable number of genes in hand to commence model building, Venter imposed a moratorium on himself until he received a green light from a panel of bioethicists. He funded the committee himself and presumably selected—directly or indirectly—its members.

The panel's report, published in 1999, contains no startling conclusions but indulges in certain provocative presuppositions. Ethics and law should not be permitted to lag behind science, it says, or science will thrust new breakthroughs upon us without adequate societal preparation and scrutiny, an unacceptable state of affairs. The ways by which ethics, law, and science are to be held in equilibrium, I would argue, is a quintessential political science question, and one gets the strong impression that Venter's Ethics of Genomics Group is rather fond of the political tool Venter had prescribed for the present state of affairs—the moratorium. The minimal genome project represents a culmination in reductionist thinking, it says, reducing life to DNA, indeed to specific genes adaptive in specific environments. There is a danger, it continues, that scientists will define the meaning of life in a narrow sense disrespectful of the Aristotelian tradition and contemporary public opinion. By way of anticipating detailed commentary in my final chapter, reductionists need not claim, and this reductionist does not claim, that to discover minimum genomes will reveal the meaning of life. The question is what DNA is necessary to life, not what is necessary to a *good* life. In reviewing ethically defensible definitions of life, it says, there is a consensus among the major Western religions that "while there are reasons for caution, there is nothing in the research agenda for creating a minimal genome that is automatically prohibited by legitimate religious considerations." That is refreshing. But I question whether ethicists are making optimal use of their time by reviewing religious texts (no matter how cosmopolitan) and taking testimony from theological witnesses. Yet, these are the only empirical sources cited. And now the panel's conclusion: "The prospect of constructing minimal and new genomes does not violate any fundamental moral precepts or boundaries."[79] I must confess that I can't think of any either.

Toward the close of 2002, Venter decided to grab his minimal genome idea by the horns. He received a $3 million grant from DOE to begin serious work, somehow constructing a thousands-of-sequences-long stretch of base pairs that could survive when implanted in a cellular home. Would the artificial genome actually function? What genes should be included in the total package? What cellular outfitting would prove viable, if any? "All this is pure basic science," Venter said, though his ultimate aim was to create a pollution-fighting microbe. The beat goes on, as I write.[80]

Over and above their intrinsic relevance to the genomics/ELSPI equation, Venter's minimal genome explorations bear witness to the larger vitality and consequence of TIGR's agenda in stark comparison to the banal, workaday Human Genome Project view of its mission. Nor am I through making this point. While he was contemplating *M. genitalium* implications, Venter joined forces with the University of Illinois's Carl Woese to sequence *M. jannaschii,* an organism that lives only in the hottest of ocean-floor environments and that, upon its discovery in the 1970s, had become the subject of intense debate. Conventional evolutionary science teaches that life on Earth divides between prokaryotes (the bacteria) and eukaryotes (all the rest). Cytologically, the archaea, as the microbe is now commonly called, resembled a bacterium. Genetically, however, Woese demonstrated that it was much more akin to a eukaryote. His startling conclusion, which received front-page coverage: the archaea represents a third form of life. Geneticists tended to ignore Woese's work because he was not really part of the rDNA revolution, while evolutionists simply scoffed at its "heresies." Venter figured that only a genome-to-genome comparison would settle the matter, so he set about applying the whole-shotgun routine to archaea. After giving Haseltine a six-month window to inspect the data, he and Woese published in *Science.* The archaea turns out to be a genetic chimera: the DNA coding for metabolic activities was more closely akin to bacteria, whereas the DNA coding for proteins was more closely akin to the higher life forms. Taken as a whole, the data tended to confirm Woese's "third branch of life" thesis.[81]

Meanwhile, what were Francis Collins and the Human Genome Project up to in the middle 1990s? Daniel Koshland, in a lead *Science* editorial, led the cheerleaders. "The project is ahead of schedule and on budget," he proclaimed. "A massive project . . . cannot be run by continual democratic votes. . . . It needs leaders [and the HGP has] provided brilliant leadership."[82] At first blush, such optimism seemed well founded. In contradistinction to Venter, project sequencers "had adopted a conservative methodological approach." They took fairly small segments of DNA, already known to be part of particular chromosomes, chopped them into yet smaller sections, then sequenced and ultimately reassembled these.[83] In April 1996, the strategy paid handsome dividends. A team of 600 HGP scientists made a Venter-like announcement: it had sequenced the first-ever eukaryote, *S. cerevisiae* (the yeast), fully deciphering all 16 chromosomes and 12 million base pairs.[84] But storm clouds were gathering. Robert Waterston and John Sulston, with their joint U.S.–UK roundworm protocol well off the ground and yielding a surfeit of data, proposed that the HGP could reach its grand sequencing goal years in advance of the deadline and save a lot of money in the bargain by simply instituting a crash program. No new elegant technologies would be needed. The idea was audacious and contentious. Koshland, speaking for the Establishment, had praised the "mom-and-pop grocery store" model of "small science" laboratories doing their own genomic things. Waterston and Sulston were saying that that approach hadn't produced the computer systems needed to cut staggering sequencing costs. Koshland had implicitly praised the HGP brain trust for its dual mapping-sequencing set of priorities. Waterston and Sulston, undoubtedly hearing Venter's footsteps, were recommending that the HGP put mapping genes to chromosomes on the back burner.[85] The British jumped on the bandwagon when the Wellcome Trust, the private charity that had already poured money into Sulston's Sanger Centre laboratory, pledged another $75 million to cover the next seven years of human sequencing at 99.9% accuracy, not the 99.99% to which Collins was wedded.[86] Collins, however, was not sympathetic to Waterston's proposal when it arrived at NCHGR, though he obviously couldn't reject the gambit outright. Rather, his shop awarded multimillion-dollar grants to six centers, which in effect would test the claim that the technology of the moment was sufficient to do large-scale sequencing, though he held firm to the pristine 99.99% accuracy rate. Waterston himself was given chromosome 22, Lander was given chromosome 17, and, curiously, Mark Adams at TIGR was given chromosome 16. The projects in toto covered 3% of the human genome and were to be completed by 1999.[87] How Venter must have scoffed at such a turtlelike pace!

Through all this, the thorny question of data accessibility refused to go away. Waterston and Sulston were releasing their roundworm sequences on a daily basis, and Merck had given the former enough money to generate a public library of 450,000 ESTs. In 1993, the Human Genome Organization (HUGO) finally obtained sufficient capital to host a meeting, the site selected being Kobe, Japan; in 1996, the group gathered at Heidelberg, Germany. I had been named by Kay Davies of John Radcliffe Hospital and Oxford to be ELSPI program organizer, so I had a bird's eye view of the proceedings and the underlying tensions. It was the most exciting professional meeting I had ever attended, as ideas shot back and forth about how to snare the euphemistic holy grail. I learned that the European Union, which had largely sponsored the yeast project, was declining to provide immediate access to the final portion (20% of the total), insisting on a multi-week delay because "[We worked so hard to get it], we cannot just give this away [without screening it]."[88] Against this background, the Wellcome Trust convened a meeting of genome leaders in Bermuda, where the group agreed unanimously on the general propositions that sequence data should be released rapidly and that such information ought not be submitted for patenting. In a more perfect political world, the "U.N. of genomics," that is, the HUGO, would have orchestrated this process; instead, leadership fell to a bursting-at-the-financial-seams charitable "corporation" motivated by *its* perception of the public interest. Venter stated his reservations: he needed three months to prescreen the output; after all, Darwin was not asked to immediately place his collections in a museum. Nor could the NIH do anything to keep Waterston, Lander, and the rest from submitting patent claims, for the entire purpose of such congressional enactments as the Bayh-Dole Act of 1980 and the Federal Technology Transfer Act of 1986 was to encourage federally funded installations and their researchers to speed commercial development through patenting.[89] As if Collins didn't have enough trouble, news hit the public in late 1996 that one of his junior assistants had falsified data in research reports to which Collins had signed his name.[90] It was not so easy to run one's own lab and manage a proliferating human genomics bureaucracy.

In June 1997, the Venter-Haseltine marriage ended, and, naturally, the gossip-mill was rife with speculation as to why. Perhaps the "Gene Kings'" egos were no longer capable of reciprocal altruism. Or perhaps the pace of discovery had overtaken, in five years' time, the viability of their arrangement. Haseltine was running a business, but Venter's sequencing programs lacked clear commercial payoffs. Venter was doing basic science, and he chafed under the terms and conditions of Haseltine's holding actions. With the divorce official, Venter released to the public all cDNA and EST data in his possession, excluding those committed to HGS by prior agreement. Haseltine was left with one of the largest EST information banks in the world and one of the largest collections of gene patents and patents pending in the world.[91]

Over the next nine months the human genomics paradigm inched forward in explication. Even HGP insiders were becoming antsy. Collins isn't spending nearly the money necessary to achieve deadline, they bemoaned. The project was "too bloody complicated, with too many groups," Eric Lander opined. What was required was a well-focused, elite cluster that would grab the sequencing agenda by the throat.[92] He was right, but he was looking in the wrong direction.

At the Bermuda conference, Venter had heard James Weber and Gene Myers present a variation of Smith's whole-shotgun strategy to the HGPers and watched as they tore it apart. He took note but knew he lacked the sequencing technology to enter the human genome sweepstakes.[93] Manna from heaven arrived in the form of the ABI prism 3700 that Mike Hunkapillar and associates had mounted at Perkin-Elmer Corporation. The 3700 was obviously going to be the fastest gun in the DNA-sequence shootout, and Venter was floored when he previewed its features. Hunkapillar persuaded Perkin-Elmer to found a genomics company, and PE, of course, sought out Venter to head it up. Venter turned TIGR management duties over to Claire Fraser, and he announced the lineaments of his new company—Celera Genomics—on May 9, 1998. Venter said publicly that he planned to get the genome in three years at a cost of $300 million, projections that boggled the minds of the orthodox. To be sure, his version would be rough and ready compared to Collins's "book of life" figment, but the idea was to get the biologically important goods as quickly as possible, wasn't it?

The next episode in the unfolding melodramatic saga was as important for social psychological analysis as it would be for genomics research. One day before Venter proclaimed his grand scheme, he and Hunkapillar went to NIH to see Harold Varmus, Bernadine Healy's successor. They told him what they planned to do, and they made him a proposition: let's collaborate and share both the data and the glory. Varmus naturally said he'd have to check with Collins. On that very day, Venter and Hunkapillar met with Collins and presented the same deal. Now it was Collins who wanted time to digest what was happening. Venter's new Celera program surfaced in the *New York Times* the next day. NIH's response came at a press conference on May 11. Said Varmus: Venter will contribute much to the challenge of sequence discovery. Said Collins: Venter's initiative is significant. Not exactly the warm words of collaboration. On May 12 at Cold Spring Harbor, Venter told genomics gurus that he would compete and he would win. At yet another press conference, with Varmus and Collins in attendance, Venter suggested that NIH should concentrate its efforts on sequencing the mouse and leave *Homo sapiens* to him, the obviously adaptive division of labor!

It did not take the HGP establishment long to regain sure footing. Watson led the charge. "He's Hitler. This should not be Munich. [Is Collins] going to be Churchill or Chamberlain?" Collins responded in print by calling the human genome Venter-style a "Mad Magazine version." Waterston labeled it a "cream-skimming approach," hence "woefully inadequate." They were beside themselves because they suspected Venter would hoard everything he found and sell it to the highest bidder. Frankly, they were being slightly irrational on this score. Venter pledged he would release almost all the data free of charge every three months, seeking patents only on a few hundred potentially lucrative genes. Celera would make money largely by selling subscriptions to those interested in Venter's annotated database. Particularly interesting was the revelation that Venter's genome would be a composite of DNA drawn from five subjects, an approach that would yield a unique set of single nucleotide polymorphisms, that is, genes that exhibit different alleles in different people and hence hold the secret to phenotypic variation. He planned to market such data as well. All this was not exactly consistent with the Bermuda accord, but since when was the Bermuda accord part of the U.S. Constitution? Watson, Collins, Waterston, et al. were also beside themselves because they knew Congress would shortly be asking them very embarrassing questions: Why should we give you any more taxpayer money when he will beat you by four years? Why is he so fast and why are you so slow? Why is he brimming with confidence if he's as "all wet" as you say he is? These fears were not irrational.[94]

The race that was now brewing would not be won by words, of course. The Wellcome Trust got things going by promising to double its invest-ment in the project to the tune of $325 million, which would permit the Sanger Centre to sequence a third of the genome.[95] Four months later, Col-lins announced a new NHGRI[96] strategy in the form of a third five-year plan: the HGP would develop a "working draft" by 2001 including a state-of-the-art set of the sequences implicated in protein expression. The mint edition of the entire nucleotide set would follow in 2003 (two years behind Venter's schedule).[97] In other words, more speed, less accuracy, getting genes rather than sequences, and preemptive publication to trump any Venter patenting machinations were the new marching orders. Sydney Brenner—no Venterite—had been outspoken in his criticism of the Watson-Collins operation, calling it "a bit like Stalinist Russia" and badly in need of competition.[98] Venter had ordered 300 of Perkin-Elmer's 3700s at a cost of $300,000 per machine. What sort of unique scientific sporting event were we about to witness?

Venter commenced the event with two strikes against him. First, he was eight years behind in know-how and just plain old experience. Second, he was using a process with no proven track record in the elaborate game of higher-order species sequencing. And before he could even get started, Waterston and Sulston published the C. elegans genome, the first full ani-mal sequence yet elucidated, after an intensive eight-year struggle. The nematode turned out to be more complicated than anticipated, contain-ing 19,000 genes rather than the expected 15,000. Demonstrating the proof of the gene-conservation theorem, 30% of the roundworm's genes had human counterparts; yet, 12,000 of them possessed no known function. The nematode clearly would make a splendid model, as Brenner had long predicted, for studying the intricacies of nervous systems, for it featured a minimal set of 959 cells and 300 neurons. All in all, the organism had 97 million base pairs, eight times more than the yeast. The authors crowed that they would now turn their attention to sequencing 50% of the human genome. Without doubt, theirs was a major achievement.[99]

The year 1998 ended with the nematode announcement. The year 1999 began with a far more unlikely announcement: Celera was joining forces with Gerald Rubin and his Berkeley team, a well-funded NHGRI operation, to sequence *D. melanogaster,* the fruit fly. Collins had nice things to say about the collaboration, but who was kidding whom? *Drosophila* was two-thirds larger than *C. elegans,* and Venter boldly asserted that he would complete the job ten times faster than it had taken Waterston-Sulston. Rubin and his European colleagues had been working on the fruit fly genome for four years, and they had gotten 20% of the sequence. They needed Venter's technology. Venter needed a eukaryotic "winner," and much to the delight of the NIH grantees, he said he'd pay the $12 million freight to get the job done. Rubin would supply the 12,500 bacterial clones containing the DNA, and Celera would do the sequencing and piece fitting. Immediate release to the public was promised.[100]

The results were dazzling, as Venter played two new cards. He now had on board Gene Myers, whose whole-shotgun variant had been rejected summarily by NIH. Myers wanted a chance—as Venter had always wanted a chance—to show Bethesda up, and he was about to succeed in spades. Venter also had at his disposal the largest, most powerful computer in the world this side of Langley and Ft. Meade. This $80 million machine could perform 1.5 trillion floating-point operations per second and would be asked to do everything Myers had in mind by way of sequence fragment synthesis.[101] In March 2000, the team proclaimed it had deciphered 97% of the base pairs; as usual, various repeat "junk" sequences resisted decoding. *Drosophila* had 13,600 genes, and 60% of known human malignancy mutations were found to have fruit fly homologues. The researchers even found a counterpart to the human TP53 gene, the most common of mutated cancer genes. Said Rubin, a leading "establishment" figure: "Working with Celera has been one of the most pleasurable scientific experiences I have had."[102] Venter now posed a clear and present danger to HGP aspirations.

Nine months earlier, when Celera loomed only as a clear and *probable* danger, the NCHGR and the Wellcome Trust put their collective foot on the "race" peddle. Armed with Hunkapillar's Prism 3700s (yes, both Hunkapillar and Venter were technically Perkin-Elmer employees, but that didn't mean Hunkapillar couldn't sell to Collins!) and invigorated by an influx of new money, NCHGR made a command decision: put the lion's share of sequencing tasks in the hands of three centers. The "winners" were Waterston at Washington University ($33 million), Richard Gibbs at Baylor College of

Medicine ($13 million), and Lander at MIT ($35 million). The Wellcome Trust upped the ante for Sulston by $20 million, and these four centers along with DOE became known as the G-5 ("g" as in genome). Waterston would concentrate on 7 chromosomes, Gibbs on 3 chromosomes, Lander on 1 chromosome plus "leftovers," Sulston on 8 chromosomes, and DOE on 3 chromosomes. The "working draft" of the entire grail was to be finished by the spring of 2000, 18 months ahead of the previous deadline. To help grease the process, Collins even broke precedent by allocating a few million to two commercial outfits, thus blurring further the science-business distinction between genomic rivals.[103] None of the Continental laboratories were in on the planning. The days of "let a thousand investigator shops flourish" were over.[104]

The internationalists demonstrated they were still major players when a consortium of laboratories, with the Japanese enjoying a prominent role, announced toward the end of 1999 that it had sequenced the first-ever human chromosome, number 22. According to Sulston, the achievement was "as important . . . as discovering that the Earth goes around the Sun, or that we are descended from apes." Here were some of the salient corollary findings: the average length of a gene is 19,000 base pairs; naturally, there are genes uninterrupted by introns (i.e., junk), but many genes are interrupted, one being divided no fewer than 54 times; knowing that "CG islands" are often situated as a prelude to genes, the researchers located 550 of these constellations and the genes they introduced. The chromosome published was a composite of several individuals' DNA, not one, and the accuracy was of a very high order, only one miscoding per 50,000 letters. And yet, 30% of the hypothesized genes might ultimately prove spurious.[105] The chromosome 22 triumph resulted from employment of the HGP standard clone-by-clone approach. At this point, entering the new millennium, no prudent person would have selected a race winner.

As the uncertainty and tension grew, the acrimony grew as well and painted not a pretty picture of science as sporting phenomenon. Each side issued reports to the media purporting to demonstrate the inevitability of its ultimate triumph. Temporarily suspended were the peer review process and the confirmation of result by objective, independent assessment.[106] But things could get worse, and they did. Confidential discussions had commenced, with Lander serving as middleman. He thought the competitors could actually help one another. And why not? The mouse genome project, a $100 million enterprise largely shared by Waterston, Gibbs, and Lander, had opted to take up the whole-genome shotgun approach in tandem with the conventional HGP methodologies in an effort to speed things up, while

Venter admitted freely that Celera was consulting the publicly available data repositories to enhance its efforts.[107] No cigar. In what some bioethicists might term an ethical impropriety, the Wellcome Trust handed a letter to the media drafted by the HGP negotiators and addressed to Celera negotiators outlining the unacceptable Celera demands, the breach of confidentiality occurring precisely at the moment of truth in the process. Supposedly Celera was insisting that all data arising from the collaboration be subject to access terms, conditions, and fees for a period of five years, whereas the HGPers would approve such controls for no more than a year. In conclusion, the letter writers, presumably with straight faces, made bold to say that "publication of other groups' primary data without consent is considered to be a breach of scientific ethics."[108]

Into this bear pit of controversy stepped—of all people—President Bill Clinton and Prime Minister Tony Blair. Probably to assist the publicly funded team in its hour of need before the bar of public opinion, the two statesmen issued the following declaration:

> [T]he human DNA sequence and its variations should be made freely available to scientists everywhere. . . . We applaud the decision by scientists working on the Human Genome Project to release raw fundamental information . . . and we commend the other scientists around the world to adopt this policy.[109]

And yet the joint announcement also says: "Intellectual property protection for gene-based inventions will also play an important role."[110] Hadn't Venter been saying right along that basic researchers would have access to virtually his entire stockpile? Nonetheless, the value of biotech stock shares, including Celera's, nose-dived.[111] Almost lost in the noise was news from Japan that chromosome 21 had yielded up its full genomic contents. The shortest of all autosomes, it featured a mere 225 genes, not the expected 800–1000. Why the poor prediction?[112]

The scientific sports competition for the ages ended on June 26, 2000. To paraphrase the bard: all's well that ends well, more or less. President Clinton is quoted as letting it be known that the bombast had become counterproductive and that rapprochement was now the order of the day. DOE director Ari Patrinos gets the credit for achieving a meeting of the stubborn minds sufficient to convene a joint coming together at the White House presided over by Mr. Clinton and, via satellite, Prime Minister Blair. Collins and Venter shook hands and announced that the HGP and Celera, each in its own way, had essentially sequenced the human genome. Clinton and Collins invoked theological references; Venter stuck to science and made the important point that genomic reductionism and the uniqueness of the "human spirit" are not at tension.[113] The fact is, however, that the public version was unfinished. Collins had promised 90% of a rough draft and had delivered 85%. Venter had assembled 99% of the sequences and, quite obviously, had won the race had this been a true Olympic Games event. Clinton and Blair, however, were not Olympic officials. They were biased parties who wanted the race terminated and a draw declared, no matter whose work product stood where on the assembly lines. As for Venter, magnanimity meant the Nobel, if indeed he or anyone else in the race would ever get one. And according to Sydney Brenner, neither party was duly qualified. "The idea that this is a tremendous scientific accomplishment is simply ridiculous. It is an entrepreneurial accomplishment, a great managerial achievement, but there isn't any new science in it."[114] Glib and incisive even if inaccurate. Collins's and Venter's data sets agreed on one highly significant point among a battery of important points: the human genome sequence contained 30,000–40,000 genes, a far smaller number than the estimate generally accepted for decades, 50,000–100,000 genes. Clearly, the manner in which genes coordinate their functions rather than their sheer number played a greater role in explaining human behavior than had previously been appreciated.[115]

Any hopes of joint publication, not even to mention serious collaboration, were shortly dashed. Collins insisted that Venter deposit his fruits in GenBank for public consumption. Venter declined, reminding him that he was running an information-processing business and was accountable to Celera's shareholders. Collins gave his sequence to *Nature* with no strings attached. Venter gave his sequence to *Science* with several strings attached. These essentially gave noncommercial scientists a free hand if they used the raw data for their own purposes but restricted access by commercial outfits unless contracts were first negotiated.[116]

The race may have ended; still, there was the little question of the final election returns. In other words, which sequence was the more accurate, the better grounded in good science? Naturally, disputation arose. An independent team asserted, based on a comparison of chromosome 4 data, that Celera's handiwork had the edge: the HGP could show broader coverage, yet what Venter's team had gotten, it had gotten far more thoroughly.[117] Lander, Waterston, and Sulston proceeded to show to their own satisfaction that Venter would have never achieved his assembly without access to GenBank. Venter shot back that if Celera hadn't been misled by faulty HGP data, his assembly would have covered a much broader swath.[118] Interestingly, when it came to addressing the mouse genome, the HGPers had no compunctions about employing the whole-shotgun procedure. Also worthy of mention was NIH's decision to permit its genomicists to pay Celera a fee in order to gain access to the latter's annotated sequence sets, particularly that of the mouse.[119]

Of much greater import to serious genomics analysis than these little power skirmishes were the rafts of data emerging from sequence comparisons. I note a few: the average gene in our bodies is 3,000 base pairs; 99.9% of these genes are exactly the same in all of us; 50% of discovered genes have no known function; less than 2% of genes code for proteins; chromosome 1 (the largest) also has the most genes (3,000), while the Y chromosome has the fewest (231); humans have far more repeat sequences than the worm, fly, and puffer fish, yet we as a species stopped accumulating them 50 million years ago, whereas rodents continue to add them; the puffer fish has roughly the same number of genes as we do (!), and comparisons between the two genomes have helped locate 1,000 putative human genes; lastly, germline mutations in men are twice that of women.[120]

In January 2002, Craig Venter left Celera. The company decided it could maximize profits not through information dissemination but by becoming a Haseltine-type drug outfit.[121] Four months later and almost a decade after putting the EST methodology on the genomics map, Venter was elected to membership in the National Academy of Sciences. He had finally become "respectable."

C. The Constitution of the Competition: An ELSPI Summing Up

There is a natural tendency when analyzing the HGP-Celera race to the finish line to forget that many other countries on the planet were taking a deep interest in human genomics during the 1990s. They may not have had the science to make critical contributions, but they certainly had policymakers and ethicists who were prepared to discuss and even act on the scientific agenda. Every nation state has policymakers and ethicists! In a very fine review, Bartha Maria Knoppers and Ruth Chadwick describe the implicit ethical and legal norms—there were no explicit rules to speak of—embodying the unfolding international consensus as to the proper governance of human genomic research circa 1994.[122] They list five core principles: autonomy, privacy, justice, equity, and quality. Fortunately, they operationalize each with precision. By "autonomy" is meant "voluntary testing based on [fully informed] choice" and a right *not* to know the results. Moreover, only "medically therapeutic" genetic testing is appropriate. The term "therapeutic" is left to each nation state to define. By "privacy" is meant "confidentiality of genetic information." Yet, "family members at high risk" should have a right of access to the subject's relevant genetic profile. Knoppers and Chadwick's report notes that specific application of privacy issues to the contexts of employer-employee and insurer-insuree relationships is largely a function of "the presence or absence of universal health insurance and social security." By "justice" is meant protection of "vulnerable populations," such as children, the elderly, and "future generations." By "equity" is meant guarantees that human genomics will not increase "social inequality" such as "creating unequal burdens for minority ethnic groups." By "quality" is meant research competence and ethical oversight of experimentation. The report's only recommendation calls for "the codification of [these] principles in an international instrument," a sort of covenant or constitution that "countries could then interpret . . . in their own domestic legislation." The report specifically rejects "ad hoc country-by-country approaches." This recommendation is refreshing simply because it *is* a recommendation, something one can sink one's teeth into. Nonetheless, I think it contrary to good political science theory construction and probably wrong. What we need less of in the dicey business of bioconstitutionalism is top-down rule drafting and implementation. What we need more of is an empirically based bottom-up process of rule drafting and implementation followed by cautious generalization. At this early stage, each country should consider itself a laboratory, building its own consensus in its own way. UN declarations in other delicate matters do not present a mantra of warming comfort.[123]

A classic example of "deductive trouble" in this policy area occurred six years later, when Knoppers, now speaking in her capacity as chair of the HUGO Ethics Committee, presented a report on genomics benefit sharing. Given the dominant role of the "have" nations in the HGP, the panel saw considerable danger that only "rich people within rich nations" would emerge as winners. Fleshing out the fourth of her "consensus principles," she labeled this outcome "unfair and inequitable." Solution? Private companies profiting from the venture should voluntarily allocate 1 to 3% of said net profits to the health care programs of poor nations. After all, argued the report, we all share the same genome just as we all share the same air and space, so in the name of "human solidarity" an extra measure of equality in result is called for.[124]

Needless to say, the chances of Haseltine and others in the genomics business, however defined, tithing themselves and then sending the proceeds to Mugabe in Zimbabwe, Khadafi in Libya, and perhaps Arafat in some new Palestine state are not sensational. And just why should only the private sector foot the bill? What about the United States government, the governments of France, Italy, China, Japan, and the United Kingdom? And why not the Wellcome Trust, whose pockets Haseltine will never match? This political scientist would enjoy knowing a little more about the political attitudes and ideologies of the committee crafters. On the other hand, if a particular nation state tried out some benefit-sharing scheme of this sort, minimal damage would transpire if my premonitions proved out. Or perhaps I'd be shown the errors in my thinking. Then, and only then, could and should we move to the context of international rule formulation.

All of which leads to a discussion of ELSI genomics policy at NIH during the Collins years. What kinds of policy laboratory parameters were at play? It turns out that the NCHGR mission in this controversial area had become quite the center of discontent. After four years and $20 million, some observers were wondering what ELSI initiatives had accomplished. As I earlier recounted, the working group had organized conferences and study sessions to which the "usual suspect" bioethicists and biolegalists had made their appointed rounds. To this phenomenon, some bench genomicists remarked: "We've had enough of this Hastings Center stuff." What was needed, these critics emphasized, were scientific policy recommendations addressing real, tangible problem areas with appropriate legislative, executive, or administrative implementation.[125]

If Francis Collins shared any of these reservations, then such was surely not evident in the early going. When I discuss in a later chapter the politics of behavioral genomics, I will show how several well-known ELSI commentators display what I call conservatively a left-leaning bias on the question of governmental regulation of genetic information. These commentators believe, among other important things, that the only proper study of DNA is to ferret out "disease genes"; any other social purpose to which DNA might be put is, to use Justice Hugo Black's famous turn of phrase, "immediately suspect." During the 1990s, several of these spokespeople found their way to the ELSI working group. One was Harvard geneticist Jonathan Beckwith, whose views I discussed in chapter I. Another was Lori Andrews, who succeeded Nancy Wexler as ELSI chair. Andrews and Collins coauthored, with others, two important reports. In the first, the working group, having determined that some insurance companies had discriminated against people because of their genetic profiles, recommended that the use by insurers of DNA data to prescribe terms of coverage and eligibility be prohibited. In the second, the working group, having found that some individuals might decline genetic testing for fear of having the results used against them by employers and also having found that "there is no scientific evidence to link unexpressed genetic factors" and job performance, recommended that employers be barred from using DNA-based criteria to judge employee qualifications and accomplishment unless they could *prove* the criteria were "job related and consistent with business necessity."[126]

These findings paint with an exceedingly broad brush. They are based on the idea, the same idea undergirding the HUGO national redistribution recommendation, that genomic information is unique. I argue that in some contexts it may indeed be unique. Consider, however, insurance coverage. If I smoke, I pay higher rates. Does the ELSI working group call this bad social policy? No. Yet, my smoking may well be a consequence of genetic malfunction. Certainly smoking itself causes mutagenesis, as do countless other environmental stimuli. Now let's consider Alzheimer's. If a medical diagnosis shows I suffer from this malady, what difference should it make by way of insurance coverage if another test shows or doesn't show that I carry a certain mutation? If, on the other hand, I don't have Alzheimer's and a genomic examination shows I will likely get the fateful disease, why shouldn't a company be permitted to consider that fact just as it considers my smoking? Well, smoking is a "choice"; Alzheimer's is "genomic lottery." But a bad heart can be "lottery," too, and companies are permitted to adjust rates accordingly. Now consider on-the-job performance. Under the working group's theory, testing to ascertain genomic susceptibility to toxins is impermissible. This means that if I contract a DNA-based pathology in the workplace, I can sue the company that, had it had an opportunity to have me tested, could have protected both my health and its pocketbook. To this, the committee responded that the employer's first responsibility is to clean up hazards in the workplace, not examine genetically its workers. Known pollutants in the factory, sure. But what about pollutants known or unknown in the air or mineshaft? Put bluntly, these reports display a profound unreality about what it takes to run a profitable business in a day and age when government regulation is omnipresent. The slightly incredible fact is that not one member of the business community sat as a member of either panel.

According to Lori Andrews, the year 1996 found her in intermittent, discomforting disagreement with Collins. I will relate her tale, providing my own assessments as I go along.[127] When she left the chair's position in 1995, Nancy Wexler advised Andrews to extract a promise of funding from Collins for her own staff person; in other words, get out from under the thumb of NCHGR. Andrews admits she blundered by ignoring this counsel. My evaluation: once again, political naïveté surfaces at ELSI. Collins, says Andrews, began to pack the committee by adding someone "whom the Working Group had not nominated nor voted on," contrary to the panel's bylaws.

My evaluation: The RAC never had any role whatsoever in nominating or choosing its membership. Collins may well have noticed that the Beckwiths and the Andrewses were picking the Troy Dusters and the Dorothy Nelkins to succeed them. So Collins was temporizing. He should have bitten the ideologically loaded patronage bullet and repealed the relevant language in the ELSI charter. When Herrnstein and Murray published *The Bell Curve*, Andrews and cohorts wanted to expose what they called its racist geneticism. Collins tacitly resisted them at every stage, Andrews claims. Finally, she and Nelkin wrote a letter to *Science* condemning the book, which Andrews says teaches people to "believe that blacks were genetically inferior to whites." The letter itself deserves quotation:

> Genetic arguments cannot and should not be used to determine or inform social policy in the areas cited [in *The Bell Curve*]. [T]he lessons of genetics . . . do not provide useful information on deciding whether or not to pursue various programs to enhance the capabilities of different members of society. These decisions are moral, social, and political ones.[128]

My evaluation: Why should the HGP consider itself responsible for anything in *The Bell Curve?* This book never discusses genomics and rarely discusses genetics. The authors neither purport to be geneticists nor to have found any genes for anything. The fact is nobody knew then (if they know now) the extent to which genetic factors influence intelligence. As I remarked earlier, the working group as a whole viewed with the utmost suspicion any argument linking genetics to any behavioral phenomenon except what the whole world would call a disease. Undoubtedly, Collins was dismayed to have on his payroll ELSI commentators who had decided all scientific questions involving behavioral genetics a priori. Lori Andrews resigned her post having served only one year of her five-year term.

With the ELSI working group's role and status obviously now a flashpoint of disharmony, Collins appointed an independent panel to review NCHGR ELSI policy. What were required, said the panel, were three separate committees, not one. The working group would be replaced by a blue-ribbon body sitting in the Department of Health and Human Services, that is, above and beyond NIH. It would address deep genetics policy questions. The other two committees would be housed on the bureaucratic organization chart where the working group had been. One of these would run the ELSI extramural grant programs, and the other would be responsible for ELSI dimensions in NIH's intramural activities. Adding to the tangle of complexity were competition from the RAC, which had on its agenda the nexus between human genomics and human gene therapy, and a new bluer-ribbon panel than the one proposed in the committee report, namely,

President Clinton's National Bioethics Advisory Committee (NBAC), which figured to talk about things genomic.[129] These recommendations have been implemented, yet, to my knowledge, ELSI policy initiatives are mere blips on the radar screen, presidential commissions easily outstripping their efforts in visibility.[130]

And just what ELSI issues were sitting on Craig Venter's doorstep? Nothing nearly so esoteric as whether white people define "genetic information" differently than black people. Celera had taken one man's genome, smashed it into smithereens, reassembled the fragments, and then sequenced parts of the genomes of three women and another man of diverse ethnic strains in order to close gaps and fine-tune the end product.[131] The Human Genome Project was also building a "representative genome," an aggregation of DNA taken from four sources. When it appeared that researchers had not met appropriate standards of informed consent and that perhaps subject anonymity may have been compromised, NCHGR decision makers agreed on some explicit guidelines. Perhaps suffering from an overdose of Andrewsitis, the group decided that genomic investigators themselves could not contribute DNA (too "elitist") and that more female DNA was required because "women have historically been underrepresented in research," a form of affirmative action that Eric Lander thought had more to do with ELSPI than ELSI.[132] As for Venter, he announced, two years after the Clinton-Blair press conference, that the principal DNA contributor to his Celera sequencing achievement was he himself. What ELSI questions did that little informational tidbit raise? Assuming we don't swallow the "elitist" pill, is there any ethical principle at risk when a Venter constructs a genome using primarily his own genes? For me, the only caveat is that he had a duty to package his participation as research subject within the compass of his protocol so that it could receive peer review from Celera's institutional review board. Implicit in Venter's acknowledgment is that he satisfied this condition to the extent of listing his identity as one of many donors.[133] Less clear is whether he honored the protocol's terms when he selected himself to be primary DNA source. Perhaps the protocol was silent as to how the principal donor was to be chosen, in which case he was free to pick himself. As a matter of fact, according to Venter's own testimony, "[t]he decision about whose DNA to sequence was based on a complex mix of factors," among which he named "diversity" and various "technical issues." Not exactly informative, but one gets the impression he had carte blanche. If, however, the process of donor selection had been specified— suppose, for example, that the protocol had used the phrase "randomized selection"—then Venter, by picking himself, would have been on shaky ethical grounds. What about Venter's waiver of his own privacy rights? Had he made a deal with Celera's ethics people to preserve subject anonymity, and he unilaterally violated that condition, an obvious problem would have arisen. Privacy is not a one-way street. It is subject, as is all our rights, to the obligations that we voluntarily attach to it, not even to mention the obligations the Venters incur as members of honored professions. No such difficulties presented themselves here, given that, according to Venter, donors retained their prerogative to waive confidentiality.[134] Media potboiling to the contrary notwithstanding, the constitution of scientific integrity seemingly remained intact.[135]

And what of the constitution of the race itself? Keep in mind how I conceive of a constitution. It is a set of rules, norms, and understandings that orchestrate the patterns of political thought and action. Constitutions, for me, govern more than certain nation states and certain organizations public or private. They govern networks of people playing political games, of friend and foe jousting for power advantage. I am suggesting that the human genome race had its own constitution. How then does one articulate it?

Gary Zweiger provides us with a possible operational definition.[136] Noting a lapse in the role of authority structures, particularly legal constraints, he argues that the watchword of the race, what I would call the overriding constitutional principle, was anarchy. His is a field theory of researchers, entrepreneurs, politicians, bureaucrats, and corporations pulling and hauling to gain advantage, each player some combination of the fiercely individualist Jeffersonian yeoman planter and the Ayn Rand enlightened self-seeker. Zweiger is careful to point out that his anarchy is not a world of violence or even chaos; it is the world of Bakunin and Kropotkin, where peace reigns without the force of power, where the original ideas of science, the strategies of economy, and the checks and balances of politics arise spontaneously. Speaking to the genesis of the phenomenon, he says: "Might an educated and well connected populace, one that is willing and able to think and speak for themselves in real time, have unwittingly created anarchy? . . . Consider the bottom-up forces at work in our society today: Artists are free to establish their own styles. . . . Athletes can rise to superstar status overnight. . . . Ballot initiatives place key decisions directly in the hands of voters."

I have no doubt that there is a certain populism abroad in the land. Television has helped increase the hype, increase the cult of the personality, and pulverize yet further the structure and function of our two major political parties. The advent of television, the impact of McCain-Feingold campaign finance legislation (for better or worse), the artist struggling to place her works at the Whitney, are not, however, the world of human genomics. That world is the world of the Beltway. There, public funding agencies dispense largesse to the favored, who tend to do "mainstream" work as defined by disciplinary leaders. There, the NIH, with its bevy of advisory committees, institutes, and staff, often operating at cross-purposes, tries courageously to orchestrate national genomics policy. There, the Congress, disarmed by a plethora of overlapping committees and subcommittees, does not even bother to try to develop a national genomics statutory framework. There, the U.S. Patent Office, inundated with DNA materials and artifacts, and under extraordinary pressure from "pro-patent" and "anti-patent" forces as strident as "pro-choice" and "pro-life" antagonists, seems incapable of mounting a consistent, rational policy. There, a biotechnology private sector, which does what business does best when smart, aggressive people have some capital behind them, was able to exploit the muscle-bound, sometimes impotent, always too conservative power structure and get the genes and get the sequences before the rest.

There is a tendency, indulged in by Zweiger, to see Venter, Haseltine, and Incyte (a California-based HGS rival) as being outside the system to the extent there is a system. Years ago, T. J. Lowi, a well-regarded political scientist, wrote about the advantages of disorder.[137] That book is the closest modern American theoretical reference point I can find that approximates Zweiger. The differences, though, are critical. For Zweiger, anarchy is a descriptive and a normative condition; he even likens it to the adaptive capacity of the genome itself, which he calls a construct steeped in anarchy. For Lowi, disorder also carries descriptive and normative qualities; however, disorder is not an end in itself. Disorder is the proper means to achieve rule by law divorced from the play of special interests. Zweiger, on the other hand, applauds the lawlessness of the moment and asks only that it be extended. Lowi, correctly, would look upon Collins-Venter as merely another example of Beltway dysfunctionalism, where the system yields fruitful results only in spite of itself. Venter's struggle, really, was John McCain's struggle. McCain could not win conventional Republican Party endorsements, so he went elsewhere for support. He eventually lost the presidential nomination to George W. Bush, but in a certain larger sense he won. As testimony to that triumph, he led the charge against his party's establishment a second time, championing successfully the campaign finance legislation bearing his name. That triumph is a classic example of Lowi's juridical democracy, enacted in spite of, hardly as a result of, the Beltway Culture. Through it all, he remained a Republican. Venter did even better, and through it all he remained a genomicist, committed to the norms of the profession. It is the Beltway—complete with its anomalies, its blemishes, and perhaps even its tawdriness—that provides the essential bioconstitutional politics parameters of the race to decode the genome.

3

Cloning

A. Disaggregating the Genomics Paradigm

In chapter 1, I traced the evolution of molecular biology from the study of genetics, through the unfolding of recombinant DNA research, to the new world of genomics. In chapter 2, I provided an in-depth review of how the human genome itself became accessible. In chapter 3, I will begin the task of characterizing the ethical, legal, social, and *political* implications (the ELSPI) of the several parameters attendant in our unique genome. I will contend that these implications—properly ordered and placed in holistic perspective—make up the constitutional framework for human genomics. But first, this question: Just what are these "several parameters" that, I argue, constitute the human genomics research paradigm?

Figure 1 fleshes out the full ELSPI nature of human molecular biological research evolution. I depict the fields of genetics, recombinant DNA, and genomics as emerging one from the other and, again reading from left to right, I show how the various lineaments of human DNA investigation have similarly evolved.

As I see it, these lineaments should be organized into four fairly broad categories. Working within the germinal paradigm of molecular biology, scholars in human genetics sought to discover, and under certain conditions manipulate, single genes. One key focus of interest was the sexual reproductive process: artificial insemination became commonplace, and genetic selection through abortion and sterilization became possibilities sometimes acted upon either by the state or by the individual. Another key focus of interest was behavioral genetics. Molecularists searched, and sometimes found, such malignant recessive alleles as the mutations for sickle cell anemia and Tay-Sachs disease. There was also talk of finding

Figure 1. The Evolution of Human DNA Epistemology from an ELSPI Perspective

Paradigm I: Genetics	Paradigm II: Recombinant Genetics	Paradigm III: Genomics
A. Sexual reproductive interventions a) artificial insemination b) genetic selection through abortion and sterilization	A. Laboratory human life creation a) IVF preimplantation diagnosis b) embryonic stem cell experimentation	A. Asexual reproductive interventions a) therapeutic cloning b) human cloning
B. [Did not exist]	B. Human gene transfer a) somatic cell therapy b) somatic cell enhancement	B. Human genome transfer a) germline therapy b) germline enhancement
C. Behavioral genetics a) malignant recessives b) "genes" for intelligence and antisocial behavior	C. [No change]	C. Sociogenomics a) SNPs b) microarrays
D. Bioethical, biosocial, biopolitical, and biolegal policy frameworks	D. [No change]	D. Consilience a) genes, mind, and culture as coevolutionary circuitry b) a coevolutionary educational constitution

genes for IQ, criminality, and schizophrenia, and studies purporting to show hereditary causality were published. These sorts of research agendas attracted attention from ethicists, sociologists, political scientists, and legalists, with each discipline bringing to bear its own tools of the trade in providing appropriate insights.

Paradigm II in the evolution of human DNA exploration features recombinant DNA procedures. As new life forms could now be created by genetic synthesization in the laboratory, so human life itself could be created in the laboratory via in vitro fertilization. Fertilized eggs were and are studied and assessed for genetic composition, with the fate of the zygote often hanging in the balance. In daring breakthroughs arising anew under the umbrella of paradigm II, scientists learned how to transfer alien DNA into the human body, sometimes to effect therapeutic results and sometimes to test various hypotheses about how the body functions. People began to talk about inserting genes in order to make individuals smarter or prettier. The behavioral genetics research agenda expanded but did not undergo qualitative change, and the various academic disciplinarians talked more and more about the larger implications of what they saw.

The paradigm of today, which embraces a global study of *Homo sapiens'* DNA, presents us with a whole host of new research agendas, some frightening, all perplexing, but, I will argue, taken together, possessing unprecedented potential to unify competing theories of who we are and how we think and behave. The first of these breakthroughs takes humans from the world of sexual reproduction to the world of asexual reproduction. In this universe scientists manipulate unfertilized eggs to replicate extant genomes either for therapeutic or reproductive purposes. In this chapter I shall provide an in-depth ELSPI treatment of these cloning procedures and their possible consequences. Paradigm III also moves us from the controversial realm of human gene transfer to the far more controversial world of human genome transfer. It is now conceivable that scientists could insert or perhaps remove DNA in the germline, thus altering the genomes of individuals yet to be and their children and their children's children. The world of behavioral genetics also takes a quantum leap forward under paradigm III, as now several investigators employ the theory, tactics, and strategies of what I shall call sociogenomics, that is, describing and analyzing the impact of an array of genetic precursors on behavioral repertoires and vice versa. Biologists are flooding data banks with thousands of single nucleotide polymorphisms (SNPs), mining the genome for alleles that distinguish human from chimpanzee and human from human. Malignancy in many forms is more and more often seen not as a genetic disturbance but a genomic disturbance, with several perturbations allied in damaging effect. Evolutionary theory itself—and especially hominid evolutionary theory—has already undergone more precise explication as we learn how the genome has adapted and perhaps maladapted to changing environmental stimuli. For our species over the past 100,000 years, the most prominent environmental stimulus has been culture. We now must study how the evolving genome has orchestrated evolving culture including political culture and, more speculatively, study how evolving culture including political culture could be working change in our genomic constitutions. All of these subparadigm shifts, some have argued and I shall also argue, have altered fundamentally ELSPI investigation of human DNA phenomena. We take the position that a grand synthesis in human understanding of social phenomena under the banner of genomics is now discussable. I shall submit that this grand synthesis, called by many of us consilience, must form the bedrock substance for the long overdue revitalization of the political science discipline. And now, on to cloning!

B. Sexual Reproduction and Genetic Ideology

D. J. Kevles has documented in precious detail the history of the eugenics movement.[1] During the period before recombinant DNA, many leading socialist theoreticians argued forcefully in favor of voluntary eutelegenesis, that is, artificial insemination, as a means to improve the human gene pool. The technique had become commonplace during the nineteenth century for animal breeding, and practitioners had achieved occasional success in the case of women whose husbands were sterile. However, the procedure was highly inefficient, resulting in pregnancies in about one of every three attempts. Evidently, it was simply too difficult at that time to keep sperm functional outside the male body. The political underpinnings of the movement were clear enough. Said one spokesperson:

> [The aims of eutelegenesis] *are* socialism, biological socialism. . . . They involve nothing less than a socialization of the germ plasm, the establishment of the right of every individual that is born to the inheritance of the finest hereditary endowment that anywhere exists.[2]

This "biological socialism," however, seemed to critics highly sociobiological and highly sexist. Its guiding principle was that through artificial insemination one brilliant man could impregnate millions of women, resulting in greatly enhanced chances of producing male geniuses. That sperm was rather more plentiful than eggs bore witness, it was held, to the inherent "fitness" advantage of those carrying the Y chromosome. George

Bernard Shaw, Julian Huxley, and C. P. Snow, among other intellectuals of the day, rallied to the cause.

After World War II, with vast improvements in sperm storage and preservation at hand, Nobel Prize winner and longtime socialist Hermann J. Muller led the charge for a revitalized eutelegenesis. According to Muller, the elevated levels of radiation arising from atomic testing and kindred stimuli had spawned in humans a plethora of deleterious mutations the totality of which he called "genetic load." As horrified as most everyone else by the Nazi coercive birth control programs aimed against non-Aryans, Muller advocated a crash voluntary program that he called "germinal choice." He formed an alliance with Robert K. Graham, a wealthy businessman, to establish the Repository of Germinal Choice in Escondido, California. Muller tried to sell the program to his friends and colleagues, looking particularly for sperm donors who ranked high in intelligence and altruism. Graham, however, was no socialist; he was a political right-wing type who wanted to produce superior minds, not necessarily nice people. Muller saw the handwriting and dropped out. Graham opened his facility in the 1970s, collecting sperm from several Nobelists, all of whom retained anonymity except for William Shockley, who proudly proclaimed his affiliation. Graham, not as sexist as the old-guard socialists, tried to screen females as well, favoring egg donations from healthy, smart prospective mothers. Aldous Huxley, who had written about state-controlled mechanisms of just this sort, thought the whole idea outstanding, while Francis Crick added his unbridled enthusiasm.[3] Twenty-five years later, we may note that there is not exactly a hue and cry to clone centers for germinal choice.

Controversial as this brand of thinking may have been even then, it paled in comparison to the tumult over state laws mandating compulsory sterilization for the "genetically unfit." To provide political backdrop, one should recall that in 1924 Congress passed the Immigration Act, severely restricting entry into the United States by citizens of Eastern and Southern European countries, not to mention citizens of non-European nations. After all, said Calvin Coolidge: "America must be kept American. Biological laws show . . . that Nordics deteriorate when mixed with other races."[4] Mr. Coolidge failed to provide scholarly citation for his observation. Speaking specifically about the feebleminded, Winston Churchill opined: "[They should be] segregated under proper conditions [so] that their curse died with them and was not transmitted to future generations."[5] In this sort of social environment, one ought hardly be surprised to learn that by the time the United States entered World War I, 17 states had enacted sterilization laws, and about 8,000 people were "eugenically sterilized" between 1907 and 1928.[6]

The constitutionality of these laws was tested before the Supreme Court in the famous case of *Buck v. Bell.*[7] The Court's opinion was written by Justice Oliver Wendell Holmes, for himself and seven others including two of the "greats" by any standard, Louis Brandeis and Harlan F. Stone. Holmes's statement upholding the Virginia statute at issue cannot possibly be understood without a close look at his tenure on the bench and his personal views on the subject before him. For twenty years, Holmes had filed compelling dissent after compelling dissent, arguing against the tendency of a Court majority to imbue the word "liberty" in the Due Process Clauses of the Constitution with substantive meaning, hence making it impossible for the Congress and the states to legislate on the policy question at issue. For example, in *Lochner v. New York* and *Adkins v. Children's Hospital,* the justices had ruled that statutes regulating such working conditions as minimum wages and maximum hours violated "liberty of contract" and hence were invalid. By "liberty of contract" was essentially meant laissez-faire economics, a particular theory of the marketplace that the Court had decided was subsumed under the term "liberty." For Holmes, statutes regulating liberty were entirely permissible if a "reasonable man" would consider them necessary and proper, assuming, of course, that government had honored relevant procedural amenities. In other words, Holmes presumed in these sorts of cases in favor of majority rule; his colleagues presumed in favor of minority rights. But, and this is critical, the right at issue in these controversies was not a right specifically set out in the document (e.g., freedom of speech), and it was not a right that anyone could reasonably say was imbedded in the traditions of our people or our law.

In *Buck,* the question was whether the state could sterilize a woman who was an inmate in a mental institution after which she would be released to make room for others. The state readily admitted she could fend for herself outside the bounds of custodial care; however, she had been adjudged feebleminded, and the state did not want her to spread her putative defective genes to offspring. Virginia had also introduced at the trial evidence purporting to show that the petitioner was an illegitimate child of a feebleminded mother and that she herself had given birth to an illegitimate feebleminded daughter.

The issue before the Court, once again, was the substantive meaning of "liberty," and Holmes had no intention of engrafting into the due process guarantees the notion that a person has a fundamental right to reproduce whenever and under what conditions that person sees fit. Holmes also saw it far beyond the Court's competence to take issue with "genetic facts" as Virginia had developed and construed those facts. Whether the petitioner was actually feebleminded and whether feeblemindedness was a hereditary characteristic were not for judges to ponder. If the law followed from facts beyond scrutiny, and the law was reasonable, the law constituted no violation of liberty. It so happens that Holmes himself was rather sympathetic to eugenic ideas, and unfortunately he compromised his opinion unnecessarily by inserting that predilection at a critical moment:

> It is better for all the world, if instead of waiting to execute degenerate offspring for crime, or to let them starve for their imbecility, society can prevent those who are manifestly unfit from continuing their kind.

Today, *Buck v. Bell* has few defenders, though it has never been overruled. And Holmes himself is no longer treated with the respect tendered him when I was learning my constitutional law. That is because we live today in the post-1960s "rights revolution," during which period the Supreme Court, under the leadership of Justice William J. Brennan, has taken upon itself the prerogative, which Holmes decried, of announcing the discovery of new substantive rights to be found within the compass of that verbal talisman, liberty. To appreciate these nuances in constitutional jurisprudence will assist us greatly in providing perspective for the cloning debates.

The Court revisited sterilization policy questions for a second and final time during World War II. In *Skinner v. Oklahoma,* the justices voided, by unanimous vote, the state's Habitual Criminal Sterilization Act.[8] This law defined habitual criminals to include three-time convicted felons whose crimes involved "moral turpitude" and who had been imprisoned on each occasion. The statute provided exemptions for embezzlers, "political offenses," and tax violators, that is, certain white-collar criminals. The case could have been—and should have been—decided on straightforward equal protection grounds. Skinner had been sent to jail three times for stealing and robbery. He was slated for sterilization. Had he been convicted three times and sent to jail for the same number of years on grounds of being a "habitual embezzler," he would not have been a candidate for sterilization. Such a classification seems quite clearly to be arbitrary and invidious, precisely the reason why the Fourteenth Amendment's equal protection guarantee is a critical civil liberty.

Speaking for the Court, Justice William Douglas, like Justice Holmes before him, could not resist *ex cathedra* commentary. Always prepared in his decision making to paint a picture rather than work a sum,[9] Douglas presented the core issue as follows:

> This case touches a sensitive and important area of human rights. . . . We are dealing here with legislation which involves one of the basic civil rights of

man. Marriage and procreation are fundamental to the very existence and survival of the race. [In our view] strict scrutiny of the classification which a state makes in a sterilization law is essential.

In other words, the burden of proof lies on the state to show that its law was necessary and proper to achieve a legitimate interest. The usual burden of proof, falling on the individual, provides insufficient protection because the law implicated "human rights." And the Court is well within its authority to ascertain what rights are human or "civil rights of man" and import these findings into the Constitution, whether the locus of analysis be the term "equal protection" or the term "liberty." It is impossible to square Douglas's analytical approach in *Skinner* with Holmes's analytical approach in *Buck,* though Douglas specifically said *Buck* stands as good law. A final point for future reference: *Skinner* does not say that procreation is a fundamental right; it says "marriage and procreation are fundamental."

During the 1960s and 1970s the judicial scene shifted from sterilization to contraception and from contraception to abortion. As we saw, the sterilization cases bore directly on disagreements as to appropriate "genetic policy." The contraception and abortion cases seemingly have nothing significant to say about matters hereditary, and yet the Supreme Court's jurisprudence has ripple effects of clear relevance to paradigms II and III. In the *Griswold* case, we learn that a right of privacy—located in the confluent penumbras of several specific constitutional liberties—protects married couples using contraceptive devices from the force of a state law criminalizing the employment of such artifacts.[10] In the *Eisenstadt* case, we learn that unmarried sexual partners possess the same constitutional rights as married partners in reproductive choice making. If the latter have a right to use contraceptives, so do the former. The state has no valid interest in attempting to discourage premarital sex, said the majority; in fact:

If the right of privacy means anything, it is the right of the *individual,* married or single, to be free from unwarranted governmental intrusion into matters so fundamentally affecting a person as the decision whether to bear or beget a child.[11]

So now, procreation *is* a personal right, a substantive liberty protected by the Constitution, and legislative bodies must meet an elevated standard of proof in order to rebut an individual's claim of deprivation. The central teaching of *Eisenstadt* is that singles have a constitutional right to have sex and therefore children.

The year 1973 saw the justices unleash not only their most far-reaching decision on this subject yet but, I think it fair to say, the most politically charged ruling since the school desegregation case of 1954. I, of course, refer to *Roe v. Wade*.[12] In this case, we learn that the right of privacy (termed here an artifact of "liberty" in the Due Process Clauses rather than in some mysterious, elusive penumbra) includes a woman's right to abort her pregnancy. This procreative choice, we are told, is absolute in the first trimester, and the state can only criminalize abortions in the third trimester (but not when the woman's health is at risk) when best medical evidence says the fetus can usually survive outside the prospective mother's womb. That is, only in the third trimester can the state satisfy the weighty burden put upon it when the Court proclaims—as it here does proclaim—the liberty interest a substantive right. This extraordinary threshold is now referred to as demonstrating a compelling state interest. But what about the argument that the embryo—not to mention the fetus—is a person? Put more modestly, what about the argument that life begins at conception? Said the majority:

> We need not resolve the difficult question of when life begins. . . . [W]e do not agree that, by adopting one theory of life, [a state] may override the rights of the pregnant woman that are at stake.

In *Buck v. Bell*, Holmes had eschewed judicial inquiry into definitions and causalities of feeblemindedness, letting Virginia carry the policymaking ball. In *Roe v. Wade*, the Court eschewed judicial inquiry into when life begins, but instead of deferring to legislative judgment in the matter, it trumped that judgment with substantive due process. As for the man who provided the sperm to make the pregnancy in *Roe* happen, the Court was unable to discover in the concept of due process liberty any right whatsoever relevant to the occasion.

Roe has been called the product of Democratic Party, liberal, feminist thinking. The claim has little support. A majority of the *Roe* Court (five of nine) were Republican, and a majority of the *Roe* supporters (four of seven) were Republican. Not a woman had as yet made her way to the American judicial pantheon. In time, however, *Roe* has obviously produced a partisan split. It is hard to find pro-*Roe* Republicans any more and almost impossible to find anti-*Roe* Democrats. This cleavage has influenced greatly the nomination and confirmation odds of prospective federal judicial appointees, not merely Supreme Court hopefuls. Republican presidents generally select anti-*Roe* nominees, and Democratic senators look for ways to deny confirmation. Democratic presidents generally select pro-*Roe* nominees, and Republican senators look for ways to derail confirmation. It has become a dirty political business unparalleled in American judicial politics. And it has had a profound impact on the cloning debate.

The final episode in the Supreme Court's reproductive liberty saga of relevance came in 1992. Eight years of Ronald Reagan and four years of George Bush in the White House had given the nation a new Supreme Court. In the *Planned Parenthood* ruling, the justices scuttled virtually everything in *Roe* except its "central holding."[13] Gone was the notion that abortion was a fundamental right; gone was the notion that the people's representatives must show a compelling interest before they could regulate abortions; gone was the trimester paradigm spelling out a woman's rights during the three-month increments of pregnancy. What was left standing was the woman's ultimate choice to eliminate the embryo/fetus during the first six months, with the state now permitted to prescribe times, places, and manner unless these constituted "undue burdens." As for men, they continued to be non-persons in the abortion policymaking arena, as states were specifically forbidden under the Court's reading of the Constitution to provide for spousal notification. What with the Court itself splitting 3–2–4 on most of the salient points, abortion constitutional law resides in a kind of limbo.

What impact does the *Roe* jurisprudence have on genetics? As screening techniques constantly improve, the pregnant female has an ever-increasing opportunity to know in advance of the third trimester all about the genetic strengths and weaknesses of the embryo/fetus in her womb. Does it carry the Tay-Sachs mutation? Does it carry the chromosome 21 trisomy? Does it carry DNA where lurks what one scientist has called the "gay gene"? Do I eliminate it or bring it to term, thousands of women will ponder in future years. And then there is this little hypothetical scenario: If all the mentally competent, adult females got together and signed an agreement that they would abort every embryo carrying the Y chromosome until their gender achieved sufficient numbers to break the back of what they call the male power structure, and they further agreed that they could not achieve their aim until those numbers constituted 75% of the voting population, then under *Roe* their choices in this matter would have full constitutional legitimacy.[14] Of course, the situation is only hypothetical. But this observation is not hypothetical: When various commentators (including yours truly) argue that the human genome belongs to all of us, that submission runs smack into *Roe v. Wade*. Under American constitutional law, the American genome belongs to pregnant women. I can't help thinking: What would Justices Holmes, Brandeis, and Stone have made of all this?

C. Assisted Sexual Creation of Embryos

A leading authority, writing 15 years ago, found that the practice of in vitro fertilization (IVF) fell into three distinct developmental periods:

"exploration, consolidation, and expansion." The period of exploration ran from 1969 until 1978, culminating in the birth of the first "test-tube baby," Louise Brown. The period of consolidation lasted from 1978 until roughly 1989. During this period "basic IVF" took hold and was virtually the only form of the new procedure in use. By basic IVF was meant fertilizing a wife's fresh eggs with a husband's fresh sperm and implanting the zygotes immediately into the wife's womb. The "expanded IVF" period of today features new technological wrinkles: egg, sperm, and embryo freezing; gamete donation by parties other than the married couple; IVF as a transaction between unmarried heterosexuals; IVF as a transaction between homosexuals male or female; surrogate motherhood, where the pregnant woman may or may not be the female genetic participant; and genetic diagnosis of embryos.[15] Right at the outset, let me make clear a terminological caveat: from now on, when I make reference to a fertilized egg sitting in a petri dish, I will use the word doctors of reproductive medicine use, which is "preembryo." I will use the term "embryo" to mean a fertilized egg that has been implanted by technology or natural process in the uterus. My definition of embryo is the dictionary definition. The word "embryo" has been expanded by pro-lifers to include zygotes, blastocysts, and all preembryos no matter what their cell counts in order to argue that to destroy a fertilized egg outside the female body is in fact to murder an embryo/human being. Science rather than political ideology should dictate scientific nomenclature.

The period of exploration was fraught with just as much disputation as has been any decade of expanded IVF. On the one hand were wives unable to have children, in a sense crippled by blocked fallopian tubes; reproductive specialists were trying to help them by learning how the process of fertilization worked in the laboratory. On the other hand were bioethicists Leon Kass and Paul Ramsey denouncing the procedure as destructive to all sorts of "natural" things: marriage, family, and the well-being of embryos and the children arising from them.[16] James D. Watson chimed in by calling IVF the initial step down the slope toward human cloning.[17]

Even before Louise Brown was born, the federal government entered the IVF field, achieving, unfortunately, counterproductive results. Fearful that researchers would apply to NIH for grants to study the procedure, HHS adopted the policy that the review process must include an ethical assessment by the department's ethics advisory board.[18] In principle, this policy made sense. The reproductive science issues were palpable, but the ELSPI issues were palpable, too. The Supreme Court had decided *Roe v. Wade*. *Roe* had talked about the rights of pregnant women; it did not say anything about zygotes produced in vitro. *Roe* could be interpreted in two very different ways. If the critical parameter was the female's privacy right inhering in her bodily integrity, then *Roe* was inapposite to the in vitro context. If, however, the critical parameter was the female privacy right inhering in the procreative process from beginning to end, then *Roe* was certainly apposite to in vitro issues. Even if *Roe* were relevant to providing a constitutional frame, it by no means followed that government had a duty to fund research on preembryos any more than it had a duty to fund abortions.[19] An ethics advisory panel could play a key role in fleshing out acceptable ELSPI guidelines for such funding.

Pierre Soupart wanted to study chromosomal functions in preembryos and applied to NIH for support. When finished with his experiments, he planned to discard the zygotes, taking the view that they might not be fit for implantation. The proposal was approved in 1977 by a committee of peer reviewers. He then had to wait, and wait, and wait for the newly appointed EAB to act. Interestingly, one of the questions that Secretary of HHS Joseph Califano put to the panel was this: "Will this research lead to selective breeding, to attempts to control the genetic makeup of offspring?" The final EAB document was eminently sensible: IVF was perfectly acceptable for marrieds; research on preembryos to ensure for improved safety was equally acceptable; the preembryo experimental subject should be destroyed in its fifteenth day of existence or sooner, in other words, before it developed a "primitive streak." "[T]he human embryo," the report concluded, "is entitled to profound respect," and yet the interests of prospective parents and scientific researchers rank higher. Incredibly, HHS never funded Soupart's protocol in spite of all these positive steps and signals.

For the next ten years, not one IVF grant proposal appeared on NIH's doorstep. HHS was flooded with letters condemning the procedure, and scientists had gotten the message. Members of Congress also offered their views, and those opinions were overwhelmingly negative. Califano's successor, Patricia Harris, disbanded the EAB. The first IVF center in the United States had yet to be established, and already the Beltway was calling preembryo study and elimination the "new abortion."

While IVF has gone relatively unregulated in this country, the United Kingdom far surpasses the United States in technologic and innovative know-how. Here, the environment is one of benign neglect; gynecologists can do their work, but research funding is hard to come by. In the United Kingdom, Parliament established the Human Fertilisation and Embryology Authority in 1990 regulating how IVF clinics are to conduct their business, but research funding is readily available. In 1993, the United Kingdom's leading experts in assisted reproduction surprised many when they noted their lack of success. Pregnancies resulted in only 12.5% of cases. Quite obviously the field then and now requires a full research plate in order to improve safety and efficacy. During this time, experimental advances had brought to the forefront a second compelling reason (the first being female tubal blockage) to essay IVF: removing from the procreative process the genetic lottery of harmful mutation dissemination. In the United Kingdom doctors had developed a tool called preimplantation diagnosis. By snipping off a blastome (cell) from a blastocyst, that is, the preembryo at the eight-cell stage, they could check for chromosomal abnormalities by employing the FISH (fluorescent detection of in situ hybridization) technology, and they could spot hundreds of gene defects by using PCR. Sex identification was also possible, an important improvement on account of the prevalence of X-linked diseases. Of course, human genomics will ultimately provide a global roster of these DNA irregularities.[20]

Throughout the 1980s, the abortion wars continued to take a heavy toll on enlightened embryonic and genetic policymaking. Congress had authorized creation of the Biomedical Ethics Board in 1985, consisting of six senators and six representatives and equally divided as well between Democrats and Republicans and representatives of pro-*Roe* and anti-*Roe* factions. This was a pretty serious initiative; we had here not the voice of a single executive branch department but the voice of Congress. And the agenda was high profile with "human genetic engineering" the lead item. There was hope for some great debate. Forget it. The board took three years to name a group of experts who would do the actual spadework, and when that panel was finally assembled, it never really got off the ground. After all, what institutional support could it count on?[21]

One of President Clinton's first initiatives in the science policy area was to announce a reexamination of the ban on federal funding of human preembryonic research.[22] The Clinton Democratic Congress took the lead in June 1993, removing from the Code of Federal Regulations the HHS-authorized language making it impossible for the NIH to award grants in support of IVF experimentation. The new provision also vested with the secretary of HHS the power to create ad hoc ethics advisory panels to assess provocative protocols. Considering that the NIH had lacked any de jure ELSI capacity in this research area, the enactment seemed a much-welcomed step forward. But what guidelines would inform the new process?

Recognizing the problem, Mr. Clinton's new director of NIH, Nobelist Harold Varmus (a liberal Democrat, who had earlier been named to replace President Bush's director, the conservative Republican Bernadine Healy), convened the Human Embryo Research Panel (HERP) and charged it to work up a set of recommendations. The committee consisted of notable biological scientists, university administrators, biolegalists, and bioethicists. Passing without notice was the fact that the committee—like the NIH genomics working group—lacked ELSPI (with a "P" for political) representation, an omission that would make itself evident in the turbulent months ahead.

The HERP's *Final Report* appeared on September 27, 1994, and on December 1, 1994, the panel met in public session with Director Varmus and his advisory committee. I attended that meeting. The salient HERP recommendation was that the NIH entertain grant proposals to study preembryos created solely for research purposes. Having published its report well in advance of the committee meeting, that cat had long since escaped the bag. The pro-life community was up in arms. The Catholic Church, an anti-*Roe* mainstay, considered IVF for *reproductive* purposes ethically unacceptable even for married couples who desired to expand the Catholic gene pool. The fall-back Catholic Church argument, certainly consistent with the Protestant fundamentalist consensus, was that once preembryos had been created, they deserved as much respect as embryos, which, in turn, deserved as much respect as fetuses and babies.

Even before the November 1994 midterm elections, 30 House members had made it clear they would do their best to block the committee report's legal force and effect. With Republicans sweeping the country one month before the Varmus-orchestrated meeting, and with key congressional leadership posts now in GOP hands, the HERP's key recommendation was in great danger.

At the December get-together, HERP chair Steven Muller excoriated the political pressures placed on his panel. In testimony before the HERP, person after person had confused preembryos with embryos, he charged; they had thought incorrectly that the states had made illegal the research practices under review; and they had no idea public funding for preembryonic experiments had long since been approved in Australia, Canada, and the United Kingdom. For Muller, the critical question was whether university scholars would be able to do the research then performed only in private nonfederally funded facilities. Science cochair Brigid Hogan added that the HERP was endorsing research on preembryos no older than 14 days, when cells had not yet begun to differentiate. Was it not better, she asked, to spot cystic fibrosis (CF) mutations before implantation than to spot them two months later and abort the embryo? Also, women undergoing anticancer therapy could have their eggs removed and frozen for later in vitro usage, thus avoiding possible damage. As for the criticism that the panel supported cloning, the committee had come out foursquare against twinning and the creation of chimeras. And yet, she pointed out, tests had shown that mitochondrial DNA in the eggs' cytoplasm might carry defects, so the HERP had agreed that nuclei from such cellular environments could be removed and placed in another woman's "emptied" blastocyst. Bernard Lo, a third committee member, noted that while the panel had approved parthenogenesis studies for funding, the transfer of parthenotes was forbidden.

During the question-and-answer session that followed, political issues predominated. Nannerl Keohane, the president of Duke University, said: We made a stupid mistake. Our committee should have called itself the Preembryo Research Panel, not the Embryo Research Panel. Steven Muller expressed the view to the director's advisory committee, which would be voting the next day: We bit the political bullet. Will you stand with us? Dorothy Nelkin, another committee member, looked Varmus's group straight in the eye and said in so many words: Don't back down.

On December 2, the Varmus advisory panel unanimously voted to support the HERP report fully and completely. Varmus responded that he would give the entire matter the consideration it obviously deserved. Eight hours later, President Clinton, who had just concluded a long transpacific flight from Jakarta to Washington and who would shortly embark on a long transatlantic flight from Washington to Budapest, issued a statement to the press. "I do not believe," he announced, "that federal funds should be used to support the creation of human embryos for research purposes, and I have directed that NIH not allocate any resources for such research." While the rest of the recommendations in the HERP report went unmentioned, the fair inference was that Director Varmus could craft guidelines orchestrating a grant policy for all other kinds of IVF research including experiments on "spare" preembryos (those not implanted) and parthenotes, the use of cell lines derived from these preembryos, and blastomere biopsy to detect pathologic genetic alleles. To Clinton's remarks, HHS officials including the NIH director made no public comment. However, Varmus quietly shelved the balance of the HERP recommendations.[23]

In the normal "constitutional" course of events, Varmus would have mulled things over, consulted as needed, and made a decision. That decision would have gone "upstairs" to HHS secretary Donna Shalala, who also would have mulled, consulted, and acted. Then, because the issues made a real difference, Mr. Clinton would have gotten involved and signed on or off. As matters transpired, the president simply jettisoned the process, short-circuiting the two public officials with statutory responsibilities for policy development in this area. Nobody on the HERP could be blamed for failing to predict the rise of Newt Gingrich to the House Speakership. And nobody could blame the HERP for not anticipating that the president would sacrifice the legitimacy that would have accompanied a pretense of deep reflection on this important question in order to cut off at the pass any possibility of Republican political gain. Where the HERP dropped the ball was in its lack of ELSPI perspective. In the murky and disputatious pre-election environment of that moment, the governing concept in the report should have been reproduction not research. Therefore, creating preembryos for solely experimental purposes should have been put on hold until clearer political signals emerged, not a large price to pay back then, and now, too, for that matter. Indeed, even in the United Kingdom, nuclear transplantation was at that time forbidden under the law. Committee member Patricia King had argued at the open meeting on December 1 that caution must be the watchword: to stand, as it were, like a gladiator in battle against the ideological enemy could result in something most unfortunate like another research moratorium being proclaimed from on high. The "on high" proved to be the president of the United States himself, who essentially was arguing that the scientific agenda here at issue was hardly worth the time of day for all the attention he was prepared to give it. How right Patricia King's instincts were.

Vigilant pro-life members of Congress could readily see the wiggle room remaining for NIH-funded researchers even with Mr. Clinton's preemptive strike in play. They also could sense that the White House had no intention of going to the barricades to protect this wiggle room so long as they observed the letter (though not necessarily the spirit) of *Roe*. In early 1996, the Congress appended to an appropriation measure a funding ban on the creation of preembryos for research and experiments in which preembryos would be destroyed, defining "embryos" (*sic*) as any living artifact created by "fertilization, parthenogenesis, [or] cloning."[24] When Mark Hughes, whose work at NIH was supported by the then-NCHGR, examined in an off-site facility a cell of a preembryo to determine the presence of the CF mutation, NIH officials were raked over the coals at a House subcommittee session, and Hughes lost his position. It turned out he was using NIH-funded postdocs in his private laboratory. So if an IVF procedure yielded seven fertilized eggs (pretty standard) and three were implanted (also pretty standard), an NIH-funded researcher could not perform a DNA analysis on one of the preembryos inevitably slated for discard![25]

When it comes to the genetics/genomics paradigms, what we were really seeing was a reenactment of the same old story. Ethicists propose, the

Beltway disposes, and then scientists come along with new discoveries and knock the entire ELSI apparatus into a cocked hat. So was the case here. In November 1998, James Thomson of the University of Wisconsin announced success in growing human pluripotent embryonic stem cells (ES cells) in culture. Put succinctly, Thomson had taken an IVF blastocyst, removed its outer layer, taken the inner cell mass, nursed it on murine skin "feeder" cells, and, lastly, disassociated the cells, after which they would produce "immortal" nondifferentiating cell lines.[26] The research promises to help achieve three major goals: 1) create new knowledge as to how embryos develop after implantation; 2) provide normal cells for drug screening; 3) produce new tissues to be used as replacements for defective human tissues. The last of these was where the money stood to be made, namely, employing undifferentiated cell lines as organ transplantation factories. Naturally, Thomson was in a race. He had beaten to the publication line John Gearhart of Johns Hopkins, who would shortly thereafter announce that he had derived ES cells from primordial germ cells, that is, cells producing sperm and egg, which had been taken from aborted fetuses.

Fearing the sweep of federal proscription, Thomson, like Mark Hughes before him, had conducted his experiments in a private facility, in his case funded by the biotech firm Geron and WARF, the University of Wisconsin's patenting agency. Geron was evidently to receive an exclusive license to commercialize the WARF patent on Thomson's work. Unlike Hughes, Thomson made absolutely certain the taint of federal subsidy could not be traced to his doorstep. But was the congressional ban on preembryonic research applicable to these stem cell breakthroughs, had they actually been U.S. taxpayer subsidized? In Thomson's case, the answer was not at once obvious. The preembryos had come from a private clinic and would have been destroyed in that clinic no matter what; still, Thomson was certainly discarding them. The cells, though, were incapable of developing into embryos. Did that fact remove them from statutory ambit? Gearhart's investigations were obviously not covered by the ban, for his stem cells were taken from fetuses, not preembryos.[27] In a certain sense, paradoxically, Gearhart's research touched upon more challenging ethical questions. Was it appropriate to abort a fetus, take eggs from that fetus, and implant them in somebody's uterus? The uterus of an egg donor's aunt? The process was guaranteed to create babies whose genetic mothers had never lived! That is why the United Kingdom permits the use of eggs from aborted fetuses strictly for research, never for reproductive purposes. On this score, at least, Gearhart was in the clear. Already, however, observers had begun to see the cloning and germline implications of stem cell production.

Those inklings required exactly one week to achieve tangible form. An obscure biotech company in Worcester, Mass.—Advanced Cell Technology, Inc. (ACT)—announced that it had created its own brand of ES cells, these derived from a cow-human DNA hybrid. The ACT protocol was designed to work this way: scientists first removed the nucleus from an unfertilized cow egg; they then removed the nucleus from a human blood or skin cell and inserted that nucleus into the cow egg; through a process called somatic cell nuclear transfer (SCNT), which I shall describe thoroughly in the next section, the scientists cajoled the hybrid cell's nucleus to revert to its pre-embryonic state, after which the cell allegedly became an ES cell factory. The ACT scientists use cows' eggs not only because they are cheaper than human eggs but because the provocative ELSPI questions surrounding human preembyros could presumably be avoided.

Needless to say, ELSPI questions were certainly not finessed in some quarters. President Clinton, for one, expressed chagrin at the news and asked his National Bioethics Advisory Committee (NBAC) to look into the matter. Nor did he stop there, for he requested the NBAC to investigate and report back to him on the entire ES cell research agenda. Many reputable scientists were also taken aback. They noted that ACT practitioners had published no peer-reviewed reports of their work, and that the company had presented no proof its people had created ES cells. Essentially, they considered the press conference a publicity stunt timed precisely to piggy-back the Thomson coup. While ACT has yet to publish a word on its chimera, it has found the time to file for patents. Perhaps also it has been the recipient of a gift or two from investors who like what they've heard.[28]

Hopes for ES cell studies received a boost when NIH lawyers interpreted the "Beltway Constitution" to permit the central core of Thomson-like research. Federal support for the employment of preembryos to make cell lines was indeed impermissible, they said, but using these cell lines to produce differentiated cell types or to investigate various cell growth procedures was appropriate. Thomson and his colleagues would have to make the ES cells (and destroy preembryos in the process) under private or state auspices, after which they could take NIH money and run cell line experiments on the grounds that stem cells were not organisms and could not develop into embryos of any type. Senators Arlen Specter (R–Pennsylvania), chair of a Senate subcommittee in charge of working up NIH's annual appropriation, and Tom Harkin (D–Iowa), the ranking member of the subcommittee, both generally sympathetic to genetic research, applauded the legal opinion.[29] To this policy pronouncement, 70 members of Congress published a letter insisting that the federal government in no way support the sort of research Thomson and Gearhart had undertaken. All that accomplished was to generate a heated reply from a diverse group of scientists including Walter Gilbert, Stephen Jay Gould, Milton Friedman, Hamilton Smith, and James D. Watson, strongly supporting the new NIH approach.[30] If all this sounds a little like pro-*Roe* and anti-*Roe* activists jousting for federal judiciary confirmation votes, then stay tuned.

The NBAC filed its report with President Clinton on September 13.[31] As I will show in my examination of the cloning debate, the NBAC had acquired in its brief tenure a deserved reputation as a "yes" committee, that is, a blue-ribbon panel that mostly did what the president wanted it to do on controversial questions. What else would one expect? If the chief executive announces ELSPI policy and then commissions a group of experts to develop guidelines on that policy, are the experts going to take issue with the boss's ELSPI predilections? Here, however, Mr. Clinton had not committed himself. The NBAC had an opportunity to be brave. So what did it do? It "overbraved" its hand.

The NBAC might have simply endorsed the NIH production-use ES cell distinction. This somewhat strained but politically prudent line-drawing exercise would not have embarrassed the president (that was clear, else he would have chastised the NIH in no uncertain terms as he had done previously with the HERP report) and would have incurred minimal congressional antipathy (NIH had already become the chief lightning rod for anti-*Roe* legislators).[32] A politically sensitive NBAC statement would have helped the cause of the new science considerably. Instead, the president's committee recommended to Congress that it amend its ban on preembryonic experiments to provide this one exception, that is, an exception with regard to stem cell production and usage, taking the view that where only "spare" preembryos consigned for discard were involved the "ethical principles of beneficence (promote good) and nonmaleficence (do no harm)" should control. As a result of brandishing these ethical shibboleths, "the NBAC found itself politically isolated." The president, his staff announced in the wake of the report's disclosure, saw no need for further legal action, was adamantly opposed to funding preembryo creation for any and all experiments, thought it clear that stem cells will be available from private IVF clinics, and had concluded therefore that NIH policy was exactly on target.[33] In August of 2000, three months before the Bush vs. Gore election, the NIH policy took legal effect.[34]

With George W. Bush's election to the presidency, nervous time ensued for ES cell advocates and opponents. While new HHS secretary Tommy Thompson was thought to support in general principle the new research program, President Bush himself was known to be militantly anti-*Roe* and had spoken unkindly about anything smacking of embryo destruction. Uncertainty increased when NIH (now leaderless as Harold Varmus had departed for Sloan Kettering) suspended its new ES guidelines pending administration review. Senators Specter and Harkin tried to fill the gap by introducing legislation to legalize the NBAC's recommendation permitting cell derivation and employment, and they succeeded in getting two conservative Republicans, Strom Thurmond and Gordon Smith, to join them.[35] The process reached crescendo proportions when Senator Orrin Hatch (R–Utah) came out in favor of funding ES cell experiments, as did the Coalition of Americans for Medical Research, while GOP House leaders Richard Armey and Tom DeLay adamantly opposed such action.[36]

On August 9, 2001, President Bush delivered a televised speech to the nation on the subject at hand. It was his first major address on a fundamental public policy issue since his hotly contested election, and all sides agreed that the president's mode of delivery was equal to the gravity of the moment. After reviewing carefully the arguments for and against federal taxpayer subsidy of ES cell experimentation, he pulled a rabbit out of his hat and proclaimed a surprise compromise solution: cell lines had already been derived in laboratories around the world from discarded preembryos; it would make no sense at all to destroy as well these cell lines; hence, NIH could entertain grant proposals enabling scientists to make good use of those ES cell materials; however, federal funding would not be available to study any cell line yet to be created. According to NIH officials, 60 of these cell lines existed in five countries: the United States, Australia, Israel, Sweden, and India. Adjusted figures enumerated 64 cell lines, with U.S. laboratories in possession of 20. It is an interesting exercise to compare the Bush approach with other policies in comparable research communities. In Germany, France, Switzerland, and Norway, human preembryonic studies, both public and private, were totally forbidden; in Japan, Canada, Spain, Italy, Sweden, Israel, Singapore, and Australia, both derivation and employment, public and private, were acceptable if the cell lines came from spare preembryos; in the United Kingdom, in privately subsidized U.S. facilities, and presumably across the board in China, scientists could create preembryos for study and then derive from them and use cell lines. Only in this country does the public-private distinction have any political currency and credibility.[37] Then again, ours is also the only country to favor separation of powers. Divided authority comes easily to us and is not a human rights violation.

At this writing, one year after Mr. Bush's speech, the ELSPI questions regarding preembryonic stem cell research far outnumber the answers. Of the 72 cell lines now believed to exist, only 16 are available, and American researchers are using a scant 4 of them. Most of the 72 have not been rigorously characterized, so their salient parameters are in doubt.[38] And all of them rely on mouse embryonic fibroblast cells for sustenance, which carry a risk of human pathogenic cross-transfer. Both the NIH and the University of Wisconsin deserve some praise for trying to make the president's policy viable. They have signed an agreement greasing the avenues of accessibility for scientists throughout the country, and Wisconsin has construed its agreement with Geron narrowly, thus minimizing monopolistic opportunities. And yet nobody knows whether ES cells are really worth all the trouble and the bombast. There is evidence that once they differentiate following insertion into the human body, they will be subject, as are all other invaders, to adverse immune reactions. Nor is the evidence firm that preembryonic stem cells are more advantageous than adult stem cells in their therapeutic potential. Perhaps only the application of genetic modification strategies to these potential vehicles will answer these questions. Then we can have some real ELSPI battles.[39]

I close this section with an assessment of the prevailing Bush policy comparing ethical criteria, political science criteria, and scientific criteria. The president drew, from all I call see, a perfectly defensible ethical line between preembryos destroyed before his speech and preembryos destroyed after his speech. His guiding criterion was fair notice or, if one will, procedural due process of law. The problem is there were other defensible ethical lines available, too. There is the ethical line between preembryos created for IVF purposes slated for disposal because they are unusable and preembryos created entirely for experimental ends. There is also the ethical line between preembryo creation and preembryo usage, not to mention the ethical line between private funding of preembryo investigation vs. public funding of preembryo investigation. Finally, there is my ethical favorite, differentiating between preembryo research and embryo research. I herewith submit that the defensible ethical standard among all these defensible ethical standards that a public official selects as a political rallying point is the option that is most politically adaptive. If by this standard we mean the "political pragmatism" scale, then Mr. Bush gets a grade of A. He satisfied most everyone, at least for the moment, except the *New York Times* editorial board on the reproductive freedom "left" and the Roman Catholic hierarchy on the reproductive freedom "right." All in all, the president put a heavy emphasis on "life" as he defines it and a modest emphasis on "scientific inquiry" as he defines it. Meanwhile, the NAS has issued a report essentially arguing that the Bush policy is unworkable and that researchers need new

ES cell lines without further delay.[40] There is a significant risk that legitimate science has come up a loser in the wake of all the ethical and political jostling.

D. Dolly and Her ELSPI Progeny

When I published *Cloning and the Constitution* in 1986, there was no such thing, even in the fantasy lives of molecular biologists, as human cloning. Standard texts and research reports, to the extent they discussed the subject at all, dismissed it out of hand: "the cloning of mammals by simple nuclear transfer is biologically impossible."[41]

Back in 1986, the critical cloning policy questions revolved around a brave new world that was very possible indeed; as a matter of fact, it was not only possible—it was upon us in the compelling world of empirical reality. As a consequence of recombinant DNA research, scientists could take genes from one organism and insert them into the genomes of other organisms. These new life forms—unique microbial mutants—were then cloned, characterized, and run through a battery of experiments. In my book, I asked the following kinds of questions: To what extent is scientific experimentation an artifact of free expression? Is cloning as we then knew it a constitutionally protected activity? If one could delineate the work ways of scientist cloners—yes, they called themselves "cloners"—then could one also delineate a set of constitutional rules and norms that guided them in their laboratory investigations? And I also asked this question in order to round out my study: If human reproductive cloning were ever possible— after all, I reflected, frogs had been cloned and birds had been cloned—to what extent would the jurisprudence of procreative liberty attach to it?

The world of human cloning did not begin as a tabula rasa with Dolly the sheep, as surprising as that may seem. One year after *Cloning* was published, a former University of Illinois student named Jim Robl returned to his alma mater and conducted a seminar on his research. That research involved cloning rabbits. He described the procedure this way: remove a fertilized embryo from the rabbit; take out the nucleus and insert it into an enucleated unfertilized rabbit egg; fuse the nucleus to the egg by administering a 160-volt shock; this shock also "causes the unfertilized egg to reprogram its own nucleus with identical genes"; the egg then commences cell division as if fertilized. In other words, "an oocyte will treat a new nucleus like that of a sperm." The result, he had demonstrated, will be numerous preembryonic twins—clones—that could then be implanted in a female rabbit for ultimate birth. Robl expressed interest only in improving livestock genetics, but he saw no reason why the procedure couldn't work with humans.[42] Had Robl applied for patents in the United States on his cloned rabbits and the method by which he had made them, his chances of receiving approval would have been excellent. The Department of Commerce had issued a statement announcing that such products and processes, if they met prevailing statutory criteria, were within the scope of the law; however, "a human being will not be considered to be patentable subject matter" as "[t]he grant of a limited, but exclusive property right in a human being is prohibited by the Constitution."[43] Precisely where in the Constitution this prohibition could be found is not clear. Perhaps the Patent Office had in mind the Thirteenth Amendment ban on slavery!

Six years later, the national media fanned the flames of what some alarmists considered a human cloning conflagration waiting to happen. Jerry Hall at George Washington University, without using any federal money and hence without being subject to any federal oversight, had created preembryonic twins in his laboratory. The trick was to arm human cells with an artificial zona pellucida so as to facilitate development, a step unnecessary in the case of lower-order animals. Hall had at his disposal 17 blastocysts that were chromosomally unfit for normal health. He then snipped off blastomeres and covered them with the synthetic substance, after which they were bathed in a nutrient solution. As a consequence, the several blastomere preembryos yielded on average 48 new preembryos, each subset displaying genetic identity of its blastocyst progenitor. None were implanted, and had they been implanted, their chances of survival would have been problematic. But Hall, much to the chagrin of many, had shown that human cloning was no science fiction reverie. And just why were the media suffering from a case of chagrinitis? Perhaps, they mused, parents might create clones to harvest organs for twins in need of transplants or would use clones as replacements for deceased children. Parents might also subject their "delayed" twin offspring to traumatizing comparisons with their "normal" twins or simply freeze the clones and sell them. Even realistic scenarios generated bioethical breast-beating. Instead of pouring over a single cell taken from a preembryo in search of genetic irregularities, perhaps using a better developed clone would yield more reliable results, some bench scientists speculated. Aha, said critics Margaret Somerville and Arthur Caplan, that would amount to creating preembryonic sacrificial lambs for the benefit of their "brothers" and "sisters."[44] Too bad we did not have the benefit of Paul Ramsey's and Leon Kass's views on this natural extension of the Robert Winston–Alan Handyside research agenda at London's Hammersmith Hospital.

In February 1997, Ian Wilmut and his colleagues at the Roslin Institute in Edinburgh told the world they had cloned the first mammal—Dolly, a sheep—through the somatic cell nuclear transfer (SCNT) procedure.[45] How exactly did they accomplish this feat? First of all, the Wilmut group used a somatic cell, more specifically, a cell from an adult ewe's udder. This was a big step beyond Robl-type experiments involving the nucleus of a fertilized egg, which by this time had become the standard medium for cloning sheep and cattle. The Roslin team removed the nucleus from their somatic cell and placed it into an enucleated unfertilized sheep egg. The prevailing theory had been that nuclei from mature cells were too far advanced to be genetically reprogrammed. Wilmut et al. showed that such cells could indeed be made totipotent by, in effect, starving them of basic nutrients, thus curtailing dramatically their genetic expression. They then applied the Robl cell fusion shock treatment to the transplanted nucleus. When cell division commenced, the full array of new DNA became functional as though in a sexually orchestrated preembryo, after which the artifact was ready for implantation. Note, of course, that Dolly the sheep is not exactly a clone because she is the product of one sheep's nucleus and another sheep's cytoplasm, the latter containing Dolly's mitochondrial DNA. Virtually lost in the emotion of the moment was the fact that the Roslin researchers had undertaken 227 attempts to engineer a clone and had managed to produce only one. Also virtually lost was the fact that one month later Don Wolf and his colleagues at the Oregon Regional Primate Research Center had cloned two rhesus monkeys, using, however, standard preembryonic cells. Still, not only the mammalian cloning barrier but also the primate cloning barrier had now been breached.[46]

If Jerry Hall's experiment had fanned some flames, Ian Wilmut's production of Dolly caused a firestorm. President Clinton, making no bones about his opposition to the use of SCNT for human procreation, at once issued a memorandum to the heads of all executive agencies prohibiting the expenditure of any federal funds for the purpose of implanting preembryonic human clones into a woman's uterus.[47] He also requested his newly formed National Bioethics Advisory Commission (NBAC) to study the relevant ethical, legal, and social implications of human cloning and report back to him with recommendations in 90 days.[48]

As had now become standard operating constitutional procedure, the usual suspects weighed into the ELSI debate. Anti-*Roe* members of Congress, who considered preembryos to be babies, introduced legislation not only barring federal funds for human cloning (no matter how defined) but also making the creation of human clones a federal crime. Both the

Biotechnology Industry Organization (BIO) and the preembryonic sciences research community came out against cloning as a human reproductive methodology, the former fearful that in the current political environment far more sweeping legislation threatening the "gene business" might be enacted, the latter fearful that their "gene research" might similarly be imperiled. The International Academy of Humanism, numbering such stalwarts in the evolutionary sciences as Richard Dawkins, Francis Crick, and Edward O. Wilson, deplored a "Luddite rejection" of cloning and urged restraint. Naturally, the bioethicists were of two minds: Leon Kass instructed the NBAC to ban human cloning because it was "radically new," while Ruth Macklin argued that a prohibition could compromise individual rights.[49]

The NBAC recommendations, announced in June 1997, can be distilled in a nutshell. Congress, the NBAC said, should pass a law making human cloning via SCNT a crime. Attached to this proscription should be a "sunset clause" of between three to five years, thus guaranteeing a formal review of the panel's unprecedented recommendation. Until such legislation was finalized, the Clinton moratorium should stay in place, and private practitioners were urged to comply voluntarily. The evidence, said the NBAC, was overwhelming that SCNT as a human procreative choice was unsafe, and hence to employ the tool for reproductive purposes was "irresponsible, unethical, and unprofessional."[50] Naturally, President Clinton was overjoyed upon receiving the NBAC's salient recommendations. After all, his own handpicked committee had provided him with all the expert cover he needed to proceed in accordance with his previously announced policy. The president stated that he would send a proposed statutory ban on human cloning to the Congress immediately, complete with a sunset provision.[51]

Very few commentators took up the cudgels against the NBAC report, particularly its proposal for legislative criminalization. I was one of them, and I desire now to restate my objections.[52] My first caveat goes to the manner in which the NBAC defined its mission. The title of its report is *Cloning Human Beings*, yet the panel explicitly found implanting twinned preembryos to lie beyond its purview and never discussed preembryonic cell nuclear transfer (PCNT) at all, including the rhesus monkey research. No doubt the NBAC was trying hard not to bite off more than it could politically chew. That is understandable. The problem is that to home in on the most controversial form of a mode of research or a mode of procreation (human cloning is both!) and eschew the rest is to rip the discussion from the larger scientific and constitutional contexts. It is easy to say, "Nazis have no free expression rights because they themselves don't believe in free expression and they have a track record of hurting people." It is not so easy to reach this conclusion when we have before us the total array of ideologically oriented political groups in this country, the various ways in which they promote their views in the marketplace, and the sundry public interests that might or might not provide reasons for legislative prohibitions. In the view I take, national commissions of any sort cannot formulate blue-ribbon reports without a full appreciation of these broader contexts.

My second caveat follows directly from my first. NBAC chair Harold Shapiro stated: "Although the creation of embryos for research purposes alone always raises serious ethical questions . . . the use of somatic cell nuclear transfer to create embryos raises no new issues in this respect."[53] That statement was highly debatable at the time, and the development of the cloning research agenda has only made it more so. Right-to-life advocates might well argue that while a preembryo created in the laboratory via the fusion of male and female gametes is inherently vested with "personhood," no such status accrues to an SCNT-manufactured entity. Correspondingly, right-to-choose advocates might well argue that *Roe* is "premised on underlying assumptions about the meaning of procreation, for example, that it is interdependent, involving the reproductive cooperation of a male and female [and] that it involves the transmission of genes vertically across a generation."[54] Clearly, to the extent that SCNT-produced human preembyros are not considered constitutionally protected, the state's right to control their destiny increases. Why, then, could not the NBAC recommend that the Congress enact legislation giving scientists reasonable access to these preembryos as proper objects of research, hence explicitly sparing them from its across-the-board regulatory hand? And once the implantation-research dichotomy is pierced, why could not the NBAC, consistent with its charge, have thrown a mantle of legitimacy around preembryonic twinning as a research tool to identify the genomic parameters of blastocysts, that is, to compare and contrast various preembryos for "fitness"? And just why, I would have asked the commission had I testified before it, is not such research constitutionally protected?

Let's be clear about one thing: never before in American legal or political history had an independent, blue-ribbon panel operating under federal governmental auspices recommended legislation in the form of penal law declaring that the *content* of biological investigation was so beyond the pale that it could not be done anywhere, by anyone, for any reason whatever. What evidence supported this unprecedented recommendation?

The NBAC candidly dismissed ethical grounds as providing the necessary factual predicate, finding that no prevailing national consensus existed with regard to such relevant terms as "self-identity" and "human dignity." Rather, as I said earlier, the governing rational was one of safety. It seemed obvious then and it seems obvious now that SCNT for human reproduction is risky business; without question, laws regulating the times, places, and manner of implementing these protocols would be eminently rational and, therefore, would survive judicial oversight. The question, however, is whether a flat-out ban can survive an elevated level of constitutional scrutiny. Hence, my third caveat.

The NBAC, as I have said, never faced the question of whether *Roe* applies to SCNT procreational processes. However, the NBAC did say that if *Roe* were the locus of analysis, then the Congress would have to show a compelling interest to justify criminalization.[55] That is incorrect. Justice Sandra O'Connor's prevailing opinion in *Planned Parenthood* rejects *Roe*'s compelling state interest standard of proof and substitutes in its stead an "undue burden" standard of proof. *Planned Parenthood* makes life much easier for lawmakers, but with the NBAC sticking with *Roe,* those who favored a statutory ban had their work cut out for them. Cloning antagonists were ready for the challenge: according to bioethicists George Annas and Leon Kass, procreation via SCNT is not "reproduction" at all; it is "replication." The NBAC never did accept or reject the Annas-Kass distinction; however, an argument for criminalizing SCNT human cloning procedures based on that formulation does not suffice. Replication may not be reproduction, but in many contexts it is scientific inquiry. Scientific experimentation, including even genetic engineering, implicates freedom of expression values. How can scientists understand the structure and function of genes without experimentation, without in some cases "engineering" DNA, without in some cases replicating genomes? Of course, genomic replication in the form of cloning humans is not an absolute right. I do not even claim that creating human cloned preembryos, implanting them in the consenting adult woman's uterus, and then aborting them in the first trimester with her permission is an absolute right. The question, as always, is one of balancing competing interests. Annas and Kass saw the SCNT version of human cloning (query: what did they think of other "versions"?) as so pernicious that there is nothing to balance, that is, there is no interest in scientific inquiry to include in the constitutional debate over the power to ban. For them, when a scientist asks about human cloning, "Can it be done?" that research question and all corollary research questions are swallowed whole by some brooding moral omnipresence deep within the general will. I submit that their theory of the marketplace of content-based ideas, practices, and experiments held hostage to their inferences about majority sentiment is itself constitutionally suspect.

So what, ultimately, was the committee's constitutional rationale? Essentially, it came to this: no matter how weighty one thinks is the researcher's freedom of inquiry interest in performing SCNT embryo creation, and no matter how weighty one thinks is the mother's procreative interest in conceiving a child in this way, both these interests combined cannot overcome Congress's authority to protect the safety of human cloned entities in utero through the device of legislation criminalizing the procedure. I respectfully dissent.

Moving on to specific policy recommendations, the NBAC might have opted for preservation of President Clinton's moratorium on federal funding accompanied by whatever action the states and professional societies considered appropriate. To test the merits of the panel's across-the-board criminalization standard, best evidence is drawn from the political circumstances and implications surrounding earlier debates over genetic engineering. When recombinant DNA became a reality, the voices of such learned antagonists as George Wald, Robert Sinsheimer, and Liebe Cavalieri were just as militant and just as strident as were the voices of Annas and Kass. The NBAC report touting molecular cloning as "form[ing] the basis of much of contemporary biomedical science" omits reference to the falsetto disputation of the early 1970s and also conveniently omits the fact that proposed federal criminal legislation banning some or all recombinant DNA investigations failed to achieve congressional approval.[56]

As I have earlier recounted, what leading geneticists agreed to do, controversial enough, was to place a moratorium on their work until the NIH had approved a set of guidelines to provide constitutional frames of reference and had established the RAC to apply these guidelines to specific cases. The debate over the next 25 years proceeded in an open forum with not only expert witnesses from all fields appearing to testify but members of the general public, too, including Jeremy Rifkin and the father of Ivor Royston's patient, who wanted his daughter to receive immediate compassionate relief. Perhaps the greatest misconception harbored in the NBAC report was the notion that human SCNT reproduction ought to be forestalled so that some great national (international?) debate focusing on its risks, benefits, and ELSI proprieties could go forward.

And so my fourth and final caveat: go forward where, and under whose auspices? These forums do not spring from Zeus's brow. They are political artifacts requiring the political wills of leadership, organization, and constitutional legitimation to achieve success. With its long and generally distinguished track record, the RAC would have made for as good a choice as was and is available. The NBAC ignored the history and achievements of the RAC in its rush to criminalize. It should have recommended for SCNT human cloning precisely the process that made recombinant DNA and, as I will show, human gene therapy viable realities. The only needed federal legislation would have been a statute compelling human cloners, even those in private laboratories and even those not using SCNT, to submit their protocols to the RAC for approval. Slowly but surely the issues would have been narrowed, and consensus on human cloning's many permutations might well have emerged. As it ultimately turned out, the NBAC report is today another yellowed pamphlet containing forlorn hopes and not very good ideas, as Congress never did adopt the Clinton NBAC proposal.

I want to turn the clock back a bit. I had made a trip to Western Europe in 1995 to gain a comparative nation-state perspective on matters genetic and genomic. I was invited to give a talk before the Gene Therapy Advisory Committee (GTAC) in London, where I shared my four years' worth of RAC experience with a panel that was essentially its opposite number in the United Kingdom. I conducted interviews with leading players on the DNA research front including Noelle Lenoir, David Shapiro, Alan Handyside, and Sir Walter Bodmer. I also went up to Oxford, where I met with Kay Davies. After a three-hour discussion, she inquired if I would want to be proposed for membership in the HUGO. I, of course, replied affirmatively. She said she would handle the paperwork, and I was elected in 1996. It means something to me to be the first political scientist to attain this distinction. That I am still, to my knowledge, the only political scientist in the HUGO verges on the farcical and speaks volumes for the current intellectual myopia of my discipline regarding which I have already said plenty but not nearly enough.

I attended my first HUGO meeting as a full-fledged member in March 1997 in Toronto. Wilmut had published his Dolly research, Clinton had expressed his displeasure, and the NBAC was hard at work. I got to meet all sorts of interesting people in the corridors, and I asked a few of them about the cloning debates. Virtually none of them had ever heard of Wilmut, and they didn't want to hear any more. "He's a reproductive scientist; we are geneticists," summed up the thinking, and particularly the thinking of HUGO officeholders. "This cloning stuff is just going to rekindle the 'regulations game,' and we are going to suffer as a result. The best we can do is pretend he and his Dolly don't exist." I didn't like what I was hearing. The DNA world, the ELSI world, and the Beltway world were in high dudgeon over mammalian cloning, and the HUGO was going to pretend the issue didn't exist?

I had organized panels at the HUGO Heidelberg conference, as I earlier recounted, but I was not then a HUGO member so I couldn't attend the business meeting. This time around I was duly qualified, and I showed up at the gathering of the inner circle expecting someone to say something about Dolly. After 40 minutes or so of perfunctory speeches, discussions, and motions, HUGO president and presiding officer Grant Sutherland asked if there was any new business to be considered. All was silent. People were getting ready to exit. I couldn't believe it. This was my first HUGO business meeting, and I, the first political scientist ever to be invited to membership, was going to have to stand up and make a pest of myself. So be it. I rose and sought recognition. While I can't quote the three-minute speech I gave, it went something like this:

> I haven't had a chance to meet many of you, so permit me to tell you that my name is Ira Carmen and I am a professor of political science at the University of Illinois. I've written quite a lot about the relationship between genetics and politics, and I've served as a member of the Recombinant DNA Advisory Committee at NIH from 1990 to 1994. I want to say a few words to you about Ian Wilmut's Dolly research and the role of the HUGO, as I understand it, in providing professional guidance and insight in special cases like the one at hand. The media have had a field day with Dolly, and a lot of everyday people are worried—even frightened—about the specter of human cloning. The politicians are making noise, and Mr. Clinton's committee is going to offer recommendations. Some members of the NBAC are friends and colleagues, but don't fool yourselves. As a political scientist I can tell you that any committee that reports to the president of the United States is a political committee. I am not opposed to the NBAC, but I don't have much faith in it. What intelligent, thoughtful people around the world are looking for is a well-considered statement addressing the ethical, legal, and social implications of this research and an objective assessment of the risks, benefits, and possibilities of mammalian cloning. They are looking for assistance from an organization like the HUGO. You are in the business of discovering genes and sequencing genomes. The genes you find are going to be placed in cloned sheep serving as unique experimental and control environments. We all know this is inevitable. So I think the HUGO has a public responsibility to exercise professional advice on controversial subjects directly related to what its members do. I am making a motion that the HUGO instruct its ethics committee to study the cloning issue and prepare a report on the matter.

The debate on my motion was fairly protracted and fairly intense. I am pleased to say that it received majority support.[57]

The HUGO Ethics Committee report appeared in March 1999.[58] It discusses the various basic cloning procedures with clarity and brevity. The panel endorses animal cloning without reservation; it concludes that, because of "the profound unease" with which the international community views human reproductive cloning, "there should be no attempt to produce a genetic 'copy' of an existing human being by somatic cell nuclear transfer"; it states that SCNT may be used to combat mitochondrial DNA diseases; it strongly supports basic research in all cloning techniques including SCNT in both animals and humans; it supports therapeutic cloning; it declines to decide if asexually created human preembryos should be regarded with the same respect as sexually created human preembryos; it opposes the deliberate production of the latter for genetic research but does endorse the creation of blastocysts for stem cell harvesting. Nothing is said about criminalizing anything; in fact, terms and conditions of enforcement are assiduously avoided. I do not agree with everything in the HUGO statement. That is not the point. In one-tenth the space, it covers more ground than the NBAC report by far, and it pulls fewer punches to boot. It takes its place as a principal source of reflection in this volatile ELSPI area.

The HUGO paper has a counterpart in a European Union pronouncement, published a few months earlier, on patenting clones. The statement says that "interventions in the human germ line and the cloning of human beings offends . . . public . . . morality . . . . [I]t is therefore important to exclude unequivocally from patentability processes for modifying the germ line genetic identity of human beings and processes for cloning human beings."[59] One can see a phalanx of official hostility outside the United States in the form of specific policy declarations opposing the cloning of human beings.[60] Also noteworthy is the link the European political leadership and undoubtedly the European ELSI community forge between human cloning and human germline gene intervention. As I said at the beginning of this chapter, the two *are* inextricably linked, falling as they both do under the paradigmatic banner of human genomics investigation. This larger connection, however, had not yet become clear to many, including those very close to the ELSPI scene.

While the Clinton-NBAC legislative proposal languished in the Congress, victimized by opposition from the American Society for Reproductive Medicine, BIO, and key lawmakers,[61] a flurry of research spin-offs from the Dolly protocol attracted a media market ready to make the most of the newest outrageous genomics play toys. Wilmut himself led the way; his laboratory had synthesized an unspecified human gene and an unspecified marker gene into lamb fetal fibroblasts. And so was born the first transgenic cloned mammal, Polly. This was not an SCNT protocol—Wilmut used fetal cells rather than somatic cells because his live birth success rate for the former was 1 in 60, not 1 in 227—but it was the sort of gene transfer experiment

I had alluded to in my HUGO speech. We would learn four months later that the marker gene was the NEO bacterial gene prominently featured in the Anderson-Blaese-Rosenberg experiments of almost a decade earlier and that the human gene was the clotting factor IX gene, the deficiency of which leads to hemophilia B.[62] Next, ABS Global, a private firm in DeForest, Wisconsin, announced the birth of Gene, a male Holstein calf. Gene was also an artifact of fetal cloning. What made him special was that a cell had been removed from the already-dividing unfertilized egg and was then fused with a second enucleated egg. This egg replicated in a dish for one week, after which it was placed in a cow's uterus for gestation. So now we had cloned cows from immortal cell lines.[63] Finally, came news from the United Kingdom that scientists had created headless tadpoles, which naturally led to intense discussion about creating via SCNT headless humans who (which?) could serve as body-part repositories.[64]

This last experimental gambit caught the attention of Arthur Caplan, a bioethicist at the University of Pennsylvania, who ranks with Leon Kass and two or three others as members of the "mass media sound bite bioethics association," meaning that whenever a molecular biologist does something new and different, they stand ready to provide instant analysis of the ethical permutations at issue or not at issue. According to Caplan, it made no sense to discuss headless human clones because "people without heads are dead." Having traversed the physiological terrain in one gulp, Caplan moved on to the moral dimension. "Intentionally creating defective human bodies would not be an acceptable use of genetic science. . . . And intentionally disabling embryos so that they would grow without heads or brains would surely be an impermissible act." Well, let's try out a counterargument for size. Suppose we agreed, by law, that human bodies deliberately programmed to be brainless can never be human beings at all, that life may begin at conception (?), but human beings begin with (among other minimal conditions) potential for brain function. For me, it would follow that if a married couple wanted to use "their" preembryo to produce a headless nonhuman for research purposes, then—consistent with both scientific inquiry and reproductive choice—they would be within their rights. Does my idea pass muster? Maybe not. Still, I don't see why his belief system is any more cogent on first impression than my own. Incidentally, coauthoring with Caplan in this essay was none other than Craig Venter, surfacing from his race with the Human Genome Project just long enough to join in the nonscientific fray.[65]

There is a touch of black humor in the fact that the debate over headless tadpoles would shortly be upstaged by a debate over Richard Seed. Seed, a complete unknown in the reproductive genetics field, had announced at a professional meeting in Chicago that he fully intended to assist infertile couples by cloning children for them and that he had recruited physicians to work on the project. Just what expertise these medical doctors possessed was unclear, and where he would get financing for his enterprise was even less clear. Nonetheless, the media had a field day interviewing Seed, who basked in the notoriety. Some bioethicists thought he should have been ignored from the outset.[66] I did not agree. One of my little beliefs, placing me very much in a minority, I think, is a firm commitment to the marketplace of ideas; and by "marketplace" I mean the views of all players with bona fide credentials and interests, not just people listed in *Who's Who*. I thought it was important to hear Seed out; it was hard to find anyone who supported human reproductive cloning, much less someone who was determined to do it. In that spirit, I invited him to participate in a panel discussion at the 1998 Association for Politics and the Life Sciences annual meeting in Boston. He accepted and sat on the same stage with George Annas, Eric Juengst, Eric Parens, and me.

Seed has a Ph.D. in physics from Harvard, and he is very familiar with the pros and cons of human cloning. In general, he held his own in the discussion. He does, however, have a penchant for the outrageous, and when that penchant is combined with his odd array of personal mannerisms, it is not so easy to take him seriously. On this occasion, he stated that he planned to clone himself, with his wife serving as carrier, before he cloned anyone else, thus defusing, as he put it, the criticism that he would be taking advantage of unsuspecting, untutored women.[67] Later in the year, he showed up at a University of Illinois forum on human cloning, and we talked further. On that occasion, he provided a fleshed-out set of reasons to proceed with his cloning research agenda, a set most would reject as controlling.[68] He also startled his audience with a new headline grabber: "People criticized me when I said I'd clone myself. They called me an insufferable egotist. Well, I'm not going to do it. I'm going to clone my wife instead." After confronting Seed at close range and taking soundings of his personal peccadilloes, some surely thought him a comic-strip character. Given the peccadilloes of certain famous scientists, that conclusion is unwarranted. The fact is that he enlivened the debate at that time and posed interesting questions others wouldn't dare entertain. But after an up-and-down examination of his ideas, it was rather foolish of serious commentators and serious politicians to consider him a clear and present danger to any values we hold dear. That did not stop President Clinton from calling anew for regulatory legislation nor did it stop Senate Majority Leader Trent

Lott (R–Mississippi) from making that legislation a high priority item. The bill, however, was written inartfully: it proscribed human somatic cell nuclear transfer period, and many scientific societies read the proposed measure as criminalizing therapeutic cloning, that is, creating human pre-embryonic clones in order to harvest stem cells. A letter signed by 27 Nobel Prize winners lambasted the bill, and a coalition headed up by Senator Edward M. Kennedy, for the liberals, and Strom Thurmond, for the conservatives, came out strongly for the original Clinton human cloning ban. Senator Bill Frist, a Harvard-trained M.D., had talked of "stop[ping] Dr. Seed dead in his tracks."[69] Now it was the Senate that was stopped dead in its tracks, as both sides blockaded one another into legislative paralysis.[70] With neither proposal generating the needed support, Richard Seed lapsed, once again, into obscurity.

While the Seed soap opera was running its course, the dreary wheels of the Beltway bureaucracy were grinding away. Deciding that there was a policy void to fill and that no one else on American soil was better prepared to fill it, the Food and Drug Administration asserted jurisdiction over human cloning. On the face of it, this seems slightly ludicrous. Cloning is not "food," and it is not a "drug." The NBAC report was long and cumbersome, but nowhere in its length and breadth can one find a single word about putative FDA competence or oversight in this area. According to FDA's rationale, "human manipulation of cells" is very much within FDA control, and where the manipulation changes "the biological characteristics" of these entities, agency leaders insisted that "such products" were subject to those regulations applicable to investigational new drugs. They announced that anybody who intended to clone a human being would have to provide data attesting to safety and efficacy including proofs of successful animal model studies. Proceeding without an FDA license would lead to fines and other penalties. This policy is still on the books.[71]

According to this logic, it is perfectly all right for the FDA to set itself up as a censorship board assessing the terms and conditions of scientific inquiry and procreative choice. I am not talking here about equipment. I am not talking here about a pill. I am talking about a scientific procedure and a reproductive process the *content* of which the FDA finds prima facie suspect. The Constitution, as the Supreme Court thankfully has construed it, frowns on such governmental heavy-handedness. I would also point out that the FDA has no ELSI capacity whatever. As we shall see, it has the RAC to rely on in gene therapy cases; however, the RAC, as I have emphasized, lacks jurisdiction over human cloning. And just why, pray tell, has the FDA never flexed its regulatory muscle with regard to IVF or any other fertility context, where, as we have seen, success rates can be discouragingly low? Perhaps the FDA will consult with the NBAC's successor panels for wisdom in these matters.

The final episode in the human cloning ELSPI story, as of this writing, commenced with George W. Bush's 2000 presidential election. We have some work to do before we get there. Important developments transpired in the two years prior to that moment, a knowledge of which is vital if we are to appreciate all that would follow. The Japanese apparently got things rolling by employing Ian Wilmut's Dolly strategy to clone cattle, but the bigger breakthrough occurred almost concurrently when scientists at the University of Hawaii employed a refined version of SCNT to clone mice.[72] Wilmut had used a cell fusion methodology; the Hawaiian team used a needle to insert the somatic cell nucleus, thus ensuring minimal transmission of cytoplasm and the mitochondrial DNA contained within it. The mouse experiments demonstrated that Dolly was no fluke: many of the cloned mice reached full maturity, were genetically normal, and produced offspring via sexual and asexual (cloning) reproduction. The first cloned mouse was called Cumulina. Not to be outdone, the Japanese employed the needle-insertion microinjection technique to clone pigs. They called their firstborn Xena.[73] However, Xena's DNA came from fetal fibroblast cells, not adult somatic cells. On a somewhat different front, the Oregon Regional Primate Research group had adapted Jerry Hall's "delayed twinning" regimen to its rhesus monkey population and succeeded in producing Tetra, the first cloned primate.[74]

These successes were tempered by the sobering realities of constant failure that dogged investigators. Wilmut's team found that Dolly's telomeres—her chromosomal "caps"—were shorter than normal, a fairly sure sign of premature aging. A later study failed to confirm the deficiency with cloned cattle; but, perhaps significantly, the subject animals in the latter protocol were derived from senescent fibroblasts and not quiescent mammary cells.[75] Then there was the nagging problem of efficiency: only 2% of cloned embryos arrived intact at birth, and some of these died shortly thereafter or soon exhibited genetic malfunction. As for genetically engineering clones to produce therapeutic artifacts, the technology of nuclear transduction remained daunting.[76] Still and all, responsible people were now asking: if mice and monkeys, then why not . . . ?

While the U.S. policy process fiddled, other countries were taking action, most of it responsible. In the United Kingdom, key players issued a report recommending a ban on reproductive cloning (naturally!), but endorsed SCNT for research and therapeutic purposes including the creation of human preembryos. At that time, UK scholars could use only sexually derived preembryos for study, and in that case for the sole purpose of improving procedures for assisted reproduction, with the Human Fertilisation and Embryology Authority insistent that all protocols be shut down before the onset of the primitive streak. Japan enacted similar legislation, while momentum in China and France moved in much the same direction.[77] The dominant message, in the form of an emerging consensus, seemed to be: human cloning is permissible under appropriate guidelines unless the result is a human cloned baby. Of course, it was inevitable that somebody would utter a grand proclamation that they could do the "forbidden." That announcement came from, of all places, South Korea. The researchers said they had used SCNT to produce a cloned preembryo that had divided twice in vitro before they had shut down their project to demonstrate how ethically responsible they were. Korean authorities were furious. They needn't have worried. People close to the scene were unified in labeling the whole thing a publicity stunt.[78] Far stranger yet were the antics of a private company called Clonaid. Founded by a group known as the Raëlians, who say creatures from another world created the human species in laboratories, its scientific guru, Brigitte Boisselier, held a conference in Montreal in 1999 proclaiming her determination to clone humans and animals for a fee of $200,000 per clone. A year later Clonaid said it had received a gift of $500,000 to clone a couple's deceased child, and that it had lined up a stable of scientists and potential surrogate mothers to make its ambitions a reality.[79] The whole charade made Richard Seed appear the acme of rationality. But the world of human cloning seemed to be inching ever closer.

Exactly two months after the Bush-Gore electoral sweepstakes, serious science again showed it had more political clout than unserious politics. Two well-known fertility specialists, Panos Zavos and Severino Antinori, told colleagues they planned to produce a cloned human child in two years. Zavos, an American, emphasized that the only legitimate reason for human reproductive cloning was to assist couples who could not have progeny any other way. Antinori, an Italian, had already established his provocative research bona fides; he had used IVF to impregnate a 62-year-old postmenopausal woman, who subsequently gave birth. Their cloning initiative was obviously no joke.[80] In a Rome conference held a month later, the group said it had "unlimited funding" and would likely conduct its experiments in either Israel or an Arab country because "the climate is more [receptive] within Judaism and Islam."[81] No doubt some observers were pondering the notion that it took human cloning to bring the warring Middle East nations together—naturally, on the wrong side.

With Rudolph Jaenisch and Ian Wilmut taking the offensive, both the reproductive and genetic science communities ventilated their expected outrage in no uncertain terms. Their arguments boiled down to three: 1) human reproductive cloning was at least as unsafe and unpredictable as any other kind of reproductive cloning; 2) the fear of these experiments would inspire a stringent regulatory response, which would put at risk not just preembryonic stem cell research with clones but *all* preembryonic stem cell research; 3) "[t]here are many social and ethical reasons why we would never be in favor of copying a person."[82] The first of these reasons is powerful, indeed; the second would prove a splendid political prognosis; the third is highly debatable, because the commentators eschew discussion of what "social and ethical reasons" they have in mind and because the phrase "copying a person" is in fact value loaded and quite unscientific. That is, only the most biased of genetic determinists believe that cloning a person is copying that person.

The scene shifted quickly to the congressional hearing room. At a House subcommittee meeting, members heard Boisselier and Zavos discuss their cloning agendas. They also heard from Jaenisch, who said that gene methylation irregularities during cloning yields unreliable and unsafe results, and from FDA spokespeople, who said their agency could grapple successfully with cloning practitioners. Committee members expressed skepticism on that score, noting that the FDA had previously asserted jurisdiction over smoking only to be slapped down by the courts. In the Senate subcommittee gathering, Senator Sam Brownback (R–Kansas), much to the consternation of scientists and biotech representatives, announced that he would introduce legislation banning not only reproductive cloning but therapeutic cloning and not only therapeutic cloning but all preembryonic stem cell research. Brownback took the classic pro-life position that these experiments entailed the destruction of preembryos, which he considered another form of abortion. The fact that a cloned preembryo was not created by a sexual synthesis of gametes made no difference to him. The fact that SCNT-produced preembryos employing the patient's own somatic cells might ultimately yield cell lines and, hence, neural cells and islets that could be transplanted into the patient's body with the hope of drastically reducing rejection scenarios also made no difference to him. Appearing at the Brownback hearing and arguing in favor of an across-the-board human cloning ban was Leon Kass, who, as it turned out, was a principal author of the Brownback bill.[83]

Over and above anti-*Roe* ideology, the arguments against human therapeutic cloning could be summed up this way: 1) if you're going to permit it, then somebody will try to implant an artifact in a consenting woman's womb, and 2) ideally, we could learn much about genetic reprogramming from SCNT, and yet the technology is lacking even to determine its research benefits including the question of whether the procedure would facilitate ES cell approaches; scientists therefore should concentrate now on animal cloning and build a body of theory for future human applications. It was pretty clear from the outset that the Bush administration saw little in human cloning of any description worthy of support, and the House of Representatives, by a vote of 251–176, enacted a bill sponsored by Congressman Dave Weldon (R–Florida), an M.D., and endorsed by prominent biolegalist and NBAC member Alexander Capron, which was essentially a replica of the Brownback-Kass initiative.[84]

By now the preembryonic stem cell controversy, which I described in the last section, was heating up, and all eyes were on the White House to see if the president would support federal funding in that research area. On August 9, 2001, Bush delivered his television address setting forth his compromise supporting scientific investigations on cell lines then in existence. During the course of his remarks, he also stated:

> Scientists have already cloned a sheep. Researchers are telling us the next step could be to clone human beings to create individual designer stem cells, essentially to grow another you, to be available in case you need another heart or lung or liver.
>
> I strongly oppose human cloning as do most Americans. We recoil at the idea of growing human beings for spare body parts, or creating life for our convenience. . . . I will . . . name a President's council to monitor stem cell research, to recommend appropriate guidelines and regulations, and to consider all of the medical and ethical ramifications of biomedical innovation. This council will consist of leading scientists, doctors, ethicists, lawyers, theologians and others, and will be chaired by Leon Kass, a leading biomedical ethicist from the University of Chicago.[85]

To my knowledge, not one commentator then or now has proposed cloning human beings in order to mine from them body parts to be used for the "convenience" of brothers, sisters, or anyone else. They have proposed using SCNT to create preembryos for therapeutic purposes such as these. The debate was now beginning to take form, and the viewing public was not off to an auspicious initiation.

I want to rewind the clock a second time. In June 1996, I joined the Hastings Center and attended a state occasion there, the retirement dinner for longtime Hastings stalwart Dan Callahan. I did not go to honor Callahan, simply because I had never met him. I went to introduce myself to the

Hastings "power network," to observe at close quarters the beehive of American bioethical discourse. I was, in short, doing what political scientists do, or should do. At the cocktail hour, I joined in conversation with two men I did not know. They were not discussing genetic engineering; they were discussing baseball. An excellent icebreaker, I thought, because I fancy myself a sports maven, particularly a baseball maven. I introduced myself, and shook hands with Leon Kass. I said, "My name is Ira Carmen." He looked at me with glazed eyes, having obviously never heard my name. Why should he have? I was a political scientist. "No one [had] explained [to him] what [a political scientist] was seeking at that altitude." Of course, I had published *Cloning and the Constitution* a decade earlier; still, I was prepared to forgive him on that score as well because we academics share a dirty little secret, namely, that nobody reads anything anymore. Kass was bemoaning the state of his beloved baseball, then recovering from the aftershock of a strike. "I miss going to White Sox games," he was saying, "but the players have turned me off." "And I really miss seeing Frank Thomas. He's the greatest hitter since Ted Williams." He was speaking with a quiet authority, which did not improve the quality of his message. "Wow," I said, hardly able to contain myself. "Better than Henry Aaron? Better than Tony Gwynn?" I might have also mentioned Stan Musial, Willie Mays, and George Brett; discretion, however, prevailed. I had made my point. He looked at me, his eyes resuming their glaze. He had no intention of answering me, of reconsidering his position in the light of the sort of data quintessential baseball fans revel in. I was like a virus perturbing his genome with my DNA. I am sure he forgot our little exchange as soon as we parted company shortly thereafter. But I never forget such exchanges. And I didn't forget Leon Kass.

Before I talk about President Bush's Council on Bioethics, I want to discuss national bioethics councils in general. During the early 1990s and even earlier, several commentators argued that the Beltway offered no forum for debating the ELSI of the "new genetics" and developing national policy.[86] They longed wistfully for a return to the days, between 1978 and 1983, when the President's Commission for the Study of Ethical Problems in Medicine and Biomedical and Behavioral Research issued important reports on such matters as human subjects' protection, fetal research, and "brain death." That commission was successful because it worked in an unstructured political climate, where the tenets of mugwumpism could surface and rule the day.[87] As soon as key interest groups entered the debate—inevitable, once abortion and its policy tentacles intruded—this national commission folded its tent. The first lesson, then, is clear: an independent, blue-ribbon body operating without the requisite political support would fall like a ripe apple should it attempt to hand down from on high some set of specific recommendations on matters of considerable public concern. The question thus arises whether a presidential commission, with the prestige and power of the chief executive sustaining the integrity of its mission, could profitably take on controversial bioethical tasks. Unfortunately, the obverse problem then emerges: the independent committee either finds itself locked into the president's preconceptions, inevitably of a political caste and character, with the result that its work product is ideologically tainted, or it makes recommendations that the chief executive disavows overtly or covertly, hence neutering the panel's impact. Scylla is de-legitimation; Charybdis is castration. All of which explains my skepticism about national bioethics commissions in the abstract, and my preference for the RAC and a properly constitutionalized ELSI working group. Absent from this assessment is the possibility that the appointing officer—in the case at hand, the president of the United States and the chair of his Bioethics Council—would simply stack the deck with experts who viewed the constitutional world of human genomics the way they did, thus making of the panel's makeup merely another example of Andrew Jackson (or Richard J. Daley) patronage. In December 1994, with Italy's national bioethics committee poised to discuss abortion policy, Prime Minister Silvio Berlusconi removed all of the lay members and substituted for them pro-life militants.[88] That sort of thing wouldn't happen here, would it?

In 1995, President Clinton, by executive order, established the National Bioethics Advisory Committee.[89] The committee issued a report on human embryonic stem cell research that I earlier criticized and a report on human cloning that I earlier criticized, these criticisms being consistent with my larger skepticism regarding the political efficacy of such deliberations and policy processes. I have made no study of NBAC strengths and weaknesses per se, though several Beltway insiders were not bashful about articulating their frustration with the committee. Said one of the commentators who had high hopes: "19 months and 21 meetings after its inception, NBAC has produced only one report. . . . [It is] distressing."[90] The NBAC ultimately did stoke its engines; however, problems remained to the end. After five years of study, the committee issued a detailed report calling for a National Office for Human Research Oversight involving human subjects. With President Clinton having departed the White House months earlier, the report's prognosis was not healthy. The president of the Federation of American Societies for Experimental Biology added insult to futility by calling the proposal essentially a waste of time.[91]

One episode in the NBAC's tenure deserves closer attention than it has received, for it gave a national commission the opportunity to sit in the shoes of the RAC. Recall that in 1998 Advanced Cell Technology had issued a press release informing the public that it used SCNT to form a cow-human chimera that had then produced embryonic stem cells. President Clinton in late 1998 had asked the NBAC to investigate.[92] The NBAC might have proceeded by conducting a RAC-like open forum: calling witnesses from various disciplines; advertising broadly to make sure those who wanted to speak could speak; debating and voting in public; and issuing a report. Instead, the panel talked by telephone to one scientist and interviewed the ACT CEO only because he showed up uninvited. In a letter to Mr. Clinton, the NBAC shared three conclusions: 1) there was insufficient evidence to conclude that an embryo/fetus/baby (take your choice) would result if doctors implanted the chimera in a woman's uterus; if that possibility existed, the practice should be banned; 2) there was insufficient evidence to determine whether the chimera was, in fact, a "human" preembryo; 3) if the chimera were a preembryo, then presumably it would have to be treated like a preembryo; if it weren't a preembryo, it would be entitled to no more respect than any other object of research.[93] I don't disagree with any of these findings. I'm not sure who would. The issue here is not substantive due process; it is procedural due process. In other words, how should a government committee review research protocols. The NBAC fell far short of patterns of practice well established elsewhere and demonstrated once again the extent to which these various bodies fail to communicate and are forever reinventing the policy process wheel.

When President Bush decided he needed a bioethics committee, he could have extended the NBAC's charter and tenure, which presumably would have included the NBAC membership among those whose terms of office had not lapsed. Had he done so, he would have perpetuated what I and others would have labeled the myth of NBAC independence, that is, the notion that Mr. Clinton's bioethicists were as politically serviceable for a George W. Bush as they had been for his predecessor. Instead, the new president, indulging in political reality rather than academic mythology, allowed the NBAC to die a peaceful death, the official date of its demise being October 3, 2001. Enter Leon Kass's bioethics advisory council. According to Kass, the committee would represent a wide array of opinion and have a "broad mandate." Chimed in Kass's Hastings Center colleague Dan Callahan, Kass "will be very careful to get a fair range of people on the council."[94]

Let us examine these aspirations on the eve of the new debate on human cloning, with the House of Representatives having passed the Weldon bill, with Zavos and Antinori telling the world about their forthcoming research gambits, and with the President's Council on Bioethics (PCB) poised to take up anew the ELSPI dimensions of the topic. I think the best way to proceed is to compare the NBAC on the threshold of its cloning investigations with the PCB precisely at the same moment. I begin with the two chairmen. The NBAC chair was Harold T. Shapiro, the president of Princeton University. Shapiro was an economist by training and by scholarly reputation. He was also a skilled higher education administrator, having served for several years as president of the University of Michigan before he matriculated to the Ivy League. Shapiro obviously knew his politics, because nothing is more polit-ical than academic politics, and university presidents are all wrapped up in academic politics. Shapiro's views on the "constitution of human genom-ics" in general and the ELSPI of human cloning in particular figured to be untutored and unformed. I am not saying he isn't a quick study; I am saying that his influence as chair, given the complexities of the agenda, figured to be more process oriented than result oriented. Leon Kass could not have presented a more dissimilar leadership profile. His writings, which I shall introduce shortly, exhibit intelligence, depth of knowledge, perspective, and an intransigent worldview, tightly bound in consistency but exceed-ingly provocative and, in the view of many colleagues, probably wrong. I am not suggesting that Kass couldn't be counted on to run a fair ship. I am suggesting that he certainly wouldn't forsake his values and his agenda, more or less the same values and agenda as the president of the United States. The NBAC had 18 members; the PCB would also have 18 members. As one scrutinizes the credentials of the respective membership rosters, a single solitary fact leaps from the résumés: the NBAC had not one political scientist, while the PCB had no fewer than four. Naturally, my biases run strongly in favor of the PCB makeup. I casually referred earlier to the NBAC's ELSI charge; this calculated choice of language was a riposte to the idea that the NBAC could feature an ELSPI capacity given the fact that no P's sat on the committee. A few sentences ago, I deliberately referred to the PCB as having an ELSPI dimension, my terminology reflecting the gener-ous selection of political scientists on board. But, query, just who were these political scientists, and what sort of political science did they represent?

I begin with Chairman Kass. In 2002, with the PCB now in place and the human cloning debate reinvigorated, Bush Republican stalwarts William Kristol and Eric Cohen published a collection of essays designed to present the anti-cloning intellectual tapestry to a putatively concerned general au-dience. Kristol and Cohen found six essays by Kass worthy of inclusion, a larger number than that of any other contributor.

The first, written in 1967, was a reply to Nobelist Joshua Lederberg, who had, off the top of his head, suggested that human asexual reproduction might provide a sanity and rationality lacking in today's social affairs, which, he felt, were an artifact of the human sexual reproductive lottery. Kass correctly berated Lederberg for his cavalier bow to the wisdom of scientists cloning humans, and he noted a few of the subtle ELSPI questions on the table. He then observed, "Does not reflection on this question suggest that the programmed reproduction of man will, in fact, de-humanize him?" The term "de-humanize" is never defined. Also, the assumption is entertained that "man" is some homogeneous genotype or phenotype for purposes of cloning policy analysis, an assumption beclouding the careful distinctions Kass endorsed.[95]

The second essay, first published in 1972, is a kind of pre-*Roe* curiosity, summing up Kass's distaste for in vitro fertilization and, if I understand his argument, even artificial insemination. According to Kass, married couples have no right, and certainly women have no right, to protect themselves and their offspring from the horrors of Tay-Sachs or cystic fibrosis by orchestrating as rationally as science permits the procreative process. He asked: "[C]an we be sure that the eradication of genetic disease (or of any single genetic disease) is biologically a sensible goal?" If he were talking about germline interventions, I could appreciate his tack. That he is talking about how my wife and I should conduct our own personal "genetic business" and that of our *immediate* offspring, where the subject at issue is a proven genetic killer, is quite another matter. What really bothered Kass was that procreation and sexual intercourse should be separated: "Man is partly defined by his origins. . . . By tampering with and confounding these origins, we are involved in nothing less than creating a new conception of what it means to be human. . . . The new technologies . . . may, therefore, mark the end of *human* life as we and all other humans have known it."[96] If this sentiment is something more than hyperbole, then the state had better step in and pass laws protecting human nature and, furthermore, how humans have traditionally viewed the world and their place in it from scientific discovery. And just who will define this fundamental humanness to be preserved by law? King Canute? Ptolomy? the pope? Creationists? Kass himself?

The third essay, originally published in 1979, is a reprise of the second essay, written seven years later, with the birth of Louise Brown and *Roe v. Wade* having occurred in the interim. Perhaps Kass has reconciled himself to "test tube babies" and abortion as representing some kind of constitutional right. Not a chance. Focusing his attention on the status of in vitro blastocysts, he found them "clearly alive." "They metabolize [and] respire";

"[They are] organic whole[s], self-developing, genetically unique." They are fundamentally akin to blastocysts in utero. While they are not entitled to a right to life because they are not human beings, they are potential human beings representing "a mysterious and awesome power, a power governed by an immanent plan." We would never eat a human blastocyst, Kass argued, for it is "protected by our taboo against cannibalism." It followed for him that experimentation on preembryos, including genetic manipulation, ought to be forbidden. His essay concluded with a last-ditch lamentation of the forbidden fruit of in vitro procreation, a technology in which "the human embryo emerges for the first time from the natural darkness and privacy of its own mother's womb, where it is hidden away in mystery, into the bright light and utter publicity of the scientist's laboratory, where it will be treated with unswerving rationality, before the clever and shameless eye of the mind and beneath the obedient and equally clever touch of the hand."[97] A scientist's laboratory is a zone of "utter publicity"? I thought it a private enclave shrouded in the "mystery" of the search for truth. The "shameless" eye and the "clever" hand of the medical doctor? I thought the doctor-patient relationship yet another private enclave where the delicate subjects of birth and death received their most sensitive and personal treatment. Eating human blastocysts is cannibalism? What about eating cockroach blastocysts?

Kristol and Cohen's genuflection to Kass now fast-forwards to the current scene. The fourth essay was written in 1997. Wilmut had cloned Dolly; the NBAC was mulling what I call the human genomics ELSI cloning fallout; and Kass was close to despair. The asexual reproduction of human progeny would be the last straw, he wailed: "[C]loning personifies our desire fully to control the future, while being subject to no controls ourselves. . . . we have lost our awe and wonder before the deep mysteries of nature and of life. . . . we are faced with having to decide nothing less than whether human procreation is going to remain human." The villains of the piece, he wrote, are not the scientists; they have warned us about the dangers of human cloning. The villains are the bioethicists: captured by "analytic philosophy," they sit on national advisory boards, worshipping the god of "utilitarianism" and "pronounc[ing] their blessings upon the inevitable."[98] By this time the reader may be excused for becoming somewhat impatient over Kass's claim of patent rights on what is "human" and what is not. But then, the Nobel Peace Prize was recently awarded to Jimmy Carter, who has his own patent on "human rights." Excuse my positivist preconceptions, but I'd prefer the humanness of our species to be either the product of scientific understanding or, where science is not equal to the task, the processes of constitutional republicanism rather than the product of Kass's and Carter's Platonism. As for bioethicists, Kass had it dead wrong: his colleagues on the NBAC couldn't wait to ban human cloning.

Kass's fifth essay, published in 2000, finds him jousting with a new and invisible foe, those who want science to find a "cure" for death.[99] "[T]ruth to tell, victory over mortality is the unstated but implicit goal of modern medical science, indeed of the entire modern scientific project, to which mankind was summoned almost four hundred years ago by Francis Bacon and René Descartes." Why debate the repugnance of genetic intervention when the real enemy is "the entire modern scientific project," perhaps even the Enlightenment itself? For Kass, Baconism is in cahoots with "liberal democracy," that is, "the desires of the majority for whom the attachment to life—or the fear of death—knows no limits." Has Kass done any public opinion polling to confirm his suspicions? And what political order would he prefer to "liberal democracy," which the Founding Fathers did not bequeath to us anyway? "[T]his is a question in which our very humanity is at stake," he noted somewhat redundantly. Reaching a crescendo, Kass invoked one of his favorite motifs—Greek mythology. Odysseus rejected Calypso's promise of immortality, saying, "Penelope can never match the impression you make for beauty and stature. . . . But even so, what I want and all my days I pine for is to go back to my house and see that day of my homecoming." For Kass, "To suffer, to endure . . . is truly to live, and this is the clear choice of this exemplary mortal." But even on this metaphysical plain of allegorical pedagogy, I think his analogy is flawed. Odysseus wanted Penelope because he knew her and loved her, had children with her, had ruled with her at his side, and all the rest. His needs, his commitments, are no true test of whether our species, unfettered by the biases of a mature human existence already lived for many years, would choose to die at 80 or 180. The test is what a person, born probably via some technology Kass deplores and shortly to enter upon adulthood, would have to say about the virtues and vices of "immortality." Or perhaps the test is what Odysseus would do if he knew Penelope, Telemachus, and his subjects all had good chances to reach 180. Who knows? I need not prolong the "defense" of an idea I personally find downright foolish. But Kass's invocation of the myths and stories of a Western Civilization he otherwise derides should not pass unchallenged.

Kass's last essay reprinted in the Kristol-Cohen reader lumbers on for 23 pages.[100] No longer is the mood one of despair; no longer are the issues broad and philosophical. Kass is upbeat and feisty. The date of publication is May 21, 2001, and I suspect he knew already that he would preside over the PCB. He is in this paper leading a charge, particularly against human cloning. For him, human cloning is the apotheosis of Huxley's *Brave New World,* a world of "homogenization, mediocrity, trivial pursuits, shallow attachments" where "[a]rt and science, virtue and religion, family and friendship are passé." According to Kass, we have come "to a clear fork in the road," a place where we now can order a course correction in what has been our reckless pursuit of technological seduction. Human cloning is a unique horror, conjuring up for Kass his most elegant array of hortatory characterizations. It will lead to "dehumanized hell"; it bespeaks "Frankensteinian hubris"; it is the epitome of "men playing at being God"; it would inevitably lead to "despotism over children and perversion of parenthood." And the remedy? "What we should do is work to prevent human cloning by making it illegal." He clarified that all these perversions resulted solely from reproductive cloning; those who favor therapeutic cloning—the use of cloned preembryos to provide cell lines to make regenerative medicine a reality—could, however, find no room for solace. "I now believe what we need is an all-out ban on human cloning, including the creation of embryonic clones. I am convinced that all halfway measures will prove to be *morally,* legally, and strategically flawed" (emphasis mine); ultimately, "[a] ban only on reproductive cloning would turn out to be unenforceable." He then threw his support behind the legislative initiatives offered by Senator Brownback and Representative Weldon.

Not just in fairness but in candor let me say that I share many of Kass's ethical concerns about reproductive human cloning as an abstract communal practice. My concerns are in the form of reservations, in the form of doubts, in the form of personal discomforts. That said, I find Kass's scholarship essentially an exercise in polemics, even when packaged in a charming rhetorical felicity. More to the point, his thinking represents a form of ideological smugness exceedingly counterproductive in the realm of either softball or hardball politics. Here, then, we have President Bush's choice as chair of his Council on Bioethics.

I do not feel the scholarly urge to track down the salient writings of most of the PCB rank and file. I have a definite urge and need to say something about the four political scientists. They are James Q. Wilson, Francis Fukuyama, Michael Sandel, and Robert George. I regard Wilson as one of the leading political scientists of his generation, as do many others. Among several distinguished books and papers, *The Moral Sense* (1993) stands out as a prominent contribution for purposes of my inquiry. I shall discuss this work at greater length in succeeding remarks; I need say now only that Wilson's argument marshals an impressive array of empirical evidence and theory to show that humans are far more than a culturally molded, environmentally

driven species. Humans also possess innate sentiments, powerful tendencies that are part and parcel of their nature and departures from which
often yield maladaptive results.[101] In the same year (1997) that Leon Kass
published his oft-cited essay on human cloning as quintessential repugnance, Wilson joined the discussion. "I instinctively recoil from the idea,"
he began; "[t]here is . . . a natural sentiment that is offended," sounding very
much like his own vision of the typical person. Why? Because "the central
problem is . . . creating [the cloned child] without parents." He called "[a]
premature ban on any scientific effort moving in the direction of cloning"
a likely impediment to "useful research." Nor could he find reason to criticize the "artificial" baby making of in vitro fertilization. For him, any antisocial results arising from reproductive human cloning could be largely
avoided in the context of parental care, affection, and bonding. "A human
mother will carry a human clone; she and her husband will determine its
fate." On the question of who is to be cloned, he simply assumed that one of
the parents would be contributing the somatic cell nucleus.[102] In a direct response to Kass's writings, Wilson remarked: "I link the meaning of children
to the existence of the family and he links it to the power of sexuality." As to
the ultimate public policy question of criminalization, he offered the following: "We do not want families planning to have a movie star, basketball
player, or high-energy physicist as an offspring. But . . . if no one can [draw
these limits], I would join Dr. Kass in banning cloning."[103] When Wilson
gently chided Kass for his preoccupation with the sexual parameter, he put
his finger on a critical Kassian weak point. For the record, I think Wilson
himself overstates the "parental" parameter as he operationalized it. In
making reference to the notorious Baby M case, he argued that when a married couple donates sperm and egg, and a surrogate mother carries the embryo to term and breaks her contract with the couple by seeking custody,
the courts should give her the child. I dissent. The woman tendered her informed consent not to be the mother. I am too much the geneticist to believe that the carrier's interest in the baby "outbonds" the interest of the
married couple who donated their gametes in a preembryonic "bond,"
watched the entity they created achieve birthhood, and then stood by and
saw their procreative dreams dashed by a cheating contract-breaker. That
little dictum aside, Wilson was a splendid choice to sit on the panel, and a
safe prediction would have been that he was a lot closer to Kass on the criminalization issue than their philosophic disagreements portended. And
given that the two coauthored the American Enterprise Institute project
cited here, one can well understand how Wilson made it to the committee.

Francis Fukuyama is the author of *Our Posthuman Future* (2002), a book I shall also discuss in subsequent remarks. For now, it is enough to examine closely two of his essays appearing in the Kristol-Cohen collection. In the first, Fukuyama applauded the "significant milestone" of sequencing the human genome, a fitting tribute, he thought, to the Baconian dream. For decades, he said, social scientists have deluded themselves into thinking that humans are "selfish, irrational and overly emotional, limited in intelligence and perception, prone to violence and aggression," because "culture and environment counted for nearly 100 percent of the variance, and biology almost none." The Celera/HGP results will permit us to trace down ultimately the many genetic behavioral antecedents involved. Such thinking puts Fukuyama in a small political science minority, but his approach comports nicely with my own. The question, at this point, is: What will we do with this new information? Citing both Leon Kass and Charles Murray with approval (provocative juxtaposition there!), he noted that the distinction between gene therapy and gene enhancement is unworkable and self-deluding, and he also noted that although the socialist states thought they could engineer people by manipulating culture, the greater danger today was that the modern liberal state would engineer people to satisfy standards of equality by manipulating their genes. I am more worried about liberals manufacturing equality by throwing out standardized test scores, encouraging grade inflation, and treating Ravi Shankar as if he were Frederick Chopin, than I am worried about genetic engineering. But if you want to get on a Bush committee, you can't do much better than cite Kass and Murray.[104]

In Fukuyama's second essay, he addressed the embryonic stem cell question, concluding that the White House should permit federal funding for research.[105] In the course of making the argument, he stated that destruction of cell-line–producing blastocysts was *not* infanticide, that adult cell lines have not been proven to be a suitable substitute, and that "American conservatives, with their single-minded focus on abortion, are missing the boat." To this, I must reply that George W. Bush conservatives may be preoccupied with abortion, but John McCain conservatives certainly are not. He concluded, oddly, by commending the Brownback bill, a surefire route to membership on the PCB; yet its ban on therapeutic cloning appears totally out of synch with his generally flexible, nuanced policy approach. Another good committee selection.

I have little to offer by way of an in-depth assessment of either Michael Sandel's or Robert George's credentials as prospective panel participants. Sandel is a political theorist who belongs to the "communitarian" school of criticism. His favorite self-selected antagonist is John Rawls, a wise choice, indeed, given that Rawls should be counted as the leading contemporary guru of rational choice "ideology," an archenemy of the biopolitical over-view. I have leafed through one of Sandel's principal studies and found it filled with detailed rejoinders to the Rawls position on this and the Kant position on that.[106] No doubt his hybrid natural law/historicist emphasis on the "good" would appeal to the Leon Kasses of this world, but solid creden-tials in conventional political theory are not enough to establish expertise in the ELSPI of human genomics. Perhaps I've missed a Sandel piece sub-stantively linked to the work of the PCB; on the record before me, his selec-tion, in the face of so many qualified scholars who were not selected, is very hard to justify. Robert George holds the most prestigious political science public law position in the country. He is the McCormick Professor of Juris-prudence at Princeton, the same sinecure held by Woodrow Wilson, Ed-ward S. Corwin, Alpheus Mason, and Walter Murphy, leading scholars of constitutionalism. Useful comparison ends there. George's writings in this field are represented by a paper appearing in the Kristol-Cohen collec-tion.[107] This essay, which might have been written by Jerry Falwell on a bad day, actually manages to out-Kass Kass. Recall that Kass, plumbing the depths of natural law, had found that human preembryos were not human beings. George, plumbing the depths of *his* version of the natural law, con-cluded that human preembryos *were* human beings. End of all discussions; end of all cases. His selection to the PCB should have inspired the Associa-tion for Politics and the Life Sciences, none of whose members were cho-sen, to picket committee meetings.

I am hardly the only commentator to cast a jaundiced eye at the ideolog-ical slant of Mr. Bush's committee. There was the usual Political Correctness critique: the panel was too male and too white. I'll stick with my criticism and simply expand on it: the political scientists reflected a Bush bias, and as one insider observed, the entire committee reflected a Bush bias and inten-tionally so.[108] There is evidence that patient advocacy groups lobbied the White House to appoint Christopher Reeve. Reeve was not selected, nor was anyone else chosen who was suffering from a genetic or some other disabil-ity.[109] Well before committee membership was finalized, an unholy coalition of genetics naysayers—Jeremy Rifkin for the left and William Kristol for the right—circulated petitions to support a ban on cloning. Sitting on Kristol's petition were the signatures of Francis Fukuyama and Robert George.[110]

As the PCB prepared to engage the cloning agenda, another round of controversial experimental exercises attracted media attention. By far the leading player was ACT, the small Worcester, Massachusetts, facility responsible for the cow-human preembryonic hybrid. Where was Harvard? Where was Stanford? Where was the Sanger Centre? Where, indeed, was Ian Wilmut? In November 2001, ACT scientists proclaimed they had created cloned human preembryos via SCNT. This was the same sort of declaration trumpeted by South Korean researchers, the difference being that ACT made no claims that it had mastered the technology to reproduce humans. Their goal, they specifically said, was strictly therapeutic. Nonetheless, the outpouring of negatives almost rivaled the chorus following the South Korean announcement. The pope and President Bush called the research unethical; German and French public officials stated that such protocols were illegal in their countries; scientists in the United States noted that none of the preembryos had reached the eight-cell stage, indicating fatal genetic malfunction.[111] At about the same time, ACT researchers reported that they had produced 24 healthy cattle, cloned from nonquiescent somatic cells. "Social interaction and behavior . . . are normal. . . . They have developed a social dominance hierarchy and the full spectrum of behavioral traits."[112] The inference beckoning was that cloned sheep may have problems, but not cattle properly cloned.

The scene shifted momentarily to the University of Missouri, where scientists cloned a pig deliberately stripped of a gene that had stood as a major barrier to xenotransplantation. It had been generally believed that pigs were the best source for organ transplants; a major problem was that the human immune system would reject a certain pig sugar and, in so doing, attack the transplanted pig cells. The optimal solution was to knock out both copies of the sugar gene; Missouri had succeeded in knocking out one.[113] Score another point for SCNT. Back to the veritable ACT cloning factory, where scientists demonstrated they could create stem cells via parthenogenesis employing monkey eggs. As the eggs were unfertilized, they were not preembryos in any sense and could not yield children if implanted. The good news was that parthenogenesis obviously finessed the "life begins at conception" or the "life begins with nuclear transfer" arguments. The not-so-good news, over and above the fact that making monkey eggs divide didn't necessarily mean one could make human eggs divide, was that parthenogenesis cannot produce all the matching organs that cloned preembryos could. After all, men don't carry eggs.[114] The very week in which the parthenote study was published, the same ACT research team informed the media it had produced functional cow kidneys from cloned cow fetuses. The scientists admitted they did not know yet how to derive stem cells from cloned cow preembryos; when they allowed the blastocysts to mature into fetuses, however, they could extract kidney cells that produced urine following implantation. We could never do this work in humans, bemoaned other leading scientists, and all ACT was doing was unleashing the wrath of Congress.[115] Finally, Texas A & M researchers, funded entirely by a wealthy eccentric who wanted desperately to clone his dog, reported that they had not been able to clone any dogs—too tough as yet—but they had cloned a cat. Named CC, cuddly initials for Copy Cat, the new addition was not a perfect replica, as random events during pregnancy help determine coat markings, the sort of phenomenon human cloning adversaries ought to keep in mind. CC was the only survivor out of 87 attempts.[116] On the day *Science* magazine announced Craig Venter was leaving Celera, *Science* also featured a story regarding a long-anticipated NAS study of human cloning implications. Confining itself to scientific and medical dimensions—perhaps given the PCB's charge—the panel's report sounded very much like an NBAC reprise. Reproductive human cloning was so fraught with risk it should be banned by statute, with the question of criminalization to be revisited in five years. Unlike the NBAC, the NAS group came out strongly in favor of human therapeutic cloning that it said should be called "nuclear transplantation to produce stem cells," a shift in terminology it thought might lower disputing voices.[117]

On January 17, 2002, the PCB held its first meeting. For those hoping to see a dialogue in which scientific research and procreative choice would share center stage with several oft-articulated competing social interests and values, the prognosis was gloomy. That morning President Bush's press secretary, Ari Fleischer, reiterated the chief executive's opposition to human cloning in all its permutations. Later in the day Chairman Kass convened the meeting, and the group expended all the natural energies one would expect to be pent up and ready for release on the threshold of this historic ELSPI moment by discussing, under his instruction, the great lessons intended by Hawthorne in his classic tale "The Birthmark." Women's groups had split wide apart on the therapeutic cloning question just as they had parted company in years past on the question of pornography regulation. Patient groups trotted out Christopher Reeve, and the liberal literati countered with Norman Mailer, who had signed on to the Rifkin cavalcade. Senator Brownback had enlisted a cosponsorship for his bill from Senator Mary Landrieu (D–Louisiana). Even some "greens" were supporting Brownback-Landrieu: cloning could well lead to genetic engineering, code language for "leaving the natural world behind." Most leading scientists and biotech advocates were digging in their heels, fearing the worst.[118] Leading the charge as the process unfolded was the president of the United States himself. While the PCB studied and discussed, Bush, who supposedly had convened his blue-ribbon team for the purpose of receiving best evidence, rallied the troops by inviting hundreds of interested parties to the White House, where he regaled them with anti-cloning rhetoric. In response, 40 Nobelists released a letter calling the Bush approach dangerous.[119] All this because the impending Senate vote was too close to call. And the contest became closer yet when Senator Orrin Hatch (R–Utah), a very strong anti-*Roe*ist, came out against Brownback-Landrieu and in favor of the Kennedy (D–Massachusetts)-Feinstein (D–California) bill, which would have criminalized reproductive human cloning but permitted therapeutic human cloning.[120]

In July 2002, the President's Council on Bioethics released its report. In a preface authored by Chairman Kass, the committee emphasized that "Human cloning, were it to succeed, would enable parents for the first time to determine the *entire genetic* [read: genomic] makeup of their children" (emphasis and bracketed words mine). This is an important and correct starting point. The panel and I agree that what is truly different about this new form of research, this new form of reproduction, is that it provides for genomic duplication; *it therefore represents a paradigmatic departure from previous related experimental protocols*. The preface then expressed the conviction that human cloning "raises for many people concerns . . . about eugenics, the project to 'improve' the human race. A world that practiced human cloning, we sense, could be a very different world, perhaps radically different, from the one we know."[121] This is even to a greater degree the stuff of a paradigm shift. It bespeaks as well a mood of foreboding, quite Kassian in tone, portending a consensus of constraint rather than a consensus of watchful, analytic scrutiny.

The report itself was sweeping in its focus and in its recommendations. Casting a broader net than the NBAC, it wrestled with ethical dimensions over and above safety issues and addressed therapeutic cloning. With the Congress already chest deep in all this, how could the PCB have acted otherwise? Beginning its inquiry with "Cloning-to-Produce-Children," President Bush's advisers unanimously urged a regulatory ban, not a ban cum sunset provision as the NBAC had touted, but a ban for all eternity. According to the entire membership, its larger locus of inquiry had led it to conclude that "*there seems to be no ethical way to try to discover whether cloning-to-produce-children can become safe, now or in the future.*"[122] This is breathtaking. Any attempt, steeped in basic research, to ascertain whether SCNT can be employed to produce humans is unethical per se and presumably ought to be proscribed. Did it not occur to the committee—to *anyone* on the committee—that environmental shifts, that even genetic shifts, might make human procreative cloning in the short term an adaptive strategy for our species? And if that assertion carries any weight at all, how can we prepare for such exigencies without the theory and data that only responsible research can provide?[123] The value judgment that there can be no such thing as responsible research in this area is dictum, and I would call it a dangerous dictum. The balance of the discussion of reproductive human cloning is old hat, enumerating the standard objections: 1) "identity confusion" for the cloned child; 2) progeny should not be artifacts; 3) cloning may lead to enhancement; 4) family dysfunction caused by intergenerational mixing; 5) societal trauma.

The committee then turned its eye to the human therapeutic cloning debate. Those who expected the PCB to opt for a ban on cloning for biomedical research to accompany the anticipated recommendation for a ban on reproductive cloning were in for a big surprise. Nobody, including Leon Kass himself, voted to go that far. In what must have been some sort of an attempt at compromise, ten committee members (the majority) supported a four-year moratorium on the practice, while a seven-member minority opposed even a moratorium, though it did endorse establishing a federal regulatory oversight mechanism.[124] The four political scientists split evenly. Professors Fukuyama and George were in the majority; professors Sandel and Wilson were in the minority. Complicating matters, the minority contingent bifurcated into two subgroups, the bone of contention being (what else!) "the moral status of the cloned embryo." The larger of these subgroups saw "serious moral concerns" arising from creating human cloned preembryos for research but believed that, on balance, the scientific benefits outweighed the costs. The minority subgroup, numbering no more than 3 members of the entire 18-member group, was not prepared to ascribe any particular moral status to these entities. The majority committee position is intriguing. I note, for example, this language: "We find it disquieting, even somewhat ignoble, to treat what are in fact seeds of the next generation as mere raw material for satisfying the needs of our own." What "seeds"? A man masturbating to produce sperm? No. Someone's somatic cell being placed in an egg that might otherwise be useless to anyone. "[H]uman individuals [have a] continuous history from the embryonic to fetal to infant stages of existence."[125] What "continuous history"? We are not talking here about a sexually created preembryo; we are talking about an asexually created preembryo. Nobody knows whether such preembryos can develop into viable embryos, then fetuses, and then babies. And nobody on the committee was prepared to permit scientists to find out! I will give the PCB this much: it did not simply bend to the president's total and complete desire, though for the next four years the difference between a ban and a moratorium would be difficult for scientists to discern, if Congress adopted the report's conclusions.

There is one final nuance regarding which I will not accord the PCB any courtesy whatever. A critical feature of ELSPI analysis in the context of public policy formation is always to ask this question: Is our solution constitutional? This is a book that takes the study of constitutions very seriously, and the most important constitution in the United States is the United States Constitution. There were and are two separate and distinct jurisprudential concerns plainly ripe for investigation. The PCB discussed the first of these in a most perfunctory way (see note 127 below) and had absolutely nothing to say about the second. I address each in turn.

Over the past decade, the Supreme Court has said that congressional control over interstate commerce is limited to economic matters. In one oft-cited case, the justices held that Congress could not make violence against women a federal criminal offense; therefore, it could not provide a federal tort remedy for alleged victims of such violence.[126] The question arises as to whether Congress could, consistent with the Constitution, enact either the Brownback-Landrieu bill or the Kennedy-Feinstein bill. If Congress cannot criminalize rape, how can it criminalize human cloning?[127] The answer to the question is that it is not necessary to wait around until the Court broadens out the Commerce Clause. The typical reproductive human cloning transaction, Congress could reasonably find, probably would involve a woman and a doctor working up a mutually agreeable contract or business arrangement covering the costs of the procedure and using the channels of interstate activity as conduits for their enterprise. It is not a prerequisite for Congress to demonstrate that reproductive human cloning clinics are procreative analogies to large steel companies or even family farms to make its "economic" case. As for human therapeutic cloning, I would only suggest that if a person can be punished for knowingly bringing obscene books across state lines even though the purpose be noncommercial, so manufacturers and scientists can be punished for producing, transporting, or procuring laboratory equipment via interstate channels for the purpose of cloning preembryos. Moreover, the notion that the ACTs of this world aren't businesses is ludicrous.[128]

The more important constitutional inquiry is whether the *Roe, Planned Parenthood* line of cases restrains the Congress from banning reproductive human cloning. Why, I have asked rhetorically before and I ask again, does not the judicially created right to "bear and beget children" include the freedom of a married couple who cannot have offspring via sexual means to have a child via asexual means? The PCB, had it faced the issue squarely, as

I feel it had a responsibility to do, could have finessed the issue fairly readily: decisions such as *Roe,* Leon Kass might have written and would have written, were never intended to cover somatic cell nuclear transfer; that being the case, the PCB declines to engraft upon the Supreme Court's procreative rights jurisprudence a generically distinct method of genetic transmission so dubious in its "morals and welfare" pedigree. Had the PCB accepted the constitutional challenge I have set before it, and even had it responded in the fashion I have outlined, it still would have had a professional obligation to conduct a literature review and construct a scholarly argument. This it made no pretense of doing. In fact, when one consults the PCB's bibliography, one does not find any reference at all to several sources that run counter to the majority's value conclusions.[129] Thus, Ira H. Carmen's book *Cloning and the Constitution* and his article "Should Human Cloning Be Criminalized?" are not even worthy of citation. I am in good company. James Q. Wilson's essay "The Paradox of Cloning" is also uncited. And on the specific point at issue—reproductive liberty—John Robertson's exhaustive study is uncited as well. Kristol and Cohen, Kass, and Fukuyama, yes; Carmen, Wilson, and Robertson, no. (On the score of committee and citation bias, I note as well that the PCB's senior research consultant was Eric Cohen.) In his article, Robertson, a leading biolegalist, roundly rejects the NBAC conviction—which would also be the PCB conviction—that one could neatly sever human cloning from the human reproduction jurisprudence of *Roe* and *Planned Parenthood.*[130] I think Robertson is incorrect. I also think the matter is not nearly as self-evident as the PCB obviously thought it was. For the record, let me say what presidential committees cannot say for obvious political reasons and what Robertson would never say: *Roe* should be overruled. In her *Planned Parenthood* Opinion of the Court, Justice O'Connor stated that *Roe* was entitled to the full benefit of *stare decisis* "because neither the factual underpinnings of *Roe*'s central holding nor our understanding of it has changed."[131] In such noteworthy decisions as *West Coast Hotel v. Parrish* (overruling cases voiding minimum wage laws and other deprivations of the now discredited "liberty of contract" doctrine) and *Brown v. Board of Education* (overruling cases upholding racial segregation in the public sector), the Court, according to Justice O'Connor, had given appropriate weight to the fact that society's values regarding these policies and practices had undergone dramatic change. I submit the same is true of *Roe,* not in the sense of what society thinks but in the sense of what science now says. In 1973 and 1992, the constitutional issue involved only abortion. Today, abortion means not just terminating the existence of an embryo or fetus, but the existence of an embryo or fetus the genetic constitution of which is now amenable to disclosure and the full *genomic* constitution of which can shortly be determined. In the next chapter, I will argue that even before the Celera/HGP joint success, a woman's right to abortion did not include the right to put upon her unborn children her preferences in the form of somatic cell gene enhancement. In the wake of the Celera/HGP triumph, I am now prepared to say that the Constitution no longer gives women alone the power to decide which human genomes shall live and which shall perish. As I earlier said in the context of patent rights and in the context of evolutionary theory: the human genome in all of its adaptive potential belongs to all of us.[132]

The Senate debate on human cloning again ended with a whimper. Not waiting to hear what the PCB had to say, probably as uninterested in its machinations as was the president of the United States, the two warring factions fought one another to a second standstill. Neither side could muster the 60 votes mandated under upper chamber rules to gain a victory; neither side could even agree as to which proposal should be voted on first.[133]

While we were waiting for the next Richard Seed to appear, the Raëlians reappeared. They announced the arrival of a cloned baby with more to come.[134] The media leaped into hyperbole, and the congressional debates were reinvigorated. Bona fide scientists fumed: show us the evidence. Brigitte Boisselier promised up and down to do so. The HUGO should have appointed a team of Watson, Collins, Venter, and Wilmut to investigate; it would have been a good public relations exercise, taking the play away from science fiction scam artists. In a matter of weeks the smoke had mostly cleared. The Raëlians weren't going to tell the world anything about who had cloned whom and how. The cold, hard facts of science dispelled the rest of the smoke and all of the mirrors: try though they might, researchers couldn't employ SCNT to make baby rhesus monkeys. There were too many "disarrayed mitotic spindles with misaligned chromosomes." "[R]eproductive cloning [in nonhuman primates was] unachievable" given today's technologies.[135] And for human primates? The vigil for the new Richard Seed continues.

E. The Constitution of Forbidden Biological Knowledge

As in chapter 2, there is a danger in my chapter 3 account that the raft of events and the convolutions of fortune will overwhelm larger patterns, larger themes. The risks in chapter 2 were less palpable because my aims were less ambitious. In those pages I needed to set down precisely how an empirical understanding of human genomics came to be; I needed to delineate the

ELSPI of human genomics discovery. I then tried to show the extent to which the larger meaning of those sundry implications required constitutional analysis in order to yield an underlying pattern. The bioconstitutional politics paradigm of human genomics investigation turned out to be what I call the Beltway Constitution, its norms, its standards of legitimacy. Here the challenge is greater. We need to appreciate how human cloning relates to the larger field of human genomics research, not in some scientific sense (I think we've done that) but in an ELSPI sense. We can achieve that end only if we descend to the depths of constitutional inquiry.

There is no Beltway Constitution of cloning research. The players who made human genomics a reality by sequencing billions of base pairs were Beltway actors, Beltway scientists, Beltway political orchestrators. Craig Venter, the classic outsider, never really left the NIH, his every action an attempt to synthesize the Beltway genome by bending it to his will. His Celera headquarters was in Rockville, not the northern climes of Scotland hard by the Firth of Forth and not the research village of Worcester. He was to the Human Genome Project what Ahab was to Moby Dick; the two could not engage their life's work without the other as foil. Cloning's intellectual roots, as I have shown, do not lie in genetics, though now cloning and genomics are joined at the hip. Cloning's intellectual roots lie in reproductive science, and reproductive science, unlike genetics, has traditionally featured few Beltway linkages. Congress has never taken seriously the option of regulating in vitro fertilization; the Supreme Court has never decided a single case involving the practice, so controversial a quarter century ago that Paul Ramsey and Leon Kass denounced it as fabrication. Cloning's rendezvous with the Beltway is a contemporary phenomenon, an artifact of the threat that someone might clone a human. To understand the dynamics of that threat, that fear, is the first step toward formulating the human genomics/cloning constitutional linkage serious analysis requires.

Anthropologists and historians of human community and culture tell us that for each civilization there is a body of knowledge that is taboo, whose content it is forbidden even for the great minds of past and present to probe.[136] For Western Civilization, much of this forbidden knowledge has traditionally involved human genetics, particularly human genetic experimentation. The race to uncover the structure of the human genome was remarkably free from forbidden knowledge taint. Some members of the political left in this country had their early doubts; yet even Jonathan Beckwith signed on as an ELSI NIH collaborator. The left contented itself with voicing storm warnings about misuse. It did not argue that the knowledge itself was forbidden, though it is still holding its breath. What we now see is that the human genomics paradigm has brought to the surface a set of forbidden knowledge taboos never before thought realistically conceivable; with the human genome unraveling in meaning as we speak, suddenly these "thou shalt nots" that earlier were only the stuff of fantasy are now the stuff of possibility. One of these forbiddens is reproductive human cloning, particularly where SCNT is the procedure of choice. This "forbidden" also includes research designed to prove that SCNT human procreation is even doable.

I herewith introduce a new concept to the bioconstitutional politics literature. It is the concept of forbidden knowledge. I am not talking about violations of the rules of political games; that turf, for political scientists, is well enough trodden. We know all about rule breaking, people for whom the ends justify the means. Nor am I talking about the larger constitutional politics literature, where, for example, book and film censorship have received close scrutiny.[137] I am talking about the substance, the content of biological science discovery. I am talking about knowledge that traditional society says must always be kept unavailable. The critical question for a nation state that adheres to a "republican form of government," then, is how will its citizenry choose to address this kind of forbidden knowledge. I opt for demystification. I opt for science. I opt for ordered liberty. I opt for a set of rules and norms that balance competing interests in a grand attempt to separate out, in some rational, patterned way, legitimate research goals and procedures from the illegitimate. We can quarantine scientific knowledge; we can even ban the pursuit of scientific knowledge. That is not the way of mature scholarship or even reflective adulthood as I think the Athenians understood those posterities. I certainly do not think it is the way of the First Amendment. Human cloning represents the first step in our dialogue of how to formulate the Constitution of Forbidden Biological Knowledge. That formulation is the apex of a pyramid called the ELSPI of Human Genomics.

4

Germline

A. Establishing a Regulatory Mentality

The history of science is a monument to human creativity. Scientists dream, ponder, tinker, hypothesize, gather evidence, experiment, and report findings if only to themselves. They carve out their little niches of inorganic or organic phenomena and seek to bring order, to bring a sense of system to the properties and to the forces they encounter. It is a tedious process of trial and error and of replication. And the entire process is driven by assumptions about how their research worlds should be studied, about what questions are worth asking and why. These assumptions are human constructs: paradigms. The entire process is also driven by assumptions about the proper conduct of research, the rules, the norms, and the understandings of appropriate scientific methodologies, standards of proof, and dissemination of information. These assumptions are human constructs as well: constitutions. All this is the quintessential stuff of free expression, of First Amendment activity.[1]

As science has become more powerful, as it seeks to explain ever more important, more fundamental aspects of human existence, it has departed the world of solitary investigation governed by the absolutes of freedom and entered a new constitutional world, the world of social responsibility and accountability. To the extent that science is a center of power influencing the attitudes and behaviors of the governing elites, of the general citizenry, so it becomes subject to the checks and balances we place on all centers of power. Certainly that has been true of science in the United States, where Americans have taken checking and balancing to new heights with dual constitutional equilibria of a national separation of political power and a division of authority between states and nation. Against the backdrop of this formal constitutional scheme has arisen an informal, "living" constitutional apparatus of scientific liberty tempered by scientific duty, a delicate set of standards and guidelines that form a critical element of this chapter's subject matter.

Scientists long accustomed to the absolute freedom of the laboratory can run afoul of societal strictures at three broad levels of endeavor. In the first instance, scientists report findings that offend the powers that be. That is, the content of their speech, of their experimentation, challenges heartfelt values. When the pope did not like what Galileo was saying, he placed him under house arrest. Attempts to regulate the *content* of the scientific enterprise constitute the most palpable affront to the research mission as a First Amendment phenomenon, and, fortunately, have occurred infrequently in the Western democracies inspired by the Enlightenment.

In the second instance, the fruits of science are taken up by either public or private centers of technological might and put to problematic social purposes. When this happens, the politics of reprisal and overkill may loom, and scientists find themselves being blamed for what technologists do. The First Amendment permits the people's representatives to regulate technology, but in the typical fact situation, the Constitution does not allow this discretion to justify direct impingements upon scientific discovery. However, scientists forfeit their accustomed freedoms when they become part and parcel of technology's designs, particularly when they serve the state for the purpose of creating weapons of mass destruction. The classic example for Americans, of course, is the Manhattan Project. The issue here is not whether the United States should have built or employed atomic weaponry. As Chief Justice Charles Evans Hughes so eloquently told us: "The war power . . . is a power to wage war successfully."[2] And when the enemy is the totalitarian fascism and communism of Hitler, Tojo, and Stalin, who is going to lose sleep over the contributions of science to the great struggle? Answer: certain prestigious scientists. And with the blessing of certain political elites, too. So it was that the Nobel Prize for Peace was awarded to Linus Pauling for his efforts to blow the whistle on the dangers of atomic testing and particularly his colleagues' silent avowal of, if not their overt participation in, his country's thermal nuclear research mission. The larger point is that already some scientists were questioning their role in harnessing knowledge to what they defined as antisocial purposes; in fact, they were also questioning the creation of knowledge that might by its nature pose antisocial consequences, thus linking content objections to technological, utilitarian objections.

In the third instance, scientists turn their attention from the "outward" study of physical and biological phenomena to the "inward" study of humans themselves, how they think and how they behave. In Pasteur's day, a researcher could study rabies prevention in human subjects without consulting oversight review panels, but no longer. Revulsion against Dr. Josef Mengele's "protocols" (though a Nazi ethics panel would have approved them!) and the notorious Tuskegee Syphilis Study conducted by the U.S. government, in which black male adults suffering from the deadly disease were given no medical assistance and instead were treated as objects of research without tendering their informed consent, have resulted in establishing institutional monitoring mechanisms.[3] The same lessons hold for DNA research. In figure 1 in chapter 3, the traditional "genetics paradigm" provides no examples of deliberate intervention at the molecular level. The field was simply too primitive. Science did have at its disposal all manner of nonhuman indirect manipulations such as plant and animal domestication and improvements in egg and milk production. At the level of our own species, one could cite such strategies as treatments for various genetic disorders, which incidentally have the effect of helping harmful genes survive in the gene pool, the manufacture of warm clothing that protects us from the elements while attenuating the role of "hairiness" genes, and perhaps the decisions smart people make when they mate with other smart people and the decisions beautiful people make when they mate with other beautiful people.[4]

As figure 1 shows, Paradigm II (the recombinant DNA revolution) changed everything.[5] Believing that they had opened a Pandora's box of potential calamities for humankind, the leading lights in the field (virtually all of them American) imposed a moratorium on their own and others' future investigations. One can cite sources extolling the Asilomar model as an exemplar of scientific responsibility. Unlike the Manhattan Project, goes the argument, geneticists were not beholden to the cause of national security, locked in deadly combat with amoral antagonists. Geneticists did what the physicists of the 1940s couldn't do and what the physicists of the 1950s and 1960s might well have done, so the argument continues. They blew the whistle on themselves and orchestrated a public debate that led to enlightened oversight, we were told then and now. One can also cite sources frowning on the new experiment in scientific accountability. Medical microbiologists had had extensive experience manipulating the most lethal known pathogens without ever declaring a moratorium on their research and without setting up an elaborate government censorship agency. Cancer epidemiologists well versed in the sorts of safety questions that nagged at the Asilomar assembly had never recommended a moratorium and a special review process on anything they did. The molecular biologists considered their research paradigm unique, because genes were involved, and genes, like atoms, are unique. Therefore, special tools of obligation and government supervision were necessary. To this, I add a political scientist's perception of the moratorium as I wrote about it twenty years ago and write about it today. The business of imposing a gag order on any subject-matter segment of the free expression marketplace is not sound public or private policy, no matter what the perceived (as opposed to the actual) danger. I emphasize, as I have done before, that we are not talking about reasonable times, places, and manner guidelines or even regulations on expression. These may well be necessary and proper—and in the laboratory venue, too—as exigencies warrant. Nor are we talking about calibrated restraints on certain especially dangerous experiments given that experimentation is part action, part search for truth; on balance, the need to regulate the former might outweigh the freedom to indulge the latter. I am inveighing against a total stoppage in dialogue; a total stoppage in communication; a total stoppage in experimentation; a total stoppage in marketplace discourse and discovery. As the years have gone by, we have seen more than once a responsible committee call for a moratorium on this and that form of genetic research. I have already spoken critically of some of these, and so I will speak with particular criticism about this initiative because it was the very first of its kind and set a bad precedent with which we still must reckon.

A critical feature of the Asilomar debates is that they focused solely on risk-related parameters. The assembly eschewed the guts of ELSPI—even ELSI—dialogue. The participants felt uncomfortable grappling with such "fluffy" value-loaded matters, and they did not see the need to invite scholars who had written on these subjects. They knew, of course, about the George Walds, that is, colleagues who wanted to ban their research. These were the same people who, later, would declare war on sociobiology and E. O. Wilson, the same people who would lambaste human behavioral genetics, and the same people who would sneer at the Human Genome Project. They ran an ideological gamut from the very liberal to the quite radical— namely, the "new left" in 1970s parlance—and the Asilomarites did not want anything to do with them. Hence, no ethical issues, no social issues, no political issues, just a legal issue or two were addressed, a smattering of law sufficient to follow up a moratorium with a viable oversight authority they could control. That authority would reside in NIH: a panel of distinguished scientists—themselves—that would decide what experiments would be funded and what experiments wouldn't be funded, that is to say, what experiments would be done and what experiments wouldn't be done.

B. The Early RAC Years

The Asilomar conference took place in February of 1975; the moratorium ended in June 1976. During the interim, the new NIH committee—the RAC—drafted a two-tiered laboratory containment system for recombinant DNA investigations. The first set of NIH guidelines, as they came to be called, dealt with physical containment, in other words, the laboratory setting. These preconditions were deemed necessary to protect against the escape of potentially dangerous microbial clones. The guidelines defined P1, P2, P3, and P4 laboratory safety criteria in prescient detail, with the last of these (P4 or "high" containment) including such exotic facilities as a special laboratory complete with airlocks and filters. The riskier the experiment, of course, the higher the level of laboratory security. The second set of guidelines dealt with biological containment. To guard against harmful consequences should a clone escape, the organisms (and, at that time, the only such experimental subject in use was the K-12 strain of *E. coli* into which genes had been spliced) had to satisfy certain minimal conditions of incapacity, depending upon the level of experimentation. The guidelines spoke of EK1, EK2, and EK3 specimens, running from the least disabled to the most disabled. The EK1 was so safe, however, that it could not survive in the human intestinal tract, while the technology did not even exist to create an EK3 variety. The two sets of prophylactic criteria conveyed an air of supreme rationality and dovetailed elegantly: for example, DNA extracted from plant viruses could be recombined into an *E. coli* plasmid under either P2-EK2 conditions or P3-EK1 conditions.

These guidelines were at the least very strict and at the most unprecedented. Thus, P2 constraints had long been considered sufficient for research involving such pathogenic bacteria as *Cholera vibrio,* and the entire concept of biological containment was unique to the gene-splicing context. Moreover, the guidelines specifically precluded federal funding for any experiments employing DNA from organisms *known* to generate hazard, such as the creation of bacterial recombinants containing genetic fragments of toxins and tumor viruses. In sum, the containment criteria were designed to regulate research where *no evidence* of risk had yet surfaced but where *no certainty* of risk-free consequences had been demonstrated. Also proscribed was the deliberate release of clones into the environment, meaning field testing.

The legal apparatus set out in the guidelines was as convoluted as the safety standards. A researcher at Harvard would, first, have to obtain approval from a campus unit called an institutional biosafety committee

(IBC), which would review all proposed projects for their standards of containment. Only after the IBC signed off would Bethesda get involved. In short order, the NSF, the USDA, and other relevant federal government granting agencies adopted the NIH guidelines. But again, we must not forget political context. During this time, Cambridge mayor Alfred E. Vellucci led a movement to ban recombinant DNA research from the city; at a National Academy of Sciences forum, Jeremy Rifkin yelled out to all concerned, "Let's open this conference up [to the people] or close it down!" a war cry reminiscent of the campus disorders and violence of those years; Robert Sinsheimer suggested that all microbial cloning should be done in a P4 facility run by the federal government; others advocated a ban on *E. coli* cloning because that creature preferred the human intestinal tract. Both President Carter and Senator Kennedy brought forward detailed legislation treating recombinant DNA research as inherently suspect and laying down elaborate licensing provisions for protocols and facilities backed up by tough penalties. Fortunately, none of these proposals was enacted, and the RAC went about its bureaucratic business as usual.

It would not be long before storm clouds gathered on the other side. James D. Watson provided legitimacy for a scientific *putsch* aimed directly at the Asilomar jugular: "I find almost universal agreement among leading molecular biologists that (our miserable) guidelines are a total farce."[6] Some of Watson's colleagues had developed new host vector systems that, they argued, eliminated all potential hazards even if the creatures could survive outside the laboratory, itself seemingly an impossible feat. By now investigators could also show that genetic exchanges among microbes in nature were far more frequent than had previously been understood. The forces for deregulation seized the initiative during 1977 discussions, training their guns on the barely dry guidelines. The burden of proof had shifted, the NIH director announced, to advocates of government oversight. The revised guidelines, as finally approved in July 1978, scrapped the rigid categorical format described above. The British "case law" approach, affording both exceptions to specific recombinations on the prohibited list and total exemptions to various other experiments, would now furnish the RAC with a discretion hitherto lacking. However, IBCs were now obligated to choose at least one member from the larger community, a constraint seemingly at odds with the salient First Amendment freedom inhering in campus academic freedom. The new provisions would also be extended to cover all cloning projects sponsored by institutions where *any* of its personnel received agency support for gene splicing. It deserves mention that the NIH had now held 19 hours of public hearings on the issue of guideline amendment and that all proposed revisions had been published in the *Federal Register* in a plea for outside commentary. As a final gesture to public participation arguments, the director, realizing that the RAC's biological sciences and public safety agenda inevitably involved ELSI questions, expanded the committee's membership to 25 and included therein seven professors of law, education, environmental studies, and occupational studies and even one ex-legislator. The only real opposition to all this came from environmental groups, which argued that no projects previously placed on the prohibited list should be approved unless a full environmental impact statement had been formulated. Now we had a new dubious constitutional wrinkle in the mix: laboratory experiments, pregnant with free expression implication, should be halted until environmental impact reports had been developed. Some of us did not know the laboratory was part of the "environment," particularly elaborately enclosed P3 and P4 facilities.

As the years inched by and as the microbial cloning field attracted a bevy of talented hands, research procedures and findings proliferated impressively. With each RAC meeting came a new call to revise the creaking guidelines, which simply could not cabin what science had to offer. The typical RAC gathering at Bethesda—there were four of these a year, with members coming in from all over the country—was not during this time a picture of protocol assessment; it was a study in administrative tinkering. All of this pleased the director, whose essential message was: these are guidelines, not regulations; we should not be fearful of fine-tuning them. But to recombinant DNA scholars, the bureaucratic turnstiles through which so much of their work still had to negotiate, with full knowledge that failure on their parts might mean a loss of financial support and professional opprobrium, certainly seemed to carry the pressure and sting of regulations. The Martin Cline case demonstrated this graphically. He will go down in medical history as the first person to attempt to transplant cloned genes into human subjects, patients suffering from beta thalassemia. A UCLA faculty member who performed his experiments in Italy and in Israel, he failed to provide proper notification to either his campus IBC or appropriate review panels abroad. NIH responded by putting him through an administrative ringer, ultimately yanking two of his grants and seriously crippling his career. In 1979, the RAC voted to free most *E. coli* K-12 work from the main thrust of guideline coverage, prescribing only P1 restrictions and requiring principal investigators to file only registration documents with their IBCs. RAC notification was deemed unnecessary, and IBC preapproval was also eliminated. Feeling heat from his HHS superiors, the NIH director modified these recommendations, according the *E. coli* K-12 research at issue P1-EK1 status and, where eukaryotic genes were in play, retaining IBC censorship powers. The important point here is that RAC decisions have always been subject to the NIH director's approval, and on rare occasions he has seen fit to revise or even reject RAC determinations. These new changes went into effect in 1980. At this point, about 50–60% of all recombinant DNA protocols were now exempt under the guidelines, so researchers in these instances had a free hand.

Slowly but surely the NIH regulatory artifice continued to unravel. The Asilomar establishmentarians on the RAC moved in 1980 to abolish prior review at both their own and IBC levels; however, a middle-of-the-road faction joined the "regulationists" (on this continuum, the political left) to retain the latter. Not only did the RAC remove itself from the censorship business, but it even removed itself from the central registration business. So while national guidelines continued to prevail, national headquarters would henceforward keep no official records of who was constrained by them. Query: How could any meaningful policy assessment of the guidelines be developed, if federal officials kept no files on experiments they considered worthy of unique legal oversight? During this period, a group of California scientists successfully instructed *E. coli* to produce the hormone somatostatin; they then contrived bacterial hybrids capable of manufacturing human insulin. Shortly thereafter, researchers located the human gene for gamma interferon and induced its production. Both human insulin and human interferon genes were transplanted into fertilized mouse egg cells, providing important steps in understanding genetic control systems.

Finally, the establishment announced that it had had enough.[7] Its members on the RAC filed a motion in February 1981 to: 1) remove all compulsory features from the NIH guidelines; 2) institute a voluntary containment regimen with P1 the generally acceptable standard; 3) retain the list of prohibited experiments to which the RAC would continue to provide exceptions. There was little debate at the April 1981 meeting over science and safety. The notion that microbial cloning was per se hazardous had long run its course. Not one calamity had arisen from this research paradigm as implemented at the bench. But there was plenty of political debate. "Are we moving too fast?" "Will the public accept what we do?" "Should we retain guidelines simply out of fear of the unknown?" "If we stop regulating ourselves, will others jump in?" Meeting followed on meeting; report followed on report. A key player in the debates was RAC chair Ray Thornton, the president of Arkansas State University and a former congressman well versed in Beltway rough and tumble. He was prepared to accept incremental guideline readjustments to facilitate important research contributions; he was not prepared to accept "cut and run" strategies. He emphasized that science capable of great good is also capable of great harm, and that the public had a vested interest, given its financial stake as subsidizer, in any *"dangerous or ethically or socially offensive experiments."*[8]

Perhaps the most important RAC meeting of these early years took place in February 1982. Again the magnet of disputation was politics, and the clarion call became one of "public consent." Said one member: The correspondence we have received from concerned parties shows that the public is not prepared for recombinant DNA guidelines as mere recommendations. Said another: Legislators don't really care about scientific assessments, only public reaction; we must tread warily in what looks like an unstructured environment. The hopes of the voluntarists were fading fast. In the end a motion was approved retaining the entire NIH and IBC coercive machinery but lowering substantially the various containment provisions. Of particular note, deliberate release of recombinants into the environment, deliberate alterations of microorganisms with respect to drug resistance, and deliberate use of various toxins lethal to vertebrates would no longer be prohibited, but they would need RAC review, NIH certification, and IBC preclearance. Also, the insertion of alien DNA into animals and plants would now be a matter solely of IBC censorship, with the RAC removing itself from the process. Not a word was uttered, nor had a word yet been uttered at any RAC meeting, about putting exogenous DNA into the bodies of human beings. The RAC assembly adjourned with the mandatory "guidelines" still in place.

C. Somatic Cell Gene Therapy

By 1985, phase 1 of the recombinant DNA debate was ending; phase 2 was about to begin.[9] The scene shifted from the laboratories at Stanford, Yale, and Harvard to the laboratories at NIH. The question now was whether recombinant DNA investigations could be harnessed in the fight against monogenetic recessive disorders and against cancer. The further question was whether human gene transfer—by which was meant transducing somatic cell nuclei, the human subject's genome, with something's or someone's DNA—was "doable." And if "doable," was it safe? And if safe, was it ethically responsible, socially acceptable, politically viable, and therefore fundable because it was consistent with the "law" of the NIH guideline constitution? The major players were no longer James D. Watson, Walter Gilbert, David Baltimore, and Norton Zinder, leading molecular biologists with essentially academic and basic research connections. (Not that these "seniors" faded away; many simply commenced new careers as cutters into another DNA policy pie: the Human Genome Project.) Now the major players were doctors of medicine, both basic and applied researchers, affiliated largely with federal government installations: W. French Anderson, R. Michael Blaese, and Steven A. Rosenberg. These M.D.s, who were caring for very sick patients, some of them terminally ill, were not merely interested in gene transfer experiments for purely scientific purposes; their concern lay in somatic cell gene therapy, that is, using transfer strategies to alleviate suffering, if not to provide cures.

Naturally, the safety issues were daunting. It had come to be appreciated that one could not "routinely" replace an abnormal gene with a normal gene in any human cell, much less thousands of human cells. Some sort of vector was required to carry the exogenous DNA to its nuclear destination, and the thinking was that the most workable vehicle was a retrovirus. But suppose the virus was still replication-competent? And what about the danger of insertional mutagenesis following nuclear integration? Also, to what extent was research using animal models a necessary precondition? Some of these questions had ethical overtones; however, I have never been convinced that ELSPI scholars have any business lecturing to biomedical specialists, or molecular biologists for that matter, about research risks and safety standards. How much science and how much medicine do we ELSPIists know, anyway? That still left us with plenty to ponder: Is human somatic cell gene transfer qualitatively distinguishable in some ethical or social sense from human somatic cell gene therapy? Are either or both qualitatively distinguishable in some ethical or social sense from other forms of transfer and therapy? If there are differences, what special rules or guidelines are required in order for gene transfer or gene therapy protocols to pass muster? On whom is the burden of proof, and what is that burden? To what extent should the public, should elected officials, be involved in the guideline-drafting process and in the protocol review process? If the patient must tender informed consent, what is meant by "informed consent" and precisely what information does the patient require? Finally, and perhaps most significantly, to what extent can we distinguish between "good" and "bad" genes in a context uncompromised by ethical and social bias, once we get beyond obvious cases? I cite but one instance of oversimplification. During this time, a well-respected geneticist wrote:

> Gene therapy . . . is . . . conceptually no different from any therapy in medicine that attempts to improve the health of the sick patient. The only difference is that DNA rather than other biologicals, drugs, or surgery is used as the therapeutic modality.[10]

The first problem is that in an essay purporting to discuss "genetic manipulation," nothing is said about human gene transfer (rather than therapy) as a biomedical tool. The second problem stems from the word "conceptually."

Experiments are not conceptual exercises; they must be judged as totalities of facts in order to determine their reliability and validity, not to mention their standing as "research expression" under the Constitution. If a retrovirus, by its nature, scrambles the genomes of patients, and if a retrovirus is the only way to deliver ameliorative DNA, then where are we with our conceptual abstractions?

I have described the unique regulatory features then in place for recombinant DNA experiments in the United States, which covered, in reality, most of the cutting-edge work. As Anderson, Blaese, and Rosenberg contemplated what would be the first-ever human gene transfer investigation, they confronted, of course, precisely that array of turnstiles and, much to their astonishment and frustration, a great deal more. As NIH scientists, their protocol would require IBC approval. It would also require approval by a "campus" institutional review board (IRB). In the 1970s, Congress had mandated all academic institutions seeking federal funds on behalf of their faculties for investigation involving human subjects to set up IRBs. Of course, the law applied to federal installations as well. These committees were required to review in advance relevant research proposals and ascertain whether the human objects of research were being placed at risk. The IRBs were given the discretion to balance the element of risk against the sum of the benefits that might accrue to participants and to the scientific community.

Let us say that a protocol received IBC and IRB go-aheads, then what? In November of 1982, one of the earliest presidential blue-ribbon bioethics panels published a report on "genetic engineering with human beings."[11] I have elsewhere discussed the commission report.[12] Here, I merely note that this committee, which sat at President Carter's elbow, was far more sensitive to regulatory nuances than either President Clinton's NBAC or President G. W. Bush's PCB. The panel specifically lauded the RAC "as the lead Federal agency in genetic engineering" with an estimable track record. It went on to support a "'next generation' RAC," armed with a broader mandate and divorced from NIH to avoid conflicts of interest.[13] No such "second RAC" ever came to be, as Beltway panels are forever recommending new genetics oversight bodies and then observing how their recommendations gather dust. The RAC itself, however, was already in place; after a decade of dexterous footwork not only had it managed to survive, but now it was about to commence a second life. In a sense, it had already entered the jurisdictional lists by placing the insertion of recombinants into human bodies on the NIH guidelines' enumeration of prohibited experiments; in the mid-1980s, with scientists beginning to talk about the prospects of gene therapy, the committee decided to take a bolder regulatory step. In 1985, the RAC created the Human Gene Therapy Subcommittee (HGTS), whose job would be to formulate special procedures and workplace criteria to govern the new field. This subcommittee produced a document entitled "Points to Consider" to guide those who would one day approach the RAC for clearance. The tacit assumption underlying these "Points" was that human somatic cell gene therapy was not per se unethical, was not per se unsafe, and was not per se without redeeming social value, but that the closest possible supervision was necessary because the work was both provocative and unique. The initial subcommittee draft was tougher than tough and exceedingly naive. The panel, featuring a robust ELSI representation, would have required investigators to speculate on whether their experiments would lead to "(a) germ-line therapy (b) the enhancement of human capabilities . . . or (c) eugenics programs." Given that that question would have sparked a lively debate between Leon Kass and Arthur Caplan, the HGTS wisely shelved the query. Furthermore, the initial draft implied that the scientists would have to provide lower-order primate experimental data. That provision was also made more flexible and reasonable; mouse or dog experiments might be just as fruitful. As finally approved by the subcommittee and the RAC, the latter voting on September 23, 1985, the "Points to Consider" asked scientists why they planned to do what they were doing, how they planned to do it, and how they would operationalize the term "patient informed consent." The "Points" specifically warned researchers not to include proprietary information or trade secrets in their applications, and they stated in no uncertain terms that the NIH would accept for consideration only *somatic* cell protocols.[14] Under the new regulatory regime, Anderson, Blaese, and Rosenberg would be obliged to go to the subcommittee first and then, if all went well, to the RAC itself. Finally, the FDA would have to go along. When the "Points to Consider" first became an official blueprint, human somatic cell proposals seemed to lurk on the horizon. The research, however, proved more intractable than anyone had anticipated, and during the next three years the HGTS received no viable protocols.

Who were Anderson, Blaese, and Rosenberg?[15] Anderson was an expert on the workings of the hematopoietic system and had for years championed the prospects and potential of gene therapy. He believed, as did many others, that the best chance for a breakthrough lay in targeting one of the rare fatal diseases caused by a single gene mutation. Anderson's candidate was adenosine deaminase enzyme deficiency. But a basic problem nagged: where in the human body does one introduce the remedial DNA substitute? Tests seemed to show that it was not sufficient to channel these agents into the bone marrow cells. A better plan was evidently to insert the alien genetic material into the blood stem cells; the technology to do so, however, resisted solution. For Anderson, all this added up to a brick wall. Blaese was an expert in T-cell production, and T-cells have long-term memories; they may divide, but the immunity they confer carries on. He had been teaming with Anderson in devising protocols for immune system pathologies, and his position at the National Cancer Institute put him in close touch with Rosenberg, one of the nation's leading oncologists. Rosenberg was of the view that the standard cancer treatments—surgery, radiation therapy, and chemotherapy—left much to be desired. His experiments, to this day, take a fourth road: biologic therapy. In this new medical paradigm as then conceived, a patient's immune lymphoid cells were removed from the subject's body, induced to recognize and destroy cancer cells, and then reintroduced into the subject as an immune reagent. So the individual's own immune defenses become the chief remedy, aided and abetted by a booster that Rosenberg called "TIL" (tumor-infiltrating lymphocytes). He had received FDA permission to begin trials in humans, where he had achieved mixed reviews with patients suffering from metastatic melanoma. So he had his own set of brick walls: (1) Why didn't all patients respond favorably? (2) How long do the TIL persist in vitro? (3) Where do the TIL lodge in the body? (4) To what extent do longevity or location correlate with clinic effect? (5) Was is possible to recover the TIL? What Rosenberg needed was a genetic marker to monitor the TIL and their offspring; what Anderson needed was a procedure to make an alien gene a functional component of a normal genome.

The Anderson-Blaese-Rosenberg (ABR) protocol included the following basic steps, proofs, and rationales: isolate TIL cells from a patient's tumor and grow them in culture in the presence of interleukin-2 as a result of which the lymphocytes will multiply and the cancer cells will die; incubate the TIL with a retroviral vector, which itself has been altered through a process of DNA recombination, to contain the bacterial gene coding for neomycin resistance (NeoR); the gene marker will become part of the DNA of each transduced cell; the vector of choice is the Moloney murine leukemia retrovirus, stripped of those genetic components that cause cancer in mice and packaged in a murine amphotropic envelope; tests on monkeys had failed to demonstrate malignant transformation by this retrovirus; the NeoR produces an enzyme capable of inactivating a substance that kills eukaryotic cells, so cells expressing the NeoR gene can be detected by their uninterrupted division when incubated in toxic levels of this antibiotic; cellular material that does not express the marker gene will be destroyed once the TIL are grown in media containing the antibiotic; inject the hybrid TIL into the subject's veins. Summing up this extraordinary set of "firsts": recombinant DNA research would permit a bacterial gene to be synthesized with rodent viral material, the mutant artifact to be inserted into a stricken cancer patient, with the gene having the capacity not only to survive but also to monitor the activities of an entirely new species of tumor fighter, the recipient's own souped-up white blood cells. N.B. The protocol was not designed to cure anyone or even to do the patients any good at all. The protocol involved transferring genes into the human body in hopes of adding to the fund of biomedical knowledge. But now would come the hard part: could the ABR team convince the political system to let them proceed?

Convening on July 29, 1988, the HGTS decided to interpret its jurisdiction broadly, to use this experiment as a case study for testing the process it felt commissioned to orchestrate even though the protocol itself was a transfer study, not a therapy study. The panel, however, also decided to err on the side of caution: it voted to defer judgment pending submission of more detailed evidentiary proofs. It took the position that the experiments weren't going to help anyone, so there was no particular rush. The ABR team was taken aback. It had already received positive reviews from not one but two IRBs plus the NIH IBC, and it had filed a Hospital Impact Statement. Furthermore, the team wanted to publish its mouse study findings in leading scientific journals; HGTS open-meeting submissions could compromise acceptance. So ABR decided to appeal directly to the parent committee.

The October 3, 1988, RAC meeting was surely one of the most important—and one of the most controversial—scientific gatherings convened in the United States. It is to this day the most important RAC meeting ever. Taking the floor to defend the protocol, Anderson noted that the RAC, like the HGTS, conducted open meetings, so he could not present pertinent raw data on the record for fear of undercutting publication opportunities. He did invite RAC members to examine his team's FDA submission off the record, and he promised to present as much information as he could in summary form via slide display. Let me stop right here. If a government review panel is going to represent the American people and not only encourage provocative, socially useful science but also guard against provocative, risky science, it needs all the information pertinent to the inquiry. A researcher's publication agenda must take a back seat, even if publication is vital to the larger cause of science. Anderson stressed that the experiments had been under review for six months, that because the recipients were terminal they would assume no risk, and that future patients might be big winners. RAC member Richard Mulligan, a prominent human gene therapy investigator in his own right, countered by saying he needed to see more animal model data and that he was quite sure his vector was more effective. Anderson retorted that the RAC was not a study section charged with making funding decisions.

Other RAC members now joined in the free-for-all. Some said the parent committee should not undercut the subcommittee, to which Rosenberg said the team deserved a vote then and there. Others believed little could be gained from more animal model research because rodents weren't human beings. A respected black geneticist, Robert Murray, stated: "If we required [mouse model work] for in vitro fertilization and sickle cell research we'd never get anywhere."[16] At this critical juncture, Harvard Medical School bacteriologist Bernard Davis took the floor. There is a relevant axiom in medicine, he began: the sicker the patient, the greater the degree of latitude a practitioner must be accorded. Also, the role of animal experiments is to do that which cannot be done with humans, a factor inapplicable here. "We have become bogged down in scholarly and bureaucratic nitpicking." Davis's statement carried the day, and, after some parliamentary wrangling, the RAC approved the ABR protocol 16–5, with the dissenters mostly those who felt HGTS responsibilities had been compromised.

It came as a shock to all concerned when the NIH director voiced his displeasure with the entire episode and insisted that the issue be reconsidered. He ordered the HGTS to conduct a de novo review based on an examination of all available data, and he warned the leading scientific journals that duly constituted NIH committees "will not be held hostage" by their editorial policies.[17] He accused the RAC of short-circuiting the HGTS, and he rejected the 16–5 vote as unduly divisive. In this delicate area, he said, the RAC should strive for unanimity. All of this conveys a preoccupation not with genetics but with politics, and the idea that serious politics requires unanimous votes would have shocked the Founding Fathers and every serious political theorist I know.

On December 9, 1988, the HGTS convened to readdress the protocol and, after a three-hour discussion, agreed unanimously to approve it. Virtually the entire discussion of note involved political questions: What was the proper review role of the RAC? Should it confine itself to safety issues or address also the merits of the research? Did it have a constitution or merely a charter that the panel could change by majority vote? Within a month, the RAC, by mail ballot (an unusual procedure recommended by the NIH director himself), had concurred in the HGTS's positive recommendation, and on January 9, 1989, the director made NIH approval official. The FDA also signed off.

The next meeting of the RAC was slated for January 30. Present for that occasion was Jeremy Rifkin, probably the most determined antagonist of genetic engineering research in the United States. He accused the RAC of gross procedural irregularities and then trained his sights on the ABR investigation. He called the committee a self-serving scientific elite; he stated that human subjects were being treated like guinea pigs; he called mouse models irrelevant because humans were involved. Then he dropped a real bombshell: "I am filing suit today in federal district court to enjoin this research." Nor did he stop there. The RAC should approve a moratorium on all human gene transfer experiments until the NIH had established an Advisory Committee on Human Eugenics that would supplement the RAC's safety expertise with a broader ELSPI capacity. According to Rifkin, the entire purpose of this research agenda was to employ genetics to improve the human condition, which he called the essence of eugenics. "I realize," he went on, "that the RAC and the scientists here aren't a bunch of Hitlers, but perhaps because I am Jewish, I worry about new notions of eugenics."[18] Here we go again! Isn't it fascinating how many American left-wing ideologues are hung up about their Jewishness! Steven Rosenberg, of the ABR team, is also Jewish, as was Harvard professor Bernard Davis.[19] Why would they be less sensitive to eugenics than Rifkin, given their Jewish credentials? Maybe because their religion (ethnic identification?) is not in their cases a rationalization for acting out against authority, especially when the authority figure is DNA.

The RAC stood fast against this last proposal, voting overwhelmingly to advise the director to reject Rifkin's plea for a new committee. On May 17, 1989, a federal district court judge announced an agreement between Rifkin and HHS. Rifkin withdrew his opposition to the ABR initiative, and HHS promised not to conduct any more votes by telephone or postal service. The first human gene transfer experiments could now proceed.

The Anderson-Blaese-Rosenberg protocol did not yield historic biomedical findings; it did yield an ELSPI biomedical genetics bonanza. It paved the way for the first human gene therapy investigations. Regulatory bodies were now prepared to judge these experiments on their own terms and not as representing a grand paradigmatic voyage into the unknown. The next round of policy decisions commenced less than a year later when Blaese and Anderson came forward with a specific plan to treat the aforementioned adenosine deaminase pathology (ADA enzyme deficiency), dubbed by the popular media the "bubble boy disorder." Those few—only about 30 in the world—who lack the enzyme suffer from a loss of T-cells, are susceptible to the most commonplace of infections, and rarely live into their teens. The BA research approach was as follows (and note how it built on the ABR protocol): take blood from the child patient; isolate the T-cells; grow them using interleukin-2; transduce their nuclei with a Moloney murine, replication incompetent, retroviral artifact altered to carry both a functional human gene for the ADA enzyme and the NeoR bacterial gene for tracking purposes; inject the now cultured T-cells into the patient's blood stream. The hope was, of course, that the transduced T-cells would produce enough enzyme to forestall immune system compromise, but a critical question was how long the recombinant T-cells would do their good work, which would, in turn, determine the schedule of reinfusions. Nobody was prepared to say that the BA protocol could effectuate a permanent cure.[20]

On March 30, 1990, the HGTS took up the Blaese-Anderson battle plan.[21] After a lengthy presentation by principal investigator Blaese, the subcommittee doubting Thomases waded in with questions. A new enzyme replacement regimen called PEG-ADA had recently become available, and several of the youthful subjects who ordinarily would be candidates in the BA trial were doing well with that drug. Living in an artificial bubble seemed clearly to be a thing of the past. Why put these children to unnecessary risks with your far more speculative, invasive treatment? Blaese and Anderson retorted that PEG-ADA required expensive weekly injections and was most certainly not a cure. On the other hand, their protocol was sensitive to PEG-ADA's virtues. Operating under the assumption that their subject patients would be making full use of the therapy, they proposed to introduce the transduced T-cells gradually; these cells should gain a selective advantage over cells carrying the aberrant, defective gene and should proliferate just as had been shown to be the case following a bone marrow transplant from an HLA-matched donor. As the patient's normal ADA gene-count increased, the schedule of PEG-ADA administration could cautiously be reduced. Others wanted to know the half-life of the gene-altered cells in the human body. No clear answer could be culled from the literature. It was agreed that a battery of peers would be assigned to review the protocol and report back their assessments. The gut unarticulated choice, a choice of much larger dimension than the fate of the particular protocol up for review, boiled below the surface: Was a good, conventional therapy the best that science could do, given that the fatal disease was written in the patients' genes, or should the experimentalist's trial-and-error procedure be given an opportunity to adapt itself to the task of human genetic reprogramming in order to find the optimal therapeutic solution?

Significant changes in their protocol format prior to the next HGTS meeting greatly improved BA's chances for success.[22] Claudio Bordignon, a

Milan biomedical professor, had a patient suffering from ADA enzyme deficiency who was taking PEG-ADA. Bordignon had removed some T-cells from that patient, transduced them with functional human ADA genes packaged in retroviral vectors, and put them into mice stripped of their immune systems. Result: the human cells displayed good survival capacity, a direct result of ADA intracellular expression that itself lasted for at least two months. Nor was there any indication that the experiments had compromised the mice in other ways. An animal model now existed that showed that the transduced cells obtained a selective advantage over nontransduced cells if only temporarily, thus conferring on the subject creatures a prolonged life span. The BA team promptly enrolled Bordignon as a coprincipal investigator and cited his important study. To drive home their newfound advantage, the team magnanimously scrapped the most controversial part of their proposal by agreeing to leave their subjects on PEG-ADA no matter what improvements in health and lifestyle would be forthcoming. They had touched all the critical bases. The question as to whether their gene therapy protocol was a better therapy than PEG-ADA, indeed, the larger question of whether their regimen acting alone could cure someone, would have to wait for who knew how long. The next HGTS meeting on June 1 carried its share of thrusts and parries; now, however, the burden of proof lay with the critics. Charles Epstein of the University of California at San Francisco, who would later suffer a severe hand injury in the wake of a Unabomber attack, deserves much credit for framing the final motion of approval, which passed unanimously. My first meeting as a RAC member took place on July 31, 1990, and my first vote was to join my colleagues in ushering in the era of human gene therapy by supporting the Blaese-Anderson protocol.

During my four years on the RAC, we spent virtually all our time reviewing human gene therapy proposals. By the time I left, in the middle of 1994, we had approved approximately 70 of them. As of January 1999, the RAC had given the green light to more than 100. That figure is misleading because throughout these years, we let loose our jurisdiction over less controversial items, taking into account the rapid progress of the field and our enhanced understanding of safety and efficacy questions. As of that same date, there were approximately 260 ongoing human gene therapy investigations in the United States, and virtually all of them since 1996 have gone directly from in-house review committees to the FDA, the RAC having waived oversight. On the international front as of January 1999, thirty-four protocols had been undertaken, the important point being that the United States even now is the dominant player on the stage.[23]

Obviously, the critical ELSPI RAC debates took place during my tenure. The typical protocol we assessed early on was not a recessive disorder of the ADA type. It was a cancer therapy protocol exemplified by a Steven Rosenberg initiative to transduce TIL cells with the gene coding for tumor necrosis factor (TNF), a protocol we approved on the same day we approved the BA proposal. Some of these were gene transfer studies, where the purpose was to track marrow reconstitution in patients suffering from neuroblastomas and leukemia. Rosenberg himself supplemented his TNF therapy research with a study involving the insertion of the IL-2 gene as a cancer fighter. Slowly the research agenda expanded: James Wilson at Ann Arbor won approval to address familial hypercholesterolemia, Edward Oldfield at NIH became the first gene therapist to experiment on brain tumors, and John Barranger's target pathology was Gaucher disease. Other researchers homed in on HIV-1 patients. One of the most provocative investigations was initiated by Ronald Crystal, who sought to employ an adenovirus, rather than the standard retrovirus, to put the normal CFTR gene into the lungs of patients dying of cystic fibrosis. The adenovirus had the advantage or disadvantage of lodging in the cytoplasm; it did not transduce a cell's nucleus. Was this less dangerous for a patient? Were the chances of efficacy enhanced? What animal models were appropriate to test various hypotheses? The vector systems would become more sophisticated yet. Liposomes would be essayed; then adeno-associated viruses; today, lentiviruses such as HIV are in vogue. And the range of genes being introduced expanded impressively: the human alpha-l-antitrypsin gene, the *Herpes simplex* thymidine kinase gene, the human glucocerebrosidase gene, and so forth.

As the years went by, gene therapists chafed under the criticism that their exotic technology had failed to cure anyone. But at least it hadn't killed anyone either, as so many critics had feared at the outset. That is, until 1999. On September 17 of that year, 18-year-old Jesse Gelsinger, a subject enrolled in a University of Pennsylvania gene therapy trial, died.[24] When a certain gene on the X chromosome malfunctions, the liver fails to produce the ornithine transcarbamylase (OTC) enzyme required to clear ammonia from the blood. The disease is usually lethal, and it strikes the very young. Using an adenoviral vector, Penn's James Wilson, a leading practitioner in the field, hoped to insert functional OTC genes into his subjects' livers. What subjects? The Wilson team had evidently decided to enroll precisely the sick people one would think at first impression ought to be enrolled, OTC-deficient babies. But because babies cannot consent to anything, Arthur Caplan, the principal bioethicist on station, advised the researchers to substitute stable "adults," one of whom was Gelsinger. In addressing the question as to why the parents of these genetically compromised children could not tender approval in the name of their youngsters, as is the case in standard life-threatening situations, Caplan argued that in the circumstance at hand the parents would be "coerced by the disease of their child" and therefore could not provide *informed* consent. I once again fail to appreciate an instance of bioethical judgment that, in reality, is a political judgment. To use Caplan's pithy language, what we have here is a terminally ill child "coercing" its parents into silence, thus constituting a per se waiver of the opportunity for, if not treatment, at least experimental benefit in the good Samaritan sense. Perhaps the parents—and the not-spoken-for child—would prefer a plain and simple death and perhaps not. I happen to prefer the politics of individual autonomy to the politics of paternalism. It should be pointed out that the RAC approved the Penn study including the researchers' choice of adult subjects. The RAC did insist on one important change: the adenoviral vector must be inserted into a peripheral vein, not the hepatic artery. The protocol then went on to the FDA, which overruled the RAC on this point. Incredibly, the FDA never informed the RAC of its decision, and Wilson never informed the RAC either. I would argue that the RAC had just as much know-how, if not more, as the FDA on the point at issue, namely, the likelihood of vector migration to germ cells. The upshot of all this was a severe setback for the gene therapy field: Gelsinger's father sued the university; the university suspended Wilson's research with humans, and he eventually resigned his position as director of the Penn gene therapy institute; Congress conducted hearings on patient safety standards; the RAC tightened its controls over the field, mandating much more paperwork for investigators; and scientists pounded their chests in mea culpa demonstrations over whether they had truly been honest with the public about the risks attendant in adenovirus infusions. I would only emphasize that the "breakdowns in the system" were no more James Wilson's fault than they were ELSPI faults. Tragically, the precise cause of Jesse Gelsinger's demise to this day remains a subject of conjecture.

Fittingly, I end this section with a Blaese-Anderson redux. In 2000, researchers at INSERM (Paris) reported a major breakthrough on the gene therapy front.[25] They removed from their two young patients not T-cells but hematopoietic stem cells, transduced these mature cells (employing an improved Moloney vector) with a functional human gene to replace its defective counterpart on the X chromosome, and introduced the package into the subjects' bloodstreams. The patients had not previously been treated with PEG-ADA. In fact, they did not suffer from ADA enzyme deficiency at all but a related form of the generic pathology called SCID (severe combined immunodeficiency). The transduced cells obtained their hypothesized selective advantage over defective cells, and both patients were doing well enough after ten months. Clearly, the Cavazzana-Calvo study marked a big improvement in results over the BA studies, but why? Anderson himself cites two big reasons.[26] The Paris team employed stem cells rather than fully differentiated T-cells. As we saw in the discussion about human preembryonic stem cells, these pluripotent entities present extraordinary untapped therapeutic potential. Back in 1990, gene therapy investigators did not have the tools to harness stem cells, much less have at their disposal the growth factors used by the Parisians to enhance proliferation in culture. Moreover, Cavazzana-Calvo was not burdened by the PEG-ADA albatross. It now seems fairly evident that PEG-ADA administration decreases the selective advantage that ought otherwise to accrue to ADA-transduced cells. To this day, BA patients continue to receive their PEG-ADA infusions.

But in 2002, the field received yet another setback. One of Cavazzana-Calvo's patients had developed a leukemia-like blood disorder, and the Paris team halted its trials. According to reports, some of the vector had not migrated to the X chromosome; it had landed on chromosome 11, where it had caused a mutagenic event by hitting a sensitive coding region (the LMO2 gene).[27] (If it had lodged in an intron, probabilistically a very high likelihood, who would have ever known?) We had debated incessantly at RAC meetings the chances of an insertional mutagenesis episode, and now it had happened. Was the whole research agenda fatefully compromised? It hardly appeared likely; too many patients over the past decade had emerged unscathed. The chances are that the investigators had enrolled a subject who should not have been there; some evidence has surfaced that even if adult patients had been fit subjects for the Penn protocol, Jesse Gelsinger ought not to have been included. In other words, the prime villain is probably human error. The upshot of the Paris tragedy was that the United States, France, and Germany suspended all roughly analogous gene therapy experiments, while the United Kingdom forged ahead.[28] Then, just when the RAC had persuaded itself that this calamity was a fluke, a second child in the same study came down with leukemia. Result: extension of the clinical hold. This time, London researchers joined in. Tests showed the LMO2 gene was again likely implicated.[29] As of this writing, gene therapy studies are back on track except those similar to the Cavazzana-Calvo protocols. C'est dommage. What was for two years the first unabashed example of a human gene therapy cure has been shown to be a cure only for the properly diagnosed. If only "proper diagnosis" was a Betty Crocker recipe.

Meanwhile, Bordignon's team at Milan reports another major success: ADA-compromised patients who had not received PEG-ADA seemed to be in good condition following the introduction into their bodies of hematopoietic stem cells transduced with a "good" ADA gene, where the cells were subjected to a nonmyeloablative conditioning regimen to enhance proliferation.[30] Maybe Blaese and Anderson had it right (at least conceptually) ten years ago. Let's fervently hope that one of Bordignon's patients doesn't become a cancer victim.

D. Enhancing the Genome

The world beyond human somatic cell gene therapy is vague and uncharted, dark and mysterious. Some would call it forbidding. It is a world that thus far has resisted the scientist's grasp. Yet it is a world often talked about: bioethicists have framed competing canons of do's and don'ts; biolegalists have developed hypothetical jurisprudential frameworks to capture the hypothesized rights and duties of major players. This is a world that I call human genetic engineering (HGE). Others have called it the world of eugenics. For me, there is a critical difference. Eugenicists seek to improve the adaptive capacity of the gene pool using whatever tools (legal, social, biological) that lie at their disposal. Human genetic engineering conjures up visions of tampering directly with somebody's DNA. People who use the term tend not to enjoy its prospects; "engineering" in this context carries very much a pejorative flavor.[31] Human genetic engineering is not a scientific concept. It is an ELSPI concept. As such, HGE must be understood as conveying an evolving dynamic. Thirty years ago somatic cell gene therapy was considered human genetic engineering. No longer. Now it is merely good research or bad research, though, unfortunately, usually not successful research. We have already encountered one species of the genus HGE: human cloning. With human cloning, however, we have a research field in which people like Seed and Antinori are actually talking about doing something as speedily as extraneous conditions permit; consequently, governments around the world are debating specific policy responses. With the various species of HGE I plan to discuss here, nobody is talking about doing much of anything. Therefore, we have no experiments to analyze and very little policy material to digest. And yet, as I say, the debates rage on. There is obviously something that strikes very close to home about even the fantasy of human DNA tampering, of HGE. What I propose to do is to provide a constitutional politics investigation of these debates. I want to understand the basic differences of opinion among the contending commentators. More, I want to flesh out the norm structures, the patterns of influence, the centers of power and authority, and especially the role of DNA, of genomics, fundamental to these commentators as they provide us with their competing worldviews. That is what constitutional politics is all about, after all, and bioconstitutional politics issues are no different.

By all odds, the least controversial form of human genetic engineering is in utero human somatic cell gene therapy, commonly known as fetal gene therapy (FGT). In FGT, the biomedical practitioner, knowing that chorionic villus sampling or amniocentesis has revealed the presence of a harmful mutation and having received the informed consent of the pregnant woman, employs the recombinant DNA tools described in the previous section to insert a functional human gene into the cell nuclei of the fetus. What makes the procedure so controversial is that the doctor and the mother are complicit in attempting to alter the genetic makeup of a "potential third party" or, as right-to-life advocates would put it, a "third party." No doubt the prospective mother would cite her reproductive liberty interests flowing from *Roe-Planned Parenthood* to justify a decision to address even by FGT the health of the fetus she carries. If she can choose to abort or not abort the fetus, then why can't she make scientific and medical decisions regarding the fetus' genetic makeup? The matter is not nearly so cut and dried. Justice Holmes, in a rather different context, once argued that Congress can abolish the post office whenever it wishes, but as long as the post office remains open Congress cannot discriminate among the forms of speech flowing through it.[32] The imperfect analogy suffices here. The pregnant woman's procreative rights do not extend to making of her fetus a guinea pig for biomedical experimentation. On the other hand, she has some discretion. A carefully targeted protocol aimed at deleting a life-threatening genetic disorder in the fetus, the only kinds of projects on the FGT horizon, would pass muster under my construction of *Roe-Planned Parenthood,* and certainly that is the case if, as most experts now believe, somatic cell gene therapy is just one of many treatment modalities. I doubt this argument would impress either the pope or Jerry Falwell and their legions of followers, even though the purpose of the procedure is to avoid aborting fundamentally flawed fetuses. So when FGT protocols eventually surface, as they obviously will, look out for more abortion-laden fireworks.

Bioethicists have not approached the matter as I have. They begin with a moral "ultimate": "ought it [the experiment] be done at all?" Only after their analysis yields a "yes" answer, do they pose what I regard as the critical questions: "exactly what is going to be done, to whom, when, and why?"[33]

The former inquiry invites worst-case scenarios to structure the discussion. Prenatal genetic diagnosis itself was initially characterized as a trapping of "Nazi science" and, when linked to abortion, a form of "infanticide." There is a fair question of why one would ever opt for FGT when preimplantation diagnosis of blastocysts is available in the larger context of in vitro fertilization. And yet a couple does not have to be Roman Catholic to insist that their baby be created via sexual intercourse rather than via test tube, if at all possible. Moreover, IVF always carries with it the intentional destruction of preembryos, a side effect not presented with FGT. The last resort of the moral naysayer is always the slippery slope. If society approves FGT, won't FGLGT (fetal germline gene therapy) be far behind? So formidable are the obstacles to the latter, as we shall see, that "There is no evidence that (FGT) would increase the likelihood of germ line genetic experiments."[34] However, advocates admit the risk is genuine; FGT, they caution, should probably be done early in the second trimester so as to provide an opportunity to assess efficacy and to give the prospective mother an opportunity to abort if the procedure either fails or causes collateral harm. The specter of the germline always adds to the HGE mystique.

W. French Anderson has taken the lead in touting FGT, playing the same role he fashioned 20 years ago in championing human gene therapy per se. He has proposed as model diseases for the procedure ADA-SCID and homozygous alpha thalassemia, as these are caused by malfunctioning genes whose regulatory regions have been rigorously characterized. Data from large animal models have yet to be garnered, however. Certainly removing blood from a fetus and reinjecting it armed with a retroviral recombinant requires elaborate dress rehearsal. It would also be risky for the pregnant woman should FGT fail to adequately redress a lethal genetic pathology in utero, and she declined to terminate the fetus. This raises the bioethical question of whether those who oppose abortion on principle should be automatically disqualified from participation. My own view is that this policy would carry its own stigma of discrimination, though many would disagree, I should think. Then there is the usual fear of insertional mutagenesis, which is more daunting in situations where the protocol is geared toward widespread vector disbursement. Finally, Anderson notes the important point that while retroviral vector injections could conceivably affect the germline, transduced cells carrying functional genes should not. This is a point in favor of selecting thalassemia-burdened fetuses rather than ADA-SCID–burdened fetuses, as the latter would be treated by direct retroviral injections into the peritoneal region while the former would be treated by directing transduced cells into bone marrow stem cells.[35]

Again, the value of ELSPI assessment is to place the experiment in an empirical context so that the totality of facts will be made part of the oversight calculus.

As we move deeper into the world of forbidden knowledge, slippery slopes abound. We have seen the phenomenon already, with the terms "therapeutic cloning" and "reproductive cloning" being brought together casually to form the dreaded concept "human cloning." As we penetrate this new terrain, we encounter *three* dreaded concepts: human somatic cell gene enhancement, human germline gene therapy, and human germline gene enhancement. Commentators are quick to agree that the last of these (HGLGE) is the most pernicious of the lot; there is disagreement as to whether HSCGE poses greater dangers to the body ELSPI than does HGLGT. Key members of the biomedical community argue simply that "Manipulation of genes for purposes of 'enhancement' should not be carried out."[36] Their attitude is that basic science produces knowledge; it is the job of medical practitioners to harness this knowledge and provide the optimal range of technological services to cure sick people. It is not the business of either science or technology, they say, to make people taller, smarter, or prettier by tinkering with their DNA. On top of these more abstract objections, pragmatic uncertainties abound. Thus, HSCGE might be close kin in final result to forms of cosmetic surgery, but the risk of cosmetic manipulations are known whereas the risks ensuing from enhancement manipulations are totally unknown. Then again, what genes would we engineer anyway, practitioners wonder?

Some of the bioethicist and social science moralists know very well that there is no true fundamental distinction between therapy and enhancement, a fact of life they bemoan, yet not a fact of life that keeps them from approving human somatic cell gene therapy while disapproving human somatic cell gene enhancement.[37] There is no scientific basis for placing all clinical abnormalities in a black box and pretending social redefinitions will not alter its contents. How short must a child be before "well-below average" becomes "dwarfism"? Are Prozac and Ritalin therapies or enhancements? A brave attempt has been made to build a wall between enhancement and "malady," where the latter is "a condition . . . such that (one) is suffering, or at an increased risk of suffering . . . death, pain, disability, loss of freedom or opportunity or loss of pleasure."[38] Increased risk? Loss of pleasure? The harsh reality is that once a procedure is approved by, let's say, the RAC and the FDA for human somatic cell gene therapy, that same procedure will be used for related enhancement purposes.[39] I believe that enhancement engineering does *not* pose more ELSPI dilemmas than any form of germline engineering so long as we are talking of human *somatic cell* gene enhancement.[40] Yes, HSCGE will be an exotic, expensive option, available probably only to a few and, therefore, raising serious questions regarding social and conceivably political equality. I also know that should Congress criminalize HSCGE, serious questions of personal liberty, absent when Congress regulates interstate shipments of furniture, would confront a Supreme Court made up of William J. Brennans, though not a Supreme Court made up of Clarence Thomases. For me, the critical points are these: 1) Human somatic cell gene enhancement implicates a relationship between the willing consumer of a service and the willing dispenser of a service; the former need not and should not be considered a human subject much less a patient, while the latter need not and should not be considered a biomedical specialist. *There are no third parties incapable of tendering their informed consent involved.* 2) Human somatic cell gene enhancement involves inserting exogenous DNA into the body of the consenting individual in a fashion that in no way implicates the constitution of the human genome. *It is a genetic not a genomic phenomenon* as I developed these terms of reference in chapter 3, figure 1.

The few relevant examples in the literature serve as useful case studies. In the 1980s, scientists invented an artificial human growth hormone in the form of a genetic recombinant as a treatment for various disabilities including pituitary dwarfness. They later arranged a clinical trial to help answer the plausible research question of whether the artifact would assist short, healthy children grow taller. No, no, no, insisted Jeremy Rifkin: science is being used to address deviations from a "social norm" rather than some sort of disease. A few years later, Ronald Crystal approached the RAC with a protocol to put an adenoviral vector carrying a bacterial gene into healthy normals. The idea was to check out immune system response in subjects who could tolerate the procedure, and if everything went according to plan he would then see how the regimen played out with cystic fibrosis patients. "Unprecedented," said one member; this smacks of "health enhancement," several opined. In both instances, the research went forward. And why not? The dominant theme of these investigations was to gather data for bona fide scientific missions, hardly the stuff of bioengineering a second Barry Bonds or Tiger Woods.[41]

And it is precisely in the area of sports competition—where so much money and glory is on the line—that the major HSCGE breakthrough may occur. Without much doubt, transgenes producing long-term physiological expression would be much safer than pill popping.[42] But again: if there is no law against baseball players taking performance-enhancing pills, why should there be a law against these athletes opting for HSCGE? By analogy, it's OK to use enhancement engineering to grow hair on someone's bald-pate, because minoxidil is available for just that purpose. Trying to construct a risk-benefit formula to judge the ethics of HSCGE as an abstract enterprise has left commentators grasping at straws. Eric Juengst says the procedure should be outlawed if it abets "invidious social tendencies." Such as? Racial discrimination, he thinks. How would he classify gender-based or gender-preference discrimination?[43] As for using the methodology to enhance intelligence, one should remember that a solitary cerebellar Purkinje cell may well possess more synapses than the total number of genes in the human genome, and these cells are specific for one and only one function, motor coordination. "[T]he finer details of assembly and intellectual development are beyond direct genetic control."[44] All this has bearing on HSCGE in both children and adults. What about the procedure in preembryos? That strategy would evidently be far more efficient than human somatic cell gene therapy has been; reaching a larger proportion of the cells would clearly be facilitated. And with preembryos, the mother's *Roe–Planned Parenthood* procreative liberty interests presumably emerge on the scale of competing values.[45] The problem is that the risk of germline gene uptake would today be a strong likelihood. Putting aside the germline issue for a moment, the RAC has never formulated a policy statement one way or the other about human somatic cell gene enhancement. That seems a wise course of action. Let the protocols come, let the nuances and distinctions unfold, and let the committee decide.

The time has arrived to focus on the principal subject of this chapter: human germline gene interventions. I begin with therapeutic aims and purposes. In the classic case, HGLGT would involve putting DNA into the germ cells of a human preembryo so that the alien gene(s) would be carried forward ad infinitum via the reproductive cycle. In the 1980s, highly reputable scientists considered the procedure "a definitive qualitative departure from other therapies since it would affect future generations."[46] Advocates of "thick" analysis, what I call members of the Department of Moral Ultimates, were considerably less restrained. A well-known theologian remarked that these experiments could implicate "what it means to be human."[47] Leon Kass, never at a loss for words or "thickness," opined that "it may . . . mark the end of *human* life as we and all other humans have known it."[48] There was even "thickness" on the other side: "Technology . . . is man's creation and man's hallmark."[49] The RAC reduced this tension between East and West to manageable thinness by approving the following language: "[We] will not at present entertain proposals for germline alterations."[50]

President Clinton's NBAC, attempting to justify a moratorium on human cloning, likened it to "the existing moratorium on the use of germline gene therapy."[51] First of all, the RAC language is applicable not merely to HGLGT; it is applicable to all "alterations" including HGLGE. Second, the so-called germline moratorium is subject entirely to the evolving debates within the RAC as to the terms and conditions under which the "moratorium" would be amended or lifted. We must, in other words, distinguish among moratoria. Its most benign form occurs when a governing body suspends discussion on a policy question subject to new information or a change of heart by some member of the group. This attempt at moderation in the face of a perhaps looming technology of unique proportions was challenged at a RAC meeting by the Committee for Responsible Genetics. This group, consisting of many of the same people who fought recombinant DNA research tooth and nail, supported the following new language: "The RAC will not review and the NIH will not approve any human genetic therapy . . . that could alter germ line cells." Note the words "will not" and "could." Protocols posing any chance of germline intrusion would be unreviewable. The RAC rejected the proposal.[52]

Even those who have something good to say about HGLGT would want us to face squarely a choice and vote a proposal or the whole subject matter itself up or down. They do not approve of "side door" germline, in which scientists performing noncontroversial work inadvertently invade the reproductive cells. Nor do they approve doing germline and calling it something else. For ten years, as I said earlier, the RAC has had its worries about inadvertent vector migration. In one experiment, a small number of clones targeted for the peritoneum wound up in a subject's liver. If exogenous DNA can gravitate to the liver, why can't it gravitate to the germline, we asked one another. In 2002, Stanford researchers, funded by Avigen, Inc., told the RAC that their very first subject in a hemophilia B trial, a 63-year-old man, displayed traces of a recombinant in his semen. The experiment was an important one: scientists had spliced the human Factor IX gene into an adeno-associated virus and inserted the package into the patient's liver in hopes that his blood would clot. And now the mutant was sitting in his semen, though, very fortunately, not in his sperm. Experts thought the chances of migration beyond the semen were highly unlikely, but the FDA and the investigators were taking no chances: the patient would have to use "barrier contraception" until all traces of the vector had dissipated. Just who would police the man's bedroom to see that he was wearing a condom was not specified. The episode provides a fascinating insight into the Constitution as a "living" document in the research context given that, according to the Supreme Court, a state has no authority to punish people for using contraceptives.[53] Do things change where the government simply says: protect yourself our way or you're off project?[54] Within three months, the recombinant had disappeared. It should be pointed out that even had the adeno-associated virus integrated the sperm cells, germline contamination might not necessarily have followed. In this respect, these organisms are unlike retroviruses.[55] The Stanford-Avigen trial eventually resumed.

In a totally different kind of procedure, researchers removed the cytoplasm from a donor's oocytes and inserted the material into the eggs of a woman who was infertile on account of cytoplasmic malfunction. This means they were transporting mitochondrial DNA from woman A's eggs into woman B's eggs. Presumably, they did not consult the RAC because they did not employ a genetic recombinant. Nor did they transduce any cells in vitro or in vivo. The scientists then had the nerve to announce that they had created "the first case of human germline genetic modification resulting in normal healthy children." I completely agree with critics that their behavior was unprofessional and should, at the least, stand as a strike against them when they come to NIH, NSF, and so on looking for subsidy. However, I do not agree with critics who argue in favor of a moratorium (i.e., a moratorium NBAC-style) on this sort of research, indeed on any research that might alter accidentally the human germline, employing as the governing standard a "reasonable foreseeability" test of possibility. Good animal model studies should suffice. Nor do I accept as a credible rationale for a moratorium the slippery-slope conviction that HGLGT would lead inevitably to HGLGE, as some have contended.[56]

There was a time, about ten years ago, when reflection on HGLGT had seemed to progress from what Alfred North Whitehead called the Romantic Stage to what he called the Precision Stage, in other words from what Evans calls the stage of "thick" debate to the stage of "thin" debate.[57] No such luck; the forbidden knowledge motif is alive and well. The spirit of a decade past is captured in an essay by LeRoy Walters, the RAC chair during those years, who said, "The time is ripe for a detailed discussion of the ethical issues surrounding germ-line genetic intervention in humans."[58] He emphasized the theme of laboratory science progress: an ever-larger array of genes was now amenable to detection in the context of preimplantation diagnosis; animal modeling had shown that HGLGT was no pipe dream, though still erratic in result. He also noted that whereas earlier policy documents and pronouncements—most of them emanating from Western Europe—had rejected germline research out of hand, the Council for International Organizations of Medical Sciences had brought forward the Declaration of

Inuyama, which stated: "The modification of human germ cells . . . might be the only means of treating certain conditions, and therefore continued discussion . . . is essential."[59] Walters peered into the future and "envision[ed] . . . a method by which the locus of a genetic defect could be precisely targeted in a reproductive cell, the defective sequence removed, and a properly functioning sequence substituted for the defective sequence."[60] A critical question was whether the methodology of choice was microinjection, retroviral transduction, or preembryonic stem cell insertion. What may seem odd to dilettantes is that each of these procedures raises its own set of ELSPI concerns. In the face of arguments such as the notion that the RAC should throw up its hands and leave HGLGT decisions to the medical profession, I jumped into the debate.[61] I urged that the RAC rethink its "at present" moratorium and open up the floor to in-depth deliberations by sending a positive signal to researchers in which we say "show us what you've got."[62] I argued then, and I continue to argue, that the RAC moratorium, which after all this time is still in place, is a prima facie infringement on scientific research as an artifact of First Amendment expression. The moratorium establishes a content-based constraint on a form of research, a form of scholarly experimentation, which is presumptively invalid, and the fact that we are dealing here with what is ultimately an NIH funding decision does not constitute a per se waiver of that presumption. There is no reason why the RAC should treat human germline gene therapy the way states used to treat obscene motion pictures—and as a matter of fact, movies were treated better! Fifty years ago, a film distributor had the opportunity to go before a city or state censorship board, show the movie, and try to convince the censors that it wasn't obscene.[63] We at the RAC weren't even permitting HGLGT scientists to make a presentation, I emphasized. Things looked promising when the RAC voted to form a subcommittee that would study germline issues, but nothing ever came of it.[64] At around this time French Anderson sounded the clarion call of caution: microinjection and retroviral insertions were too scattergun in their consequences. "Until the time comes when it is possible to correct the defective gene by homologous recombination . . . , the danger exists of producing a germline mutagenic event." He also conceded—and this is a big concession—that "infants [might] have the right to inherit an unmanipulated genome," so it was up to philosophers, ethicists, and theologians to work out that little dilemma.[65] The root problem in the United States was that the federal government, as I reported in the previous chapter, refused to fund preembryonic research of any sort, so the abortion controversy had a direct impact on the sort of experiments necessary to get the HGLGT agenda off the ground.

Lack of empirical research in the field emboldened the Continental Europeans, who, as in all things political, love their "moral ultimates," to repair once again to the high ground.[66] For example, some writers called for a "principled framework in the form of an international normative code" to govern germline issues. This code would "transcend cultural and jurisdictional differences."[67] My political science colleague Andrea Bonnicksen took on these commentators and, in my view, put them in their place.[68] She noted the wide disparity in national regulatory approaches that would have to be harmonized: the Germans, egged on by the Green Party, believed that preembryos should have the same legal rights as born humans; the British allowed preimplantation diagnosis and had set up a licensing board to govern it, but they had specifically banned therapeutic cloning, and their rules, it could be interpreted, had banned HGLGT as well; the Americans, she argued, adhered to a policy of laissez-faire because Congress and most of the states had enacted no legislation criminalizing HGLGT, though I certainly wouldn't call the RAC moratorium an exercise in laissez-faire; the French deplored the research on principle; and the Spanish position seemed mired in ambivalence. Bonnicksen advanced the position that these, and other, discrete national policies should be permitted to bloom without prejudice, and that eventually, through a process of bottom-up incrementalism, a consensus would emerge.

Her thoughtful essay spawned some very useful responses. Robert Winston predicted that preimplantation diagnosis would not enhance the need for germline therapy; rather, it would enhance the implantation of normal blastocysts. David Shapiro rebutted the argument that conflicting national policies would promote medical tourism. Those who emphasize the Rights of Man, he pointed out, ought to enjoy the choices available to them with regard to competing nation-state cultural norms and research guidelines. Ruth Chadwick pondered why parents shouldn't have a right to cleanse the Huntington's dominant from the genes of all their future progeny. Robert Cook-Deegan drew a distinction between the Nuremberg Code, which had arisen in response to well-defined noxious practices, and proposed international guideposts for HGLGT, an evil perceived by some but not others and lacking rigorously specified parameters to all.[69]

This debate seemed constructive enough. And yet we still can't get HGLGT out of the "forbidden knowledge" cave and into the bright light of scientific discourse. I would like to be able to say that the lack of good data, either because of adverse government policy or sheer theoretical/experimental ignorance, is the chief stumbling block. That conclusion would be myopic, because it would ignore the deeper ELSPI dimensions of the conundrum. I will close this discussion with a checklist of these difficulties.

Some highly reputable biologists do not believe human germline gene therapy should be done, and their convictions are almost an article of faith. In an appearance before the RAC, noted population geneticist James Neel, after indulging in thirty minutes of literature review and scientific evidence, threw up his hands and stated glibly that HGLGT constituted "intellectual arrogance" on the part of those who would seriously entertain the research as a bona fide scholarly agenda. Neither "shotgun" DNA injection nor homologous recombination could ever work, he pronounced. The risk, now and forever, of maladaptive modifications affecting untold future generations was too great, he thought, ever to permit the use of germline therapy to address the plight of consenting adults suffering from life-threatening mutations.[70] Such genomics luminaries as Eric Lander and Jean Dausset favor outright bans against modifying the human germline.[71] Query: When does DNA study become DNA tampering?

Another factor is that the lay community never ceases to be frightened by HGLGT-related science as it proceeds in collateral contexts, a process that, of course, refuses to abate. A prime example is sperm research. A decade ago, Ralph Brinster et al. astounded the molecular biology and biomedical communities by taking sperm precursor stem cells from mice and injecting them into the testes of infertile mice. The spermatogonia developed into sperm, and the mice went on to produce offspring. You could conceivably collect a human male's spermatogonia, alter them genetically, and return them to the male. Presto: germline! You could also conceivably take the human male's spermatogonia, splice in genes from other species, and return the package to the male. Presto: a germline chimera! And then there was the prospect of human genes in mouse spermatogonia.[72] So guess what? Ten years later, researchers transplanted key portions of pig, goat, and horse testes into mice. The subject mice produced the animal sperm of choice. "Oh My God," said bioethicist Arthur Caplan, a comment he may have made with one tone of voice while Leon Kass would likely have used another tone.[73] At about this time, scientists also created an immortalized spermatogonial cell line that should facilitate the creation of transgenic mice and germline gene procedures.[74] Perhaps the topper is a project wherein University of Milan researchers cloned the human DAF gene into swine sperm that was then employed to fertilize pig eggs. Presto: "spermline" gene transfer, in which 20–50% of progeny exhibited the human DNA.[75] It is experiments like these that prompt Francis Fukuyama to write about our species' "posthuman future," thus alarming the untutored even more.[76]

Yet another deep reason why HGLGT cannot escape forbidden knowledge status is the fear posed by responsible discussants that the procedure, once unleashed, will compromise our evolutionary posterity. In an earlier section, I argued that *Roe v. Wade* ought to be overruled because, in theory, it put potential mothers in charge of our genomic constitution. That assertion, however, is relevant only to the question of what formal jurisprudential theory or what division of political labor we should legitimate as governing the evolutionary process to the extent it is governable. The issue before us now has to do with how appropriate decision makers—whether it be a RAC-like body, a Supreme Court–like body, or a congressional-like body—balance the right to perform HGLGT experiments against the risks of perturbing the Darwinian paradigm of genomic propensity. All of the evidence points toward weighting the scales in favor of the former set of interests. Preimplantation diagnosis, abortion, even immigration, tax, and environmental laws and regulations already work their various wills on the makeup of the future gene pool to little effect.[77] Even if scientists created new genes and proliferated them via the germline, "any effort to enhance the human species experimentally would be swamped by the random attempts of Mother Nature."[78]

We come now to the last, and most philosophical, deeply rooted objection to HGLGT: it will compromise our "genetic patrimony." Here are a few quotations capturing the flavor of the sentiment: "[T]he protection of the human genetic patrimony in its individuality, complexity, and universality is a question of respect for human dignity from which all fundamental human rights derive."[79] "The gene pool is a joint possession of all members of society."[80] "[T]he human gene pool is a joint possession belonging to all members of the human species."[81] Almost 20 years ago, I wrote: "Our genetic constitution belongs to us."[82] That statement was narrowly tailored to rebut the proposition that scientists could discover the content of the genome and patent that content. It was never intended to justify blocking research in, and modifications of, the human genome in the context of reproductive liberty and scientific investigation as free expression. According to Bonnicksen: "Genetic patrimony indicates a collective genetic heritage—'the collective assets of a community (or of mankind)'—that is 'both irreplaceable and of enduring worth, and therefore subject to specific forms of social protection.' . . . [Genetic patrimony] touch(es) upon something with a nearly mystical aura."[83] To provide the argument with its best face, HGLGT would lead our species to the ultimate tragedy of the commons: if all mothers or parents or scientists exercise rationally the full play of their germline options, everyone will be the loser. And yet if the genetic patrimony vision suppresses HGLGT in all its manifestations, for all its purposes, we would have before us just another eugenics statute, freezing in the genome, until the blind hand of evolution decreed otherwise, each and every life-threatening mutation.[84] Why must humankind accept on metaphysics alone such a legacy?

Human germline gene therapy—indeed, human somatic cell gene therapy—is a form of enhancement because both are examples of genetic improvement or, in the case of the former, germline genetic improvement.[85] HGLGT, controversial though it is, is not the heavy-duty stuff of enhancement; it is demarcated in people's minds from HGLGE, by which is meant manipulating the reproductive cells to work changes in the germline having nothing to do with therapeutics. I do not know of a single advocate of HGLGT who has anything good to say about HGLGE. Most treatments of enhancement as an ELSPI phenomenon do not even bother to discuss it.[86] The aforementioned Declaration of Inuyama, noted for its moderation and dispassion, says: "Attempts to enhance cognitive abilities or behavioral characteristics . . . are inherently suspect, and there is no consensus that such applications are acceptable."[87] This statement was made in the context of HSCGE, not HGLGE!

Human germline gene enhancement is the métier of science fiction books *(The Boys from Brazil)* and movies *(Gattaca)*. These scenarios often combine HGLGE with human reproductive cloning in order to intensify the message that our twin lusts for perfectability and perpetuity will overcome us all. The final message of *Gattaca*—"the birth[s] that may never have taken place [are] your own"—did not sit well with the viewing public, which hardly enjoyed being held up to the obloquy of mortality.[88] I am still going to stand my ground. Although the chances of engineering the germline to enhance such polygenic/polycultural propensities as intelligence, memory, golf ball striking, and cooperation are slim to none, I think it is bad public policy—and that it fosters the rudiments of a maladaptive mindset—to ban practices that might conceivably, in highly unusual circumstances, be worthy at least of discussion. Keep the marketplace open even to Prometheus.

Tampering with the germline draws enemies from all quarters. The ideological right tends to call it "playing God"; the ideological left tends to see it as widening the disparity between the "haves" and the "have-nots."[89] Reading the mind of God is chutzpah; widening the gap between economic classes has resulted from countless technological improvements, which only the affluent can fully enjoy. The great unwashed hardly see germline issues as salient (why should they, given the crude state of the science), but what they see they don't like. The result is that governments around the world can proclaim propositional ultimates about hypothetical horrors and even impart to them the force of law.

There is a small cluster of concerned citizens who do not share this paradigmatic fear of forbidden knowledge in general and the taboo associated with human germline and cloning interventions in particular. Some of them are liberals, and some of them are conservatives; some of them are Catholics, some are Protestants, and some are Jews; some are Americans, some are Europeans, and some are Asians. The cluster represents all races, all religions, all continents. These people are members of what I call the constitutional elite. Anybody can be a member, but few desire to be. To obtain one's card of admission, the person must truly believe in the Enlightenment, in Western Civilization, and especially in the understandings and canons of science that have freed all of us to think objectively and creatively. Most fundamentally, card-carrying members have freed themselves from the trappings of ideology, the normative noise that seemingly forever has stood against the flow of science.[90] And yet, to make of this comment a little mystery, one more element or ingredient is needed to attain full membership in my elitist club of true constitutionalists, of those who play the most adaptive game under the most adaptive rules available to our species. In other words, accepting at face value the credo of the Enlightenment is not enough. That is because the Enlightenment itself speaks not with one voice but with other (too many other) voices. How can we manage to speak with just one? That is the essential subject of my last two chapters.

5

Sociogenomics

Molecular biologists and their colleagues have been studying behavioral genetics for many decades prior to the recombinant DNA revolution. By "behavioral genetics" is meant here the search for those genes that act as precursors to human thought and action. One of the oft-cited benefits arising from the mapping and sequencing of the human genome is that it will facilitate the discovery of these genes. This chapter begins by reviewing the literature of behavioral genetics, emphasizing the extent to which selected parameters of political behavior can be ascribed to genetic causes and circumstances. Conventional wisdom among natural scientists is that human behavior—across a spectrum of repertoires—is a 50–50 blend of genetic and environmental influences. If so, a valid delineation of the political behavior landscape requires the specification of those DNA sequences and their physiological consequences that give shape and expression to relevant genetic influences. From this appreciation of the role of genetics in the behavior of humans in their sundry political roles and capacities flow many far-reaching policy consequences, some of which have already inspired profound differences of opinion and receive here considered formulation and assessment.

In the second part of the chapter, I move from the world of genetics to the world of genomics. Precious few political science research studies are relevant to serious behavioral genetics investigation. I discuss some of these (others crop up in chapter 6), and I show how the genomics paradigm will enable political scientists to move far beyond the current frontiers of investigation. This commentary will feature not only abstract theoretical points and counterpoints but also empirical reports culled from the lower-order species literature demonstrating the efficacy of the new sociogenomics research agenda. Nor would my treatment be complete without a demonstration that sociogenomics qualifies fully as a representative of the forbidden knowledge scholarly genre.

A. Disease Genes

The nexus between a person's health and political behavior must be self-evident; how else can one explain the fact that political scientists spend almost no time discussing it?[1] The significance of the former to the latter did not escape the notice of Lasswell and Kaplan, who found the health and safety of the human organism key "welfare values" that set the stage for influence patterns and relationships.[2] The issue before us is not the extent to which the typical individual's well-being effects political orientation and activity;[3] the critical question is the extent to which pathologies and infirmities, genetic in cause and consequence, undermine the performance of key political actors and, employing as a laboratory the context of American power relationships, threaten the stability and integrity of the constitutional order. I begin with physical maladies.

Once a year, the U.S. president's personal medical doctor issues a report to the public, invariably concluding that the chief executive enjoys superb health. No doubt this evaluation covers the standard litany of bacterial, fungal, and viral infections, not to mention broken legs and chronic lower back pains caused by overjogging. And certainly it covers such "exotic" diseases as glioblastomas, carcinomas, and AIDS. Perhaps we can stipulate that it includes a determination that the subject is HIV negative.[4] But of one thing we can be sure: the assessment will not include a survey of the president's genome, specifying whether he carries mutations that, down the road (read: his second term, if reelected) could produce serious medical problems compromising his performance. And what of the genomic prognoses for federal judges, including, of course, Supreme Court justices, who serve lifetime appointments checked only by the credential of "good behavior"? As to their physical serviceability, there is not even the custom of a perfunctory annual report.[5] What genes and what diseases might present such clear and possible risks?

Even before the major gene-sequencing, gene-mapping, and gene-cloning exploits of the 1990s, twin studies had pointed toward powerful proportions of heritability contribution for several common medical disorders. For example, close to 100% of idiopathic epilepsy, approximately 80% of rheumatoid arthritis, and approximately 50% of peptic ulcer phenotypic variation in the population is ascribed to genetic precursors.[6] Of course, a heritability score of 100% would automatically result where a pathology is transmitted via standard Mendelian processes. Such is the case with Huntington's disease (HD), among others, which, as we have seen, was known to be a dominant mutation decades ago, was mapped to chromosome 4 in the early 1980s, and was finally isolated in the early 1990s.[7] To develop further the discussion in chapter 2, HD is a late-onset, irreversible, degenerative disorder that ravages both body and mind, just the sort of gene-gone-wild that poses the dilemmas of concern here. Fortunately, it is rare; and fortunately, tests are available to ascertain whether one is at risk. But unfortunately, there is no cure. In these circumstances, a "tragic choice" arises: to know or not to know when the fruits of knowledge are seemingly indigestible.[8] But if potential candidates for high political office know that one of their parents was an HD victim, don't these aspirants have an ethical obligation to be tested and to reveal publicly the results of these tests? It is easy to agree that a right of privacy, even a right of constitutional dimension, attaches to the contents of one's genome, but it is almost as easy to conclude that one waives one's right to such privacy when one enters the arena of high public responsibility and where the search for "genetic fitness" reasonably constitutes a bona fide occupational qualification. Even traditional searches and seizures are constitutional under the Fourth Amendment when reasonable, and I do not think a standard of elevated scrutiny is called for in the context of health risks.

Questions of this sort received enhanced national visibility with the discovery and cloning of the BRCA1 susceptibility gene, located on chromosome 17, which hosts mutations increasing greatly the risk of breast cancer.[9] Twin study data had long shown minimal heritability for this disease, though, of course, *all* cancers, including breast cancer, are properly classified as genetic disorders in the sense that some stimulus, usually environmental, has perturbed one or more somatic cell genes to bring about the malignancy. But I adopt here, as I have done throughout my commentary, a strict construction of the term "genetic pathology," employing it to mean malfunction caused by an inherited DNA mutation.[10] So while the BRCA1 gene accounts for only 2–5% of the breast cancer cohort or 1 in 200 women, the fact that breast cancer has had such traumatic effects on the female population plus the fact that American Jews are much overrepresented in that cohort has brought the question of genetic testing into larger relief. As with HD, there is no straightforward cure; yet, while mammograms are of questionable utility, mastectomies remain a viable prophylactic. Weighing and assessing these and other considerations in making a decision to permit testing rely upon a dependable, accurate predictive methodology, but with BRCA1 there are hundreds of possible harmful mutations, that is, deleterious letter substitutions in the base pairs of the gene's double helix code, so that extensive testing of family members is sometimes a prerequisite. Moreover, 15% of those who carry a mutant-flawed BRCA1 remain cancer-free during their productive years. Most experts think the costs of BRCA1 testing currently outweigh the benefits.[11] Clearly, the question whether the public can insist that female candidates for, and incumbents in, high political office submit to BRCA1 gene scans represents a closer case than compulsory testing for the HD gene.

The discussion can be sharpened with an example plausibly drawn from real life. Marfan's disorder is caused by a dominant gene on chromosome 16. This mutation produces an elongated limb structure and a compromised circulatory system at great risk of sudden collapse. Many experts hypothesize that Abraham Lincoln was afflicted with Marfan's and that if John Wilkes Booth had not killed him he would quite possibly never have finished out his second term. The harmful polymorphism has yet to be cloned, but a simple PCR examination of Lincoln's DNA drawn from his blood or hair in 1859 probably would have told the electorate what it needed to know if geneticists then had possessed the theory and data they possess now.[12] Or did the electorate need to know? To my mind, mutations that kill or disable are very serious business; moreover, performing the high duties of constitutional office is exceedingly taxing and exceedingly important, requiring substantial expenditures of physical and mental resources. Therefore, the public has a right of access to the "genetic basics" at issue here, assuming there is an isomorphic relationship between the tests employed and the predictions of phenotypic fitness and unfitness drawn, what I regard as the essentials of procedural due process relevant to this policy context.

It seems as though each hereditary infirmity has its own public policy story to tell. Let's take the still common practice of smoking. It is what I would call a "hereditary infirmity" in that roughly 60% of the statistical variance among those who are characterized clinically as regular users is gene-based, according to the most authoritative twin studies.[13] Smoking is a "sort of" addiction, meaning some people can kick the habit and some can't. Smoking may or may not cause cancer in the particular subject. One thing is for sure, however: if somebody smokes and develops lung cancer, then in 90% of cases the individual has a defective TP53 gene on chromosome 17. And if you are born with a malfunctioning TP53 gene, your chances of developing early-onset cancer are 95%.[14] The fragile entity we call a healthy human body is in serious danger when oncogenes don't shut down and when tumor-suppressor genes don't turn on. Here is where TP53 enters the fray, expressing a protein that causes cancer cells to commit suicide. I'd certainly like to know if the candidate for Congress in my home district who happens to be a smoker has a damaged TP53 gene. And if I were a shareholder in a big company, I'd like to know if the smoking CEO had a damaged TP53 gene. In fact, why wouldn't various constituencies want to know whether their key decision makers had compromised oncogenes, tumor-suppressor genes, and TP53 genes, putting to one side their smoking appetites? Apoptosis (cell suicide) evolves as an adaptive strategy through the "group selection" phenomenon: what is maladaptive for a cell may be highly adaptive for the body as a whole. This does mean that we are stuck with the genomic parameters of our individual apoptotic systems, unless we decided to harness germline interventions to improve our posterity. Query: Would these experiments be germline gene therapy or enhancement?

The tale of BRCA2 mutations differs in important features from that of BRCA1. As with BRCA1, BRCA2 pathologies afflict the Ashkenazi population in markedly disproportional numbers, but with BRCA2, which was tracked to chromosome 13 in 1994, the precise mutational pattern is known: the 6,174th letter of the gene is absent. Various Jewish organizations in the United States, mindful of advances in modern genetics and recognizing the ravages of Tay-Sachs (chromosome 15), cystic fibrosis (chromosome 7), and the breast cancer pathologies on the rank-and-file membership, have taken special steps in implementing customary marriage policy to obtain blood samples, notify couples at risk, and either advise against matrimony or recommend abortion in cases where the embryo is compromised. Although compliance is voluntary, few can doubt the efficacy of "elder persuasion," particularly when the elders may be a consortium of parents, doctors, matchmakers, genetic professionals, and, in some cases, rabbis. Both the above noncancerous disorders have been virtually eradicated in this population. Interesting is that the *New York Times* denounced the project as eugenic.[15] Is aborting a Down's trisomy embryo also eugenic? Some may not approve, but is it eugenic? Perhaps the *Times* would condone it if the woman simply relied on her *Roe* procreative rights rather than admitting she took the advice of theological elders. Is the latter—in broadest compass her right of religious "free exercise" under the First Amendment—not as legitimate a rebuttal to the eugenics charge as would be a rebuttal predicated on *Roe*? The Ashkenazim are now training their guns on other recessive genetic pathologies, such as Canavan, Gaucher, and Fanconi disorders, especially hurtful to their group. They appear unimpressed by Lori Andrews's slippery slope objection that "this [might be the start of something very troubling because it] has an impact on self-concept and on relationships with others."[16]

The BRCA2 phenomenon also strikes another inbred population—Icelanders—with much higher frequency than a normal distribution would allow. In this population, letters 1000–1004 are missing. The government of Iceland has gone way beyond the American Jewish policy response to deleterious recessives: on December 16, 1998, the national parliament, by overwhelming vote and with the correspondingly enthusiastic support of public opinion, entered into an agreement with deCODE, a private company, to establish a national data bank genomic repository of all living Icelanders. The legislation gives deCODE what amounts to a twelve-year franchise permitting it to market drug development rights to pharmaceutical companies. Medical practitioners around the island of 270,000 people provide their patients' genotypes, which the company then correlates with Iceland's immaculately preserved genealogical charts and hospital records. Although Icelanders see the scheme as a sound investment in health care and as a vehicle for employment diversity (the firm has hired several hundreds of skilled personnel) in a single-industry (fish) economy, American and European observers well understood from the outset that the unique initiative displayed all the trappings of forbidden knowledge one associates with the human genomics paradigm. No longer was society faced with policy questions arising from this odd mutation or that odd mutation. Now the government—and even worse from their standpoint, a for-profit company—would have access to the full and complete genomes of all citizens, with permission to hawk the global array of their DNA wares to the highest bidders. During the debates leading up to the final vote, key questions regarding informed consent and donor privacy were hotly debated. The legislation, as approved, provided for "presumed consent"; given the fact that the country as a whole, through democratic processes, had tendered its endorsement, supporters felt free to entertain the presumption that everyone would cooperate unless they specifically refused, and as a matter of fact, only about 6% have refused. The legislation also wraps genotypic data in a cloak of anonymity; the problem is that in a country as insular and homogeneous as Iceland, it is not difficult to identify people if one has only two or three pieces of basic demographic data. DeCODE argues that its elegant computers are so programmed as to resist such invasions. Another thorny matter, that of handing over a treasure trove of unique data to one group of scientists, was finessed by providing a right of access to noncommercial researchers. The brains behind deCODE is Kari Steffensson, who has all the earmarks of a successful modern molecular biologist: his visionary data bank protocol was rejected for funding at NIH; he left Harvard to found a commercial biotech venture; he courts competition with the Human Genome Project, whom he says he will trounce in the race to unlock the locations of disease genes; and he holds news conferences announcing that his team has homed in on critical and sensitive sequence regions of interest while rarely submitting his findings to peer-reviewed outlets. No doubt the concept "informed consent" is evolving in nature. Up until very recently, principal investigators would tell their subjects what they planned to do and how they planned to do it. The genomics paradigm in the context of public data bank repositories seemingly requires subjects to sign blank checks or opt out. Only if the fundamental choice to participate remains in play will the essential constitutional contours we associate with personal liberty and autonomy be preserved. The price, however, may not be worth the candle for people at risk where those who do opt out deprive themselves of badly needed information.[17]

Estonia has followed Iceland's lead, putting in place a new wrinkle: genome contributors are permitted to inspect the contents of their profiles. The project is slated to begin as I write. Far more ambitious is a UK genetic census, sitting at this time on the drawing board and requiring a decade to achieve full implementation. The new creation, called U.K. Biobank, will be jointly funded and operated by a government-sponsored charity, the Wellcome Trust, and the Medical Research Council (the UK's NIH). The idea is to obtain DNA profiles from a random sample of general population volunteers and ask them as well to supply all manner of biographical information. The 500,000-person cohort will then be tracked for ten years. The ultimate goal is to "quantify the roles of genes and environmental influences like smoking, alcohol, viruses, pollution, exercise, and diet in unrelated people for all common diseases." Iceland's unique feature is a homogeneous population, splendid for isolating disease genes; the UK population is highly diverse, splendid for extrapolating "the nature-nurture, environment-genes secret" to the worldwide stage. So far, "opposition has been muted and polite."[18]

This dialogue about "hereditary infirmity" mutations would not be complete without discussing certain diseases that are not hereditary at all but that have powerful hereditary implications because certain environmental stimuli compromise certain susceptibility genes that are inherited. The classic case study is asthma. The commonest cause of asthma is the dust mite dropping. Without doubt, asthma runs in families, though the precise dynamics of "asthma families" may well differ appreciably given that disparate susceptibility genes are at play. There is evidence of eight candidate genes on chromosome 5, two on chromosome 1, and seven scattered among five other chromosomes. The variance also contains ethnic and gender biases: for instance, men have problems with environmental stimulus A but not B, while women display the opposite reactions. When a physical malady of consequence is as complicated as asthma—when pleiotropy is clearly the nature of the beast—theory tells us that the Icelandic data bank approach ought to be particularly useful. Data amassed by deCODE link 104 living asthma sufferers to the same family tree going back ten generations. Tracking down their shared DNA sequences of relevance should lead to isolating the critical susceptibility genes. Thousands of miles to the south lies the island of Tristan da Cunha, many of whose inhabitants are also asthmatics. Scientists have now collected blood samples from almost all the population universe of 300, so that an islander genomic repository is now essentially in place. Again, the chances of tracking key susceptibility genes should be enhanced. Realistically, the odds that Icelander asthmatics and Tristan da Cunha asthmatics exhibit the same constellation of compromised genes are hardly good, and the chances that either set will be anything like the compromised clusters among sundry urban dwellers are far dimmer. Thus does the slog of laboratory genomics inch along, in certain respects no faster than the pace of earlier paradigmatic investigation.[19]

What is the comparable landscape for "behavioral" disorders? Twin studies yield the surprising conclusion that on average heritability accounts for an even higher share of causation for this cluster than is the case for medical malfunction. Striking examples: virtually 100% of autism, approximately 88% of major affective disorder, and approximately 54% of Alzheimer's disease result from genetic antecedents.[20] The Alzheimer's instance is particularly instructive. In 1992, Allen Roses and his team at Duke University announced that they had discovered the first late-onset gene associated with this dreaded disease, the APOE susceptibility gene located on chromosome 19. It is estimated that 30% of Alzheimer's cases feature a breakdown in the function of this gene. More specifically, the gene presents at least four possible alleles (versions) of which the most common is APOE3 and the deleterious mutation is APOE4. In the general population 90% exhibit one copy of the former, while 30% exhibit one copy of the latter; in that same population, 60% exhibit two copies of the former and are "safe," while 2% exhibit two copies of the latter and are at "high risk." A research distinction must also be drawn between patients in families with a late-onset history and "sporadics," the larger cohort, patients who are members of non-late-onset families. If one is a member of an afflicted family and features no APOE4, the chances of having Alzheimer's are 20% by age 75; for those with one APOE4 the chances are 45%, and for those with two APOE4 alleles the chances are 90%. If a sporadic, the odds of having one or two copies of the

APOE4 are about 2 to 1. And yet many people with the "dreaded couplet" die of old age and never get the disease; conversely, many come down with late-onset Alzheimer's and have "clean" APOE genes. Interestingly, all these data are for whites and hold neither for African Americans nor Hispanics.[21] Also interestingly, boxers who are "punch drunk" display the APOE4 allele to a statistically significant degree.[22]

The plot thickens with recent news. In 1998, Rudy Tanzi reported a high correlation between a mutation in the A2M susceptibility gene on chromosome 12 and late-onset Alzheimer's.[23] Even more promising was a Tanzi investigation published in 2000 indicating that a breakdown in the as yet elusive IDE gene on chromosome 10 is quite likely an Alzheimer's contributor.[24] Tanzi's work at the Massachusetts General Hospital received confirmation at the Mayo Clinic in Jacksonville, Florida, where researchers, noting that relatives of late-onset patients featured the same elevated amyloid peptide levels as early-onset familial patients, also reported genetic linkages in the former group to chromosome 10.[25] Enter the Human Genome Project. A team at Washington University performed a full-length scan of affected Caucasian sibling pairs and declared that a susceptibility gene on chromosome 10 was about as likely as the dreaded APOE4 chromosomal 19 allele to cause Alzheimer's.[26] Now to find the culprit!

Ronald Reagan's long-term bout with late-onset Alzheimer's has brought to the surface more than one critical policy dilemma; it is quite likely that the disorder manifested itself during his second presidential term and conceivably even earlier. A Stanford blue-ribbon advisory panel has argued that general testing for the breakdown is unacceptable given the twin facts that there is no effective remedy and that knowledge of the subject's APOE allelic structure (the only DNA constellation that had for sure been linked to the malady) could be devastating to all concerned.[27] Moreover, unlike a similar test for Huntington's, an Alzheimer's test could not at this stage be dispositive as to the ultimate course of pathological events. Libertarian Matt Ridley finds such recommendations a clear instance of bioethical paternalism. Why is it OK to test for the HIV but not OK to test for the E4 allele? Ridley also thinks a Nuffield Council study reaching similar conclusions is biased, in that it puts genetically based diseases in a taboo category not accorded other sorts of pathologies for which testing is commonplace.[28] I think his criticism of the report is well made; however, until the fruits of genomic research kick in, a test for Alzheimer's, except for those at clear, objectified risk, seems entirely too problematic. Now let's get back to Mr. Reagan. A few sentences ago, I referenced the argument that Alzheimer's testing could wreak unfortunate results to all concerned. But suppose the "all concerned" are the citizenry of the United States, and suppose the test for the E4 mutation is 95% accurate, as indeed it is? The totality of facts surrounding the complexities, nuances, and burdens of the disease counsel that the public should have exercised, presumably through appropriate legislation, its need to appreciate at some point before or during Ronald Reagan's terms of office the composition of his APOE gene in order to assess better his competence for executive leadership. In no way would this have compromised his civil liberties. As a matter of fact, such a policy, applied uniformly and predictably to all prospective Officers of the United States (to use the language of article 2, section 2, clause 2), might make a nullity of noxious and no longer improbable scenarios in which partisans surreptitiously obtain DNA samples from hair (and other bodily materials) in order to get the "goods" on their rivals.[29]

Why does the APOE4 persist in the human genome, and why don't we start thinking about eradicating it? Along with such other "nasties" as the sickle-cell and Tay-Sachs mutations, which in heterozygotes confer some advantage against malaria and tuberculosis respectively, there is evidence that the E4 allele serves to fend off trypanosomes, the African sleeping sickness messengers. As for E4's fate in the world, evolutionary and genomics theory provide the answer: close scrutiny of the critical 5500 base pair regions on chromosome 19 reveals that while the E4 locus predominated in *Homo sapiens* of old, E3 and E2 competitor loci have been proliferating over the past 300,000 years. Assuming pertinent environmental conditions remain favorable, the E4 allele may well be history in another 300,000 years.[30]

Tempering the relative success of Alzheimer's research is a mixed, even erratic, record in tracking down other suspected genes for behavioral disorders, a record that helps round out prognoses for future policy agenda proposals. Take manic depression. How does it affect political behavior, particularly the display of leadership qualities? More than a decade ago, scientists announced they had found a gene for this mental illness, but follow-up studies have shown that the report was invalid.[31] As I was preparing to file this manuscript in final form with the University of Wisconsin Press, what for the past several years had been pretty much a research brick wall suddenly showed signs of give. In one breakthrough, scientists demonstrated a strong association between a flawed version of the GRK3 gene and bipolarism, presumably accounting for about 3% of all cases.[32] In a second and far more elaborate study, researchers who followed a cohort of almost 1,000 New Zealanders for 25 years found a clear genetic influence in how individuals coped with stress. The DNA in question was the promoter sequence triggering the 5-HTT gene on chromosome 17, a key vehicle in serotonin carriage across synapses. If respondents presented two "short" alleles, they were twice as likely to be depressed as those who presented two "long" alleles. No matter how often the latter group came under stress, they were no more likely to experience depression than people who reported a total absence of stress-related episodes. Note that in order to succeed, the investigators were compelled to combine both genetic and environmental independent variables.[33]

What about schizophrenia, especially paranoid schizophrenia? How many notable political actors have been afflicted with it? In the heyday of Freudian psychology, experts said that bad parenting ("nurture") was the culprit, but twin studies now show that the heritability quotient for schizophrenia is 60%.[34] So where are the genes? An initial study concluded that a dominant gene on chromosome 5 was linked to the disorder, but these findings could not be replicated.[35] Then researchers announced that paranoid schizophrenia was at least in very small part a function of the HLA-A gene on chromosome 6;[36] however, linkage studies did not confirm the association. Neuroscientists did a little better: the brains of schizophrenics seem to feature an abnormality in the thalamus,[37] and the left inferior parietal lobes of male schizophrenics are 16% smaller than normal to the unusual point of being smaller than their right IPLSs. (For women, who are twice less likely to be saddled with the disease than men anyway, these aberrations could not be observed.) Said one authority: "[W]e may be able to split [schizophrenia] into a series of diseases."[38] Enter once again the genomics scientist. Conducting in-depth scans of 22 extended families cluttered with this behavioral nightmare, we have now learned that chromosomes 1 and 13 are likely repositories of the salient genes at work. It is "merely" the task of positional cloning to isolate particulars.[39]

Now let's consider alcoholism. Political science biographers have cited drinking abuse as a critical variable in decision making.[40] The conventional wisdom, shared by several Supreme Court justices, is that alcoholism is a disease.[41] If by disease is meant "in the genes," twin studies debunk this conviction.[42] But the search for causal DNA goes on, perhaps because "five times the normal number of the male relatives of alcoholics are themselves alcoholic." Also, if male alcoholics put their sons up for adoption, these children "had a four times higher risk of becoming alcoholics . . . than the children of nonalcoholics." It is quite possible that certain chronic drinkers lack the physiological armamentarium to tolerate alcohol.[43] In 1990 came the report that alcoholism was linked to the A1 allele of dopamine receptor D2 on chromosome 11. Confidently, other researchers waded in, publishing findings that this same region featured strong associations with such related antisocial behaviors as drug abuse and gambling addiction.[44] As I write, this array of conclusions is the subject of an intense debate between boosters and critics.[45] The current research agenda is eclectic; investigators are examining "hot spots" on chromosomes 1, 2, 4, and 7, as well as the ever-popular 11, this last retaining particular attention in part because it houses genes with implications for the brain's "pleasure centers."[46] If ever scientists could use a genomic-wide scan, this would be a subfield of choice.

Finally, we have obesity. Eating—unlike smoking and alcohol consumption—is obviously highly adaptive, and the species has probably conserved many genes influencing us to take in food and store it for future use. Our genome is an age-old set of constitutional behavior norms, very much the product of times when humankind had to work overtime to survive and when nutritional scarcity was a constant consideration. Today, when their descendents sit behind desks and indulge in poor eating habits, excess poundage has become a public enemy.[47] As recently as 1994, investigators thought that the heritability variance for human obesity was 60%;[48] in 1998, the upper limit was pegged at 70%.[49] That same year, one of the leading researchers in the field announced that the true quotient was 88%, the highest rate of variance in the literature associating genes with behavior.[50] Certainly murine models yield that high a figure, and the human LEP gene is known to be a homologue of the cloned mouse obesity gene. The LEP gene produces the protein leptin, and leptin deficiency in the hypothalamus was at once proposed upon its discovery as the principal cause of obesity. That obesity is a salient mover of "death precursor disorders" is well known even to laypeople, so there is a lot of general interest in ferreting out the genetic roots of fatness. As quickly became evident, "obesity in most people . . . was not caused by a mutation in the leptin gene,"[51] and the $25 million Amgen had paid to develop a leptin-enhancing drug is probably a dubious gambit. Recent investigations have focused on the MCR receptor gene, which turns out to be the commonest monogenic antecedent of obesity. Mutations in this gene display a high correlation with both being overweight in childhood and adult binge eating. These projects took a big step forward when researchers determined the complete nucleotide sequence of the key gene.[52]

If we agree that complex human behaviors implicate marked genetic activity, as twin studies tell us again and again, then we must be prepared in the best spirit of science to track down the clearly polygenic root contributions. How can we ever do it absent a viable, operational genomics paradigm? Ultimately, we will get a handle on these dynamics, at which point we can apply my "officeholder genetic index" to monitor obesity. Let me precisely state the constitutional principle on which I rely: if we test the William Howard Tafts for the LEP gene but not the Woodrow Wilsons, if we test the Abraham Lincolns for Marfan's mutation but not the Steven A. Douglases, if we test the Ronald Reagans for Alzheimer's but not the Walter Mondales, we will have created for ourselves more than an unpalatable policy. The equal protection principle in our Constitution mandates congruent testing for everyone in our sample of political notables or no one. Candidates for important positions should undergo the same acceptable tests for the same agreed-upon genomic constellations.

B. Personality Genes

The locus of analysis for political scientists is power in all its sundry dimensions. The dynamic of seeking power is one such dimension. Are power seekers "different," and if so, how does one explain the difference? In a controversial study, Douglas Madsen linked whole blood serotonin (WBS) to power seeking in humans. His research did not purport to show causation but did purport to show—and, indeed, did show—association, if one accepts how Madsen defined power seekers. In that study, power seekers were self-defined *dominance* seekers, whose responses to written questionnaires demonstrated they were of the type A personality configuration.[53] In subsequent work, Madsen provided two improvements. First, he substituted for questionnaire responses an actual competitive environment, and second, he compared reactions to that environment among clusters of high WBS subjects, average WBS subjects, and low WBS subjects. Again, a strong association between power seeking as defined and WBS prominence emerged.[54]

The Madsen research agenda has attracted criticism, some of it vitriolic. His work has been labeled an example of the "new biologism," a wastebasket school of opprobrium into which scholars are deposited if they imply that political behavior has significant organic roots.[55] Constructive skepticism would have focused on whether competition is a satisfactory proxy for power seeking plus the very great need to elucidate the genetic mainsprings of serotonin in general and WBS in particular.

Important steps in achieving the latter have now been taken. Serotonin is a brain chemical, a monoamine. The serotonergic system is the most elaborate neurotransmitter system in the human brain. The raphe nuclei are found in the midbrain and house the bodies of the serotonin nerve cells, from which axons branch out to the hippocampus, the hypothalamus, and the pituitary. When the system is functioning properly, unused serotonin is evacuated by a transporter; when the system breaks down, serotonin overload causes cranial disturbance and personality shifts. After years of experimentation, Eli Lilly marketed Prozac, the ultimate generic serotonin vacuum cleaner. In the early 1990s, the gene for the serotonin transporter function (the 5-HTT gene referred to above) was cloned. The gene is clearly a factor in midbrain raphe serotonin (MRS) function. Yet, scientists could find no correlation between the gene's structure and either manic depression or schizophrenia. What they eventually did discover was that a critical difference lay not in the gene coding sequences but in the promoter regions located "upstream." It turns out that the system can display a "long" version and a "short" version. The long or normal system (57% of the general population) effectively clears serotonin deposits; the short mutational system (43%), which is dominant, permits serotonal accrual. The genetic disparity? The normal version contains 16 sequence repeats approximately 20 base pairs per repeat, while the shorter version contains 14 of these repeats, a difference, then, of 44 nucleotide sequences.[56] Evidently, the next step on a Madsen-type research agenda would be to construct samples based on these genetic profiles and then measure over time WBS and MRS carriage in controlled, power-oriented environments, defining power in several ways to compare the play of each behavioral constellation in each genetically distinct cohort. Lasswell's "political man" may well turn out to be "political men."

Fascinating, against this backdrop, are possible ties between power orientation (seeking?) and neuroticism. There is a high correlation between the serotonin transporter "short" version regime and neuroticism, but no statistically significant association between this mutational display and such personality dimensions as extroversion, openness, agreeableness, and conscientiousness. Of the several behavioral strains known to be manifestations of neuroticism, four—anxiety, angry hostility, depression, and impulsiveness—are significantly related to the short version.[57] As regards anxiety, the key elements of association are tension and suspicion. Yet another high correlation was reported for harm avoidance (but not novelty seeking), especially worry, pessimism, fear of uncertainty, and fatigability (but not shyness, though the shyness-boldness parameter is 50–60% heritable and 70–90% in the cases easiest to classify).[58] (Note: In the neuroticism studies, males appear to be much overrepresented.) Summing up: Investigators believe they have located a gene that "influences a constellation of traits related to anxiety."[59] This finding received a powerful boost when researchers reported a statistically significant correlation between subjects carrying at least one copy of the "short" serotonin transporter promoter mechanism and elevated activity in their right amygdalas following exposure to anxiety-producing pictures. Control groups made up of "long-long" individuals recorded significantly lower levels of response.[60] Acting instinctively in re-action to messages received from the thalamus, the amygdala sets up the first line of defense against perceived dangers, and now we appreciate a linkage between that system and the serotonin carriage system, though the causation trail remains unmarked.[61]

Still and all, it is generally accepted that serotonin receptors responsible for altering transduction, ligand affinity, and amino acid repetition may also contribute to personality expression, a recent proof of principle being the link between serotonin disruption and bulimia.[62] Twin studies conclude that approximately 50% of anxiety-related traits, that is, neuroticism, are heritable.[63] No wonder then that geneticists say the short version allele ac-counts for a maximum of 50% of the heritability quotient for this syn-drome, that at least ten other genes are likely involved, and that 50% of the variance is environmental.[64]

From everything I have been saying, one can see that the empirical task of teasing out both genetic precursors and power-relevant behavioral or-derings is a formidable but necessary task of the genomics paradigm if the truly scientific study of politics is to proceed. During this search, political scientists should always be mindful that genes are just as likely to be effect as cause. For example, while there may be a strong positive correlation between leadership position and high serotonin levels, post-Madsen re-search shows that the former, not the latter, is the prime mover. And people who perceive themselves as ranking at the bottom of pecking orders not only exhibit low serotonin counts, but they also exhibit high cortisol counts. Stress acts on the hypothalamus, which triggers the CYP17 gene on chromosome 10, which produces an enzyme, which instructs cholesterol to release cortisol, a steroid, to sketch the process simplistically.[65] Darwinian theory explains that increased cortisol can help us cope with danger just as short-versioned low serotonin carriers (probably in the heterozygous con-dition) are more interested in sex, thus enhancing their reproductive op-portunities. Also, it is one thing to prescribe Ritalin in modest amounts to classroom children suffering from attention-deficit hyperactivity disorder in order to boost their serotonin levels;[66] however, it is quite another thing to recommend extreme therapeutics even for genetically based pathologies producing deviant adult behavior patterns. What we now call "bipolar dis-order" could somehow correlate in certain instances with creative genius, as conceivably was the case with Poe, van Gogh, and various public figures. But DNA is always relevant because the mixture of political responsibility with either genetic or environmental "imbalance" could lead to dire straits for millions.

The quest for the genetic antecedents of power provides clues in charting yet-to-be launched quests for the genetic antecedents of other salient political science variables. Consider traditionalism. In his classic study of American political theory, Clinton Rossiter contends that "the first and most visible of . . . conservative principles is traditionalism." Traditionalists are constitutionalists, he argues, who presume in favor of stability and "the institutions and values of the contemporary West." But according to Rossiter, there are also "pure traditionalists": "enemies of change as well as reform . . . [who] live in a state of acute cultural schizophrenia."[67] Twin studies report that the heritability score for traditionalism is 45%, but it is not at all clear whether the Multidimensional Personality Questionnaire used in these protocols captures Rossiter's "traditionalists," "pure traditionalists," or some combination of the two.[68] More fundamental for our purposes is the question whether traditionalist subgroups disaggregate along heritable-environmental lines. Rossiter himself might be chagrined to learn that certain core conservative predispositions have genetic roots; political scientists at century's beginning would be guilty of gross myopia if they thought otherwise. The trick, as always, is to locate the genes; and it is precisely here that collaborative research involving political scientists and molecular biologists is most viable. The former would construct the necessary questionnaire instruments tapping various features of traditionalist and nontraditionalist mainspring and then would devise game theoretic behavioral formats to compare and contrast the play of "traditional" vs. "innovative" decision rules by selected experimental and control groups. They would also compile family tree and voting record data. Rigorous DNA investigations employing RFLP, EST, and linkage disequilibrium analysis could then be brought to bear on the genetic constitutions of high and low scorers. As I shall elaborate later in this chapter, the genomics paradigm provides us with even more sophisticated theoretical and methodological tools for plumbing the depths of subjects' genomic constitutions in these and similar contexts. If Rossiter's theoretical profiles are sound, we should not be surprised to find genetic proofs of hard-core conservative and liberal attitude and action. Various models of decision rule complexes could also be subjected to computer simulation scenarios in order to evaluate their evolutionary potential.

The fact is that geneticists have already given political scientists a leg up. Dean Hamer has investigated the dynamics of what he calls the novelty or thrill-seeking syndrome. High scorers on questionnaire items enjoy the play of new ideas; they are predisposed to openness in thought and often action. Low scorers are cautious and conventional, prudent and orderly. He speculates that John F. Kennedy belonged to the former group, while Dwight D. Eisenhower belonged to the latter. How exactly novelty seeking or novelty avoidance manifests or expresses itself is probably a question of environmental circumstance, but that the personality characteristic itself is genetic can no longer seriously be doubted. The misnamed "novelty gene," actually the D4DR gene, is located on chromosome 11. This gene makes a dopamine receptor protein. Rather like serotonin, dopamine is one of those brain chemicals that needs to be at equilibrium in the typical case, or personality problems and worse arise. As is well known, a severe loss of dopamine can lead to personalities without physiological nuance, that is, Parkinson's disease. Dopamine overload correlates with highly risky behavior: too much gambling, too much sex, too much drinking. The D4DR gene contains a series of 48 letter repeats. The average number of repeats runs from 4 to 7; those with 2 or 3 are extraordinarily effective in clearing dopamine, whereas those with 8 or more (the ceiling is 11) are not very effective at all. If a subject has two "longs" or a "long" and a "short," the correlation with novelty seeking is far greater than for a subject exhibiting two "shorts." In other words, people with less acute pleasure centers have a genetic impetus to develop compensating behavioral propensities. Hamer's modest conclusions are that while novelty seeking is 40% heritable, the D4DR gene accounts for only about 4% of that estimate. He thinks there may be ten other equally significant genes.[69] So now our quest to understand the genomic constitutions of "unambiguous" liberals and conservatives, traditionalists and innovators, must include D4DR comparisons, the same sorts of full-sequence scans earlier described, and, as I shall show, even more newfangled research regimens. Of course, it would be wrong to assume that these "pure types" are forever gene-driven, that they can do nothing to counterbalance their firmly established, well-ordered mindsets. We take action contrary to our genetic pulls and pushes all the time. Risk takers can force themselves to do more planning and analyzing; compulsive types can read more widely and introduce fresh interpretation into their otherwise closed systems. All of this, however, is corollary to the constitution of the genomics impulse.

Twin studies show at least 50% heritability quotients for a number of other personality dimensions;[70] still, political scientists will likely remain skeptical as to the probative value of these data for their own research unless and until clear connections present themselves. A good example is happiness. Everyone experiences emotional "ups" and "downs," which turns out to be another way of saying that environmental stimuli trigger short-term (but only short-term) joy and despairs. That is because psychologists now know that each individual possesses a baseline level of happiness, and this baseline is an incredible 80% heritable.[71] In this instance, the relevant genes are as yet unknown, but quite clearly it would be straightforward to segregate the "truly happy" from the "truly unhappy" and embark upon the sorts of investigations outlined previously. The question is: What theoretically interesting information would we recruit? If our universe be political leaders present and potential, then we might do well to recall James David Barber's distinctions between "positive" and "negative" presidents. Barber believes that some chief executives have been happy and satisfied with their use of power; for others, the game of political thrust and parry yields feelings of irritability, perhaps even self-doubt, but certainly generates unhappiness and lack of fulfillment. Barber goes on to argue that America's better presidents have tended to be happy in their work, and by "happy" he surely means the baseline happiness described above.[72] Though some of Barber's conclusions appear overdrawn (Lyndon Johnson is classified as a "negative," yet it seems to me that as Majority Leader Johnson he was a "positive"), who can doubt the validity of Barber's central point? And if valid for presidents and prime ministers, then why not for judges, mayors, and interest group leaders? Quite obviously, the DNA of baseline happiness and unhappiness is critical to our understanding of political behavior.

One personality characteristic of quintessential political significance—and presumably therefore of political science significance—is sexual orientation. Reports in the early 1990s demonstrated that the brain structure of some male homosexuals differed from the formation typically presented by male heterosexuals.[73] Another study showed that female homosexuals were more likely to display left-handed skills than female heterosexuals to a statistically significant degree.[74] Twin studies have long shown that male homosexuality has a powerful heritability index; recent research pegs the figure at 50% and probably higher.[75] Then, in 1993, came the startling announcement that Dean Hamer and his colleagues at NIH had found a unique stretch of five genetic markers in the q28 region of the X chromosome for 33 out of 40 pairs of homosexual brothers.[76] Hamer himself has replicated this finding, but a comparable DNA linkage for female homosexuals has yet to materialize.[77] However, in 1998, researchers detected the first physiological marker for lesbians: receptor cells in the cochlea of the ear.[78] Hamer has no doubts on the matter: women become homosexuals largely for cultural reasons, and the biggest cultural reason is the girl's relationship to her lesbian mother.[79] At first blush, this is a startling conclusion to say the least, because twin studies show convincingly that shared environment accounts for precious little in the transmission of behavioral traits. But the truism about shared environment, one must remember, speaks to the impact of home life on siblings and says little about the one-on-one intergenerational dynamic. Thus, political scientists appreciate that Democratic parents generally produce Democratic children, and Republican parents usually spawn Republican children. Meanwhile, a new study challenges directly Hamer's central thesis, finding nothing on the X chromosome to correlate with male homosexuality.[80] Hamer sticks to his guns, noting that his critics' sample was strikingly dissimilar from his own in that the X chromosome hypothesis led him to investigate families in which male homosexuality abounds in the brothers' maternal line.[81]

We may assume—and the responsible literature does assume—that there are many roads to homosexuality. We may also assume that if there is a gene linking to male homosexuality on the X chromosome or somewhere else, that gene or those genes will yield up their identities to the genomics onrush. When that day happens, should it happen, we will have to confront the difficult policy questions arising from genetic testing. Of these, two seem especially noteworthy. First, is there justification for the state *ever* to insist that a person be screened for homosexual orientation? Even Freud recognized that homosexuality was not a mental illness; obviously, homosexuals are just as fit as anyone else to perform assorted professional responsibilities, including those associated with public office. But what of genetic screening for service in the armed forces? A Sam Nunn or a Colin Powell has been known to argue that the "homosexual lifestyle" is incompatible with military rules, regulations, and norms. How does one define "homosexual lifestyle"; indeed, how does one define "heterosexual lifestyle"? Even more to the point, is a piece of DNA a satisfactory proxy for a lifestyle? Unless we are to indulge in a determinism not shared by experts in the field, the answer must be no. Second, will the inevitable improvements in preembryonic, embryonic, and fetal testing trigger some impetus for women to ascertain the "genetic sexuality" of their future progeny and induce them to abort or otherwise discard suspect potential homosexual offspring? Under the *Roe–Planned Parenthood* line of cases, the state could prohibit this option in roughly the third trimester but not before. This prospect fills gay commentators with horror, it being for them, and for others, too, a form of genocide; and one can almost visualize the term "civil liberty," for 30-odd years the preserve in this context of the women's movement, being turned on its head. In the face of this dilemma, it seems a violation of any theory ever before labeled "democratic" to contend that such issues are for judges and not for elected officials to decide.

Beyond issues of normative political theory are issues in Darwinian political science. Harking back to the themes addressed in chapter 4, before we go around enhancing somebody's genome, we had better understand why it is that various DNA loci have persisted for hundreds of thousands of years and more. Every society on Earth, whether more or less civilized, has its share of homosexuals. I have yet to encounter a truly plausible argument that same-sex propensities enhance species fitness, but who knows? Maybe women carrying a mutation on an X chromosome have an expanded window of reproductive opportunity.[82] Maybe male homosexuality was adaptive during those long stretches when hunter-gatherers were "on the road" foraging, which would mean that today a mutation(s) could be classified as another one of those DNA relics.[83] As usual, the answer is more research penetrating to the structure and function of the human genome, a constitutional norm hopefully to be preferred to some competing, emerging norm entitled not "DNA enhancement" but "DNA elimination."

Genetics may have elucidated to a small degree and therefore complicated the dynamics of (male) sexual orientation, but genetics has only complicated the even more fundamental question of what is a female. Biologists used to say that if a person presented the XX constellation, then surely that person was a woman. Now we know that there are individuals whose primary sexual characteristics are female, but who display the XY configuration. The issue of gender classification has given Olympic Games officialdom headaches in recent years;[84] the more critical policy challenge is providing "best evidence" in order to optimize socially and scientifically acceptable implementation of the Constitution's equal protection provisions. The current estimate of "best evidence" is that the TDF gene on the Y chromosome makes the "real" difference. If the gene is functioning properly, the person is a male "fully and completely." If the gene is not functioning properly, the person is a female though not "fully and completely" because "she" cannot conceive normally. So under the U.S. constitutional system, who makes the final call? Who decides if XY women get drafted, or receive the same affirmative action "booster" points XXers get, or are tapped for membership in the Augusta National Golf Club? I don't mean to trivialize the critical question, which I take to be this: In what decision-making forum are the fundamental biological science policy issues, the fundamental human genomics policy issues of this and future times, to be thrashed out? In a very real sense, it's the old William H. Rehnquist vs. William J. Brennan war all over again, except now the stakes have never been higher. Throughout my commentary, I hope readers can see the very great need to balance the value judgments of what Holmes called the "dominant forces of the community" against the themes of liberty and equality, which themselves are often at tension. Notice, then, the many fact situations in which I ultimately conclude that both Rehnquist and Brennan had it wrong. And so my relatively easy choices, at least this time around: the military is Congress's call; affirmative action is the judges' bailiwick; Augusta is Augusta's business.

C. Cognition Genes

Let's not kid ourselves: when we consider the efficacy of genomic scans for leading politicians, we have commenced a gradual (rapid?) descent into the pit of forbidden knowledge. Outside the doctor-patient hospital setting, most people get squeamish when it comes to anybody knowing anything about human genetic constitutions, especially their own, of course. The pace is about to quicken.

The most hotly debated subject in human genetics 50 years ago and today is the extent to which intelligence is inherited. On this matter, twin study data speak loudly, unambiguously, and convincingly. General intelligence (IQ or *g*, the latter being today's "politically correct" scholarly denotation) is approximately 52% heritable; verbal reasoning is approximately 50% heritable; and spatial reasoning is approximately 40% heritable.[85] Moreover, investigators have demonstrated beyond doubt the counterintuitive finding that environment does *not* mute IQ heritability as time goes on. Quite the opposite is the case: the older one gets, the higher one's general intelligence heritability quotient is likely to be, and it can reach as high as 62% in later life.[86] To be sure, the data sets typically developed in these studies are from the United States and Europe. Also, these sets represent "normal" ranges; it is odd that scientists spend so little time analyzing the tails of this bell curve, particularly when the two most prevalent causes of low IQ (the chromosomal 21 trisomy and the Fragile X syndrome tracked to chromosome X) are manifestly a function of DNA. An excellent surrogate for *g*, we are told, is scholastic achievement. The two "overlap almost completely," and "if a gene were found that is associated with intelligence, we would predict that [it] will also be associated with the determinants of scholastic achievement."[87] This focus on factor *g* has been challenged by Robert Sternberg, who presents as an alternative a "triarchic" theory of intelligence in which the brain expresses three quite different skill functions: analytic intelligence (the putative domain of *g*), creative intelligence, and practical intelligence.[88] It is quite possible that Sternberg is correct, in which case political scientists ought to encourage a search for the genetic precursors of practical intelligence, given his argument that this cranial system likely best explains the whys and hows of George W. Bush's and Al

Gore's political successes. On the other hand, Sternberg's two intellectual add-ons may constitute the nonheritable parameters of cognition.

The notion, once entertained with delectation by eugenicists, that there exists some supergene for intelligence is nonsense. There do not even appear to be "intelligence genes" transmissible via standard Mendelian principles. This is not to say that geneticists are still standing at home plate, however. Researchers have developed the concept of QTL, namely, quantitative trait loci, which refers to what is now deemed the standard genetic recipe for complex behavioral disorders: many genes, each contributing small and different effects, are responsible. During the 1990s, investigators refined their methodological craft, devising such tools as allelic association to home in on QTL.

Recently, these and related efforts have provided two important fruits in the search for cognition genes. In 1996, University of Utah scientists announced the discovery of the *LIM-Kinase 1* gene on chromosome 7, called almost immediately the "jigsaw puzzle gene" because those who lack it cannot perform the spatial processes required for problem solving of this nature.[89] In 1998, Robert Plomin's group at the London Institute of Psychiatry announced a discovery of far greater social consequence. Having tackled head-on the unanswered question of "nature" and superior intelligence, Plomin's research team found a DNA marker for high *g*, namely, the IGF2R receptor gene on chromosome 6. The study, replicated shortly thereafter, reported that a sample of gifted children were twice as likely to carry allele five of this marker than a sample of average children. The group's conclusion: the marker is closely associated with an as yet unknown gene that could well account for approximately four IQ points.[90]

Some Americans have long found it hard to accept the role of heredity in intelligence, essentially fearful that the "new genetics" will undermine the nation's commitment to the equality principle. "Equality" is a term pregnant with constitutional implication, so it is useful to frame the debate along constitutional dimensions broadly defined. I think we need not tarry over the precise language of American fundamental law: "No state [read: instrumentality of government, according to the Supreme Court] shall . . . deny any person . . . the equal protection of the laws."[91] We are not worried about the onset of legal orders favoring genetically endowed geniuses and disfavoring genetically limited dullards. I acknowledge our past experience with restrictive immigration policies, forced sterilization statutes (discussed in chapter 3), and other Galtonian flirtations and misadventures. Still, extensions of Robert Plomin's research will hardly inspire legislatures to give subsidies to high IQ adolescents. (Query: Would such grants trigger strict judicial scrutiny?) Moving toward a more dynamic conception of equality, even should the term be defined to mean "political equality" such as, for example, the notion that the typical adult should be permitted to vote or run for high public office, I see no flash point of potential controversy. In most legal and political contexts here and elsewhere, money and education are surely more significant factors making for inequality than intelligence, whether genetically orchestrated or not. The bone of contention, then, really centers around social equality and, to a lesser extent, economic equality. Americans, however, do not believe in either social equality or economic equality, as political scientists have many times shown; and yet there is in Western culture the basic value sentiment that if some people succeed and others fail, it is mostly because the former work harder while the latter just don't want it badly enough to make the necessary sacrifices. Sure, some people are born "behind the eight ball" because they have lousy parents, and sure, the Kennedys and the Rockefellers have big advantages. The former can still make it, while the latter can still (and do!) screw up. Nonetheless, speaking in generalities, we have difficulty abiding the thought that some of us are born smarter, which would mean that some of us are born dumber, even though most of us are pleased to believe that our children may be gifted because, of course, they carry our DNA.

All in all, Americans would ordinarily be perfectly happy with their prevailing constitutional mythology of meritocracy/equality-of-opportunity-modestly-underwritten-by-government-intervention-as-in-the-"leveling-the-playing-field"-metaphor without losing any sleep over the fact that the equation, should it ever be seriously lived up to, would be a prescription for genetic enhancement. In an open society where environmental conditions are a relative constant, what else would determine status and power more than DNA?[92] What they do lose a little sleep over is the fact that the term "social equality" sometimes means "racial equality." They look around them and they see what most would prefer not to see: blacks have less education, less money, less of everything tangible than whites. And well-schooled Americans are familiar with an oft-cited, oft-criticized 1960s–1970s literature, exemplified by the scholarly writings of Arthur Jensen and the rantings of William Shockley, which essentially argued as follows: 1) intelligence is largely a function of heritability; 2) black Americans, on average, have substantially lower IQs than white Americans; 3) these group disparities are primarily a function of genetics; 4) compensatory efforts, such as public welfare programs, can't make a dent in this predetermined cycle of events.

In the 1990s, two substantial research studies appeared elaborating and enlarging upon that literature. The more newsworthy by far, and the *less* controversial in scientific circles by far, was the bestseller *The Bell Curve*.[93] As I read *The Bell Curve,* the authors offer two major conclusions, the first scientific, the second normative. The scientific finding is that IQ counts for more in the world of social behavior, in the world of power relationships, than does what sociologists have long labeled SES (parental socioeconomic status). The normative finding is that the empirically documentable failures of New Deal–type social welfare initiatives stem from the tight linkage between heritability and IQ; hence, what America needs is less government redistribution of resources and more freedom for the better endowed. Based on the twin study data I cited earlier, the first of these contentions may well be correct, though much more rigorous investigation is needed to pin down key parameters and dynamics. The terms "IQ" and "SES" themselves require tighter operationalization. The second of these contentions is not the stuff of serious social science scholarship, much less the stuff of serious biological science scholarship. The statement "They have inferior IQs, so we can't help them" is nothing but a political value judgment, belied incidentally by the fact that people suffer all the time from physical and mental disabilities, and yet medical research has done a great deal to alleviate these systemic pains and deprivations. *The Bell Curve* received a big play in the *New York Times* and on Sunday television talk shows because not only did it have the temerity to challenge Democratic liberal orthodoxy with regard to the value of government subsidies to the "disadvantaged" (read: the less intelligent) but, worse, it dressed up its libertarian (conservative?) bias in scientific terminology and methodology. Agreeing with liberal commentators that the two prongs of *The Bell Curve* argument should be neatly severed, I am advancing in this book a perspective that begs for adoption as well as a different critical tack. The biggest weakness in *The Bell Curve* is that it trifles with genetics. The authors have built an edifice of social structure predicated ultimately on what they say genes do and don't do, without ever searching for, much less finding, a single mutation to support their grand theory. They could not even bestir themselves to conduct independent research on heritability by constructing their own twin studies.[94] The Social Darwinists of Herbert Spenser's ilk were guilty of the same sorts of shortcomings. No wonder *The Bell Curve* possesses precious little scientific currency.

Much more honest is J. Philippe Rushton's research on race and intelligence, research generally ignored by the media because it does not directly challenge the politics of Franklin Roosevelt, Harry Truman, and Lyndon Johnson, to speak only of the dead. In fact, Rushton acknowledges that his findings cannot be translated into any social policy agenda.[95] What exactly does Rushton say? He begins with the literature on IQ: no matter where and how they are tested in the Western world, Orientals exhibit an average IQ of 106, whites exhibit an average IQ of 100, and blacks exhibit an average IQ of 85. When education and income are folded into the mix, the categories close somewhat but not appreciably. Naturally, he rereports the hereditability indices derived from twin studies conducted around the globe. He then focuses on his central assertion: no matter how one measures cranial size and capacity, Orientals are better endowed physiologically than whites, who are better endowed physiologically than blacks, and not even controlling for body size will wash out these differences. Orientals have, on average, approximately 102 million more brain cells than whites, and whites have, on average, approximately 480 million more brain cells than blacks. Finally, he attempts to blow away competing environmental explanatory models. If such-and-such behavioral trait is 0.50 heritable for whites, it is 0.50 heritable for blacks. If white middle-class parents adopt a black kid, he says, the child's IQ will be slightly lower than a natural-born sibling raised in the same home; ten years later, the black kid's IQ will have reverted to 89, or four points higher than the average black of that age. He concludes by mocking the Jesse Jacksons: "There is no special factor like the history of slavery or White racism that has made cultural influences stronger for one race than for another."[96] Rushton also mocks Herrnstein and Murray when he says: "*The Bell Curve* is soft on race."[97] I agree with him on both counts. Rushton is a serious scholar whose efforts require a serious level of engagement. To whom can we turn for enlightened dialogue? Certainly not to the academic "left"; the best it can muster is to call Rushton and his hereditarian allies "pro-racist[s]" and champions of "scientific racism."[98] We can do much better.

Rushton himself points the way: "Because Whites and Blacks come from different races, they have many different genes."[99] The clear inference to be drawn—in fact, an inference central to Rushton's entire argument—is that there is something out there in the real world called the "Black Genomic Constitution," an empirical assessment of which yields reliable and valid information on what the average member of this group has the capacity to say, to think, to do. With regard to the name of this game, *thinking*, let's consult the data ourselves.

What exactly does the term "race" mean from a genetic point of view? Craig Venter says: "Race is a social concept, not a scientific one." A Venter archrival in the race to sequence the human genome, Eric Lander, says: "There's no scientific evidence to support substantial differences between [racial] groups."[100] Naturally, there are genes that code for skin color and a few other physical features, and, yes, we respond to these features in laboratory tests. When the typical white sees a picture of the typical black, the subject's amygdala fires off an emotional response, a loose translation of which might be: here is someone different. As I said in chapter 1, we carry around the genomic baggage of prehistory when facial recognition developed as a key monitoring device for sorting in-groupers and out-groupers. Environmental variables can alter the equation, however. If I see a picture of Michael Jordan, then chances are my amygdala won't respond. I will have unconsciously coded him as an insider.[101] The "race as social concept" argument unfolds as follows: evidence gleaned from the human genome, through the application of such methodologies as mitochondrial DNA similarities in women, Y chromosomal similarities in men, and noncoding microsatellite short tandem repeat sequence similarities in both sexes, establishes that *Homo sapiens* originated in Africa; several tribes commenced migratory journeys 100,000 years ago and eventually colonized the planet; our species has not had sufficient time to develop unique genetic subspecies or "races"; skin and eye color, which can be linked to discrete races, are orchestrated by a few genes in response to extreme temperature variations; putative genes influencing a complex behavior such as intelligence must be manifold and work in synergistic patterns; they would require many more generations to produce significant differences among groups; "only 3 to 5% of genetic variation is due to differences among the major population groups."[102] Rushton's critics also cite Neanderthal Man, a creature with a big head but far dumber than those who have displaced him.

Of course, the "social concept" folks are prey to infiltration from the "political correctness" brigade, thus making for a kind of silly season in terminology. Racial clusters should be replaced with ethnic clusters, we are told, though for scientific analysis the term "Hispanic" hardly has more value than the term "black." Also, some aver, the term "race" has a plethora of nuances, so researchers had better be consistent in word usage.[103] Unfortunately, the silly season can become the serious season. Blacks are twice as likely as whites to die from heart failure, and they do not respond nearly as well as whites when they take standard drugs. So a reputable, politically naive doctor has contrived an alternative artifact; the FDA even went so far as to test it on black subjects, a step it had never before taken given today's climate of opinion. "This is a social time bomb, dangerous even in the hands of well-intentioned persons," warned Troy Duster, a black bioethicist and a member in good standing of that discipline's environmentalist wing. Blacks can't have it both ways. They want, quite understandably, to retain the words "black race" (what other term would they invoke?) to capture and keep alive a tradition—I call it a constitution—of trials and tribulations, of unique felt attitudes and values, a unique societal view and even worldview, an identity. And yet some black leaders want medical science to ignore fatal maladies afflicting their own people more than others to a statistically significant degree in the name of some abstraction called "color-blindness."[104]

This serious season implicates pure science as well as therapeutics. One of the most important goals of human genomics is to discover single nucleotide polymorphisms (SNPs), those genes that present different alleles and that, therefore, are responsible for the genetic differences among us. In the late 1990s, Francis Collins led an NHGRI effort to organize a data bank in which SNPs would be stored and made available to researchers. Incredibly, these SNPs will be derived from demographically anonymous sources; all geographic, gender, and racial identifiers have been removed. Why? "Research on alcoholism or schizophrenia, for example, could cause offense if linked to a specific group, and NHGRI wants to avoid any 'group stigmatization'?" We are talking here about important contextual data without which a full scientific understanding of human behavioral patterns might be impossible.[105] Rushton must be howling with derision as censorious ideology plays right into his hands.

This is not the first time the enemies of behavioral genetics have invoked the tools of censorship rather than the tools of the idea marketplace to engage their adversaries. For example, a prominent publishing house in the mid-1990s rejected two book-length manuscripts, one by Arthur Jensen himself, seemingly because the "race and IQ" theories endorsed therein were deemed "repellent."[106] For another example, as I reported in chapter 2, sociologist Dorothy Nelkin and legalist Lori Andrews published a letter in *Science, in their official capacities as NIH-DOE ELSI Working Group officers,* lambasting *The Bell Curve.* They wrote: "Genetic arguments cannot and should not be used to determine or inform social policy in the areas cited by Herrnstein and Murray."[107] "The areas cited" involve the length and breadth of what government can, does, and should do over and above the domain of disease. To these critics, genetics, by its very nature, has no place in a reputable dialogue addressing the causes and consequences of political behavior. I couldn't disagree more, and I find appalling their manipulation of taxpayer legitimacy to flaunt their value judgments in an area where they have no expertise whatever.

What, then, is a final assessment of Rushton, who, remember, is more extreme in his racism than Herrnstein and Murray in their most offensive (to their critics) moments? The major geographical groupings the world over can be distinguished from one another by allelic frequency, and practitioners can determine a subject's geographic (read: racial) identity by trolling the genome and collecting information on about 100 markers at random or at

30 specific locations. Such are the value of SNPs. You can e-mail a Sarasota, Florida, outfit and find out if you are black and therefore eligible for affirmative action consideration.[108] In the wake of these stipulations, concerned scientists have offered a compromise: identify the groupings by SNP frequency but not by geography. Others dissent, saying geography (i.e., race; whom are we kidding?) will provide the researcher with phenotypic as well as genotypic information.[109] The distinction would ordinarily be Talmudic but for this: no matter how you slice it, white Americans have an average IQ of 100 and African Americans have an average IQ of 85, and we need to know why. I (unlike Nelkin, Andrews, and Duster) don't pretend to have an answer. I also think, much to Rushton's chagrin, that it doesn't make a lot of difference anyway. I asked Rushton at the 1995 meeting referenced above why Jews did so well on all his criteria, and yet Israelis were bogged down in endless rivalry by competing religious divisions and political parties, groups galore, while Arab enemies made the most of these opportunities, a scenario hard to square with Darwinian insights. His answer, in front of about 100 people in the panel meeting room, was broad, diffuse, and uninformative. Let's face it, I took advantage of him; he isn't a political scientist. As if we political scientists have any answers either! My point is that people may be a little "smarter" or a little "dumber"; still, they are at the mercy of their limitations, 50% of which are genetic and 50% of which aren't. These limitations test the brain power of us all in the real world of political thrust and parry.

D. Criminality Genes

Why do some people break the law over and over and over again? This question has nagged at students of antisocial behavior for decades. Twenty years ago, the battle lines were drawn as rarely before with the publication of James Q. Wilson and Richard J. Herrnstein's massive examination of the voluminous data. Criminals, they contended, often exhibit similar "constitutional" profiles (their well-chosen term), referring to certain physiological commonalities such as mesomorphic body formation, low intelligence, and above-the-norm aggressiveness. The authors were prepared to cite contributing environmental factors including dysfunctional family ties and antisocial peer pressure particularly in the context of inferior schools, and they were very careful to say that "constitutional" causes may not necessarily be genetic in character.[110] These qualifiers were insufficient to escape the wrath of most sociologists, as ever committed to the "culture paradigm," and the ever-watchful "left," determined, as usual, to keep biological theory from having any social relevance and content whatever. Wilson and Herrnstein were not prepared to rule out genetics explicitly; they even cited with approval scholarship arguing the role of genetics in partially explaining the violent predatory acts of criminal recidivists. Clearly, they were the enemy, and it did not take the opposition long to produce its own in-depth treatment: environmental conditions such as poverty constituted the overwhelming, if not the sole, causal determinants, and it was undemocratic—perhaps even racist—to blame biology.[111] One need not hue to any ideology but respectable social science to ask these questions: Where were Wilson and Herrnstein's cross-cultural data? Is there any correlation between the constitutional factors informing blue-collar crime and those informing white-collar crime?

In the decade of the 1990s, these battle lines formed anew when the NIH's ELSI office agreed to subsidize a conference at the University of Maryland entitled "Genetic Factors in Crime: Findings, Uses, and Implications." This initiative was at once denounced by a coalition of blacks and whites. Said political scientist Ronald Walters: "[T]hese studies [are] looking at a connection between violence and race."[112] Said psychiatrist Peter Breggin: "[T]here isn't any scientific evidence that violence is genetic."[113] NIH officials thereupon canceled the conference. On appeal before an in-house advisory panel, the university argued that its protocol had won out in an error-free peer review process and that the grant had therefore been appropriately tendered; it was nothing but content-based censorship for the NIH to renege on its promised largesse. The NIH counterargued that the meeting hosts had changed the focus of the conference after the award had been announced, asserting in their promotional literature that there was already scientific evidence showing a nexus between DNA and violent crime. The review panel specifically rejected NIH's argument on this point and ordered the parent agency to help the university prepare a revised format, but, in a dictum pregnant with implication, it also concluded that the initial project should not have been supported in the first place because it could reasonably have been perceived "as an affront to the black community."[114] The meeting was eventually held under a revised, unique set of rules insisted upon by NIH: "for each psychologist or behavioral geneticist researching predispositions to antisocial behavior, the conference also featured a scientist, philosopher, or historian leery of suggestions that crime is causally linked to genes."[115]

The episode is of essential relevance to a key message of this book. If scientists cannot convene assemblies to share theory and data on human genetics and its consequences, how then can we reexamine with profit the political science teaching and research agenda along the lines suggested here? On behalf of those who opposed the gathering tooth and nail, let it be noted that in unveiling the project a highly visible NIH official likened the inner-city environments of America to jungles, a metaphor that to many observers conjured up images of black males being compared to monkeys.[116] It certainly is curious—and highly unscientific—that whenever genes are hypothesized as triggering agents for criminal behavior, it is invariably mugging or rape that are mentioned and never embezzlement or other impersonal derelictions. For empirical example, one need only cite, as I have already cited, the notorious Oklahoma Habitual Criminal Sterilization Act, voided by the Supreme Court in *Skinner v. Oklahoma* (1942), which had authorized sterilization for incarcerated three-time perpetrators of "predatory felonies" but which exempted a whole host of "nicey-nicey felonies." Why are Willie Hortons hypothesized by some to be genetically driven while a Michael Milkin, an Ivan Boesky, or an Enron executive (people who wantonly steal millions) is hypothesized by everyone to be environmentally driven?

Nonetheless, the merits of this particular case surely lie with the advocates of free expression. The Supreme Court may be correct when it says Congress can withhold subsidies for "indecent" art while underwriting meritorious "decent" art, but by no means does it follow that Congress or the NIH can open the door to the funding of research on criminal behavior yet decline to sponsor any investigation aimed at testing for genetic antecedents.[117] Viewpoint-based prior restraints are prima facie suspect in the world of federal grantsmanship, just as they are suspect in other contexts. Nor does the First Amendment countenance a role for minority veto groups deciding which public research projects are "friendly" and which are "hostile." As for NIH mandating a 50% representation of "outsider opinion" at the conference, this is reminiscent of Louisiana's Balanced Treatment Act, which obligated public school biology teachers to instruct their students in creationism whenever they instructed them in evolutionary theory. That law and this agency ruling "stack the free speech deck" in favor of a state-ordered equality-of-idea participation and representation, hardly the stuff of uninhibited, robust dialogue that is the centerpiece of free expression.[118]

What, then, is the truth of the matter? When some commentators say, "there isn't any evidence that [criminal behavior] is genetic," are they stating scientific fact? Mainstream opinion among geneticists themselves is quite to the contrary. Confining the discussion to the biased sample of violent, chronic lawbreaking, there is at least one genetic marker that looms extraordinarily large in the profile of serious offenders: the Y chromosome. On the Y one finds the SRY gene, which makes people carrying the XY tandem men. The Y also attracts other genes particularly useful to males, a process entirely consistent with Darwinian survival theory. If gender specialization for our species was critical to successful adaptation, then males and females needed all the genetic assistance their respective behavioral repertoires could provide. So if there are genes coding for "offspring nurture and protection," individuals carrying the XX constellation tend to have them. If there are genes enhancing aggressive behavioral modes in the rough and tumble world beyond the "nest," those possessing the XY pair tend to have them.

Recent research has proven exceedingly edifying. There appears to be a high correlation in the male population between violent behavior and *low* serotonin levels. Serotonin deficiency also is strongly associated with depression, aggression, and impulsivity; each and every one of these traits displays high heritability quotients according to the latest twin study data. Elevated serotonin level, we have seen, is an artifact of a deletion in the relevant nucleotide sequence promoter region. Similar alterations may well be responsible for low serotonin levels and, perhaps, account for at least some of the masculine mayhem on our streets.[119] I would also like to see white-collar criminals undergo serotonin testing. Here is a smattering of related results already on the table: higher levels of the neurotransmitter vasopressin appear in the cerebrospinal fluid of aggressive subjects, exactly where lower levels of serotonin metabolite are found; mice lacking the gene for nitrous oxide synthase are more aggressive than your average field mouse; and in a study comparing convicted murderers with a control group, the former displayed a diminution of glucose metabolism in the prefrontal cortex while another experimental panel had 11% less gray matter than controls.[120] Working toward a unitary theory of criminal behavior in which genetics is accorded the respect it deserves would require major funding, commitments the NIH has declined of late to entertain. Years ago, Harvard Medical School researchers tried to study children exhibiting the XYY chromosomal pattern. To what extent, they wanted to know, was there a correlation between XYY and various personality and behavioral traits putatively deemed antisocial. Even after their Harvard peers gave them the green light, Jonathan Beckwith and his allies hounded them until they threw in the towel.[121] It was an early, precedent-setting example of ideological censorship in this arena. Thirty years later, the tactics and strategies of the naysayers and nay doers have become only slightly more varnished as the Maryland crime conference episode made painfully clear, frightening an establishment prepared always to bargain away the scientist's constitutional craft and drying up research opportunities in what should be a most fertile domain of inquiry.

Meanwhile, the DNA results trickle in, and the battle goes on. Geneticists have ascertained that male members of a Dutch family who had committed such "abnormal behaviors" as "impulsive aggression, arson, attempted rape, and exhibitionism" all suffered from MAOA enzymatic deficiency. MAOA is principally known to metabolize serotonin. Analysis of the subject males' X chromosomes revealed that each carried a mutation in exon number 7 of the MAOA double helix, this alteration changing a glutamine to a termination codon. When one of the authors of this investigation was offered a post at the Max Planck Institute in Berlin, he was denounced in certain media outlets as laying the groundwork for possible eugenics experimentation.[122] In a recent, much larger update, scientists studied 1,000 New Zealand boys from birth, focusing their attention ultimately on about 450. They divided their universe into two groups, those displaying antisocial behavior and those who did not. They then demonstrated high correlations between a) parental abuse and hyperaggresive behavior, and b) the genotype for low MAOA activity and hyperaggresive behavior. Those boys suffering from both conditions committed four times as many violent crimes than the others. To this, Beckwith countered that genetic causation had not been shown, while Duster cautioned that antisocial behavior might also vary with social context.[123] They downplayed the obvious fact that some maltreated boys can conform and others can't, and one reason why some can't is genetic. But at least, in this case, they were not in a position to abort the project! Clearly, we need to know as much about the political psychology of those who fear the fruits of genetic knowledge as we need to know about the political psychology of those who think "better genes" make "better people."

E. Political Genes?

"The central problem for social science is to explain social order."[124] It was refreshing to hear the president of the American Political Science Association, at the very beginning of his presidential address, frame his colleagues' major research mission not in the parochial garb of disciplinary discourse but in the larger current of social science exploration. Today's scholarly agenda requires many hands and skills. By way of constructive criticism, I find the call to arms sufficiently broad but not sufficiently deep. Social science—drawing sustenance from a biological sciences program now deeply committed to the molecularist perspective—must provide heavier emphasis on the reductionist turn of mind. More fundamental than social order and social disorder is sociality itself, and here I do not necessarily mean human sociality. Scientists who study the nematode can learn a lot by consulting scientists who study the fruit fly, and scientists who investigate the honeybee have much to learn from scientists knowledgeable about the nematode. Key genetic structures and functions are conserved up and down the phylogenetic tree. But, says the typical social scientist, the genus *Homo sapiens* is different: we build cultures that take on lives of their own. What and how we build anything is an empirical question, not a theological imperative, and the weight of evidence now runs strongly contrary to the notion of some "high wall of separation" between one big-brained organism and the rest of the planet's life forms. Put fairly succinctly, when a set of organisms finds sociality, that is, interaction beyond the reproductive act itself, an adaptive strategy—a vehicle enhancing either reproductive opportunities or survival in a highly competitive Darwinian world—then sociality will spread over time throughout that set's genome. Assuming environmental conditions remain relatively stable, the species' fitness—or should I say, the genome's fitness—will receive all the advantages natural selection has to offer.[125] In sum, the "central problem" for social science is sociality; the central problem for political science is sociality as it affects and as it is affected by power relationships; and the locus of analysis at the deepest explanatory level is DNA in its full and complete context, namely, species genomics.

And so has been born only in the past decade the term "sociogenomics." Officially baptized by Gene Robinson, my colleague at the University of Illinois, it elaborates a new area of research: the social behavior of organisms as an outgrowth of genomic cause and consequence. The difference between James Q. Wilson's pregenetics explanatory angle of vision and what I shall call the Robinson-Carmen postgenomics approach to a study of social behavior deserves precise specification. Whereas Wilson says humans possess a "moral sense" arising from an "attachment response" or an "affiliative trait" that has been "selected for" by evolutionary processes,[126] we search for the global genetic determinants of such cooperative behavior, first in lower-order species and then in humans themselves employing a comparative species perspective that is sufficiently robust to appreciate the play of environment, the play of culture, in genetically relevant scenarios.

What are the lineaments of this new line of inquiry called sociogenomics? Before we can address that question, we need to appreciate the fruits of the short-lived, fading paradigm—sociogenetics. That research approach took hold in 1997 when an NHGRI scientist knocked out a gene called *disheveled*-1 in ordinary laboratory mice following which the rodent subjects failed to perform an important social duty for their peer creatures called whisker trimming. Dominant mice normally attend to the barbering needs of social secondaries in this particular strain. According to influential media observers: the "first gene for social behavior" in mammals had been isolated.[127] Two years later, Emory University scholars removed a gene from a prairie vol, a highly sociable rodent, and inserted it into the genomes of considerably less sociable mice. The mice adopted the salient exercise in sociality, for which the prairie vol is well known: constant romantic, monogamous behavior. As matters transpired, the coding DNA itself was not the causal agent. The prairie vol and the montane vol share exactly the same relevant gene, yet the latter animal is a conspicuous loner. The difference lies in the promoter sequences, the instructional software that the investigators had prudently also transferred to the mice. (Recall the critical role that promoter DNA can play in human serotonin release.) The parameters of this particular promoter region have a significant impact on brain structure, and the brains of the experimental mouse subjects developed from the embryonic stage to resemble prairie vol brains. Closer inspection has shown that the affected proteins are vasopressin receptors, and the location of these receptors varies enormously across species including the human species. So now we have a small opening into the dynamics of *Homo sapiens'* sociality.[128]

The groundbreaking work thus far in sociogenetics, not surprisingly, has come from the world of social insects. This universe is largely composed of bees and ants, species whose community lives are rivals in complexity even of our own interpersonal network arrangements. Of critical importance to fire ant organization is queen number. Some colonies have a singleton queen; others have multiple queens, as many as 200. A single gene makes all the difference. Ants residing in a monogyne community carry the BB allele of the Gp-9 gene, but a significant minority in polygyne communities carry the Bb allele. The whole process is orchestrated by chemical signals emanating from colony workers, with Bb carriers detecting and eliminating potential BB queens. Conventional evolutionary theory had frowned on the prospect of simple allele shifts determining intricate social behaviors. No more. A search of public data repositories shows Gp-9 most closely resembles in comparable function certain moth genes.[129] Along the same line, the entire caste system upon which is predicated social relations among red harvester ants is genetically programmed. The single queen mates with members of one lineage to yield reproductives and members of a separate and distinct lineage to yield steriles. Here, the consensus among experts was that environment determined caste permutations, a second theoretical musing now coded as insufficiently sensitive to the power of DNA transmission.[130]

Gene Robinson, who is an entomologist with a specialty in honeybee behavior, has made important contributions to sociogenetic empirical investigation. The question he addressed in a notable breakthrough was not the play of allelic variation but disparate levels of gene expression, an alternative predicate for explaining social behavior. In the honeybee, worker females spend their early days in the role of hive nurses and their later days as food foragers. Fruit fly researchers had discovered the *for* gene and had gone on to demonstrate that allelic variation in that gene explained why some fruit flies were "sitters" and some were "rovers." The fruit fly, however, is not a social creature by any stretch of the imagination, and the two organisms are separated by an evolutionary distance of 300 million years. Robinson and colleagues isolated the *for* gene in honeybees, found no comparable "active" and "passive" allele sets, and decided to test the hypothesis whether level of expression determined social role. And so it did. The *for* gene turns on, and the bees become foragers. To show causality, Robinson's team constructed a hive with nurses only. Almost at once, some of them became foragers as their *for* genes swung into action, an excellent example of how environmental change can trigger genetic effects. Neuroscientists would be pleased to learn the precise location of elevated gene expression in the bee brain. Robinson and colleagues provide an answer: in the coronal section, particularly the lamina of the optic lobes and the Kenyon cells, the latter of great functional significance to the mushroom bodies, the critical center for large-scale sensory processes. Do humans have a *for* gene? Indeed, they do, though how its place in the behavioral scheme of things compares with those of other species has yet to be worked out.[131]

Sociality is ubiquitous, showing up in the oddest places, even in the behavior of normally nonsocial organisms. Sociogenetics explains why. Consider the nematode *(C. elegans)*. These soil worms eat bacteria either alone or in groups. The difference is accounted for by genetics. Each cluster possesses the *npr-*1 gene, but the solitary consumers bear one amino acid (valine) at a certain juncture while groupie consumers bear another amino acid (phenylalanine). The latter group lacks *npr-*1 expression. Researchers pondering the hows and whys of this "social feeding" repertoire made a checklist of possible relevant neurons and knocked them out one by one. They settled on two sets, one in the anterior and a second elsewhere. Exchanges between the two determine feeding patterns. Only after the anterior neurons sense such aversive conditions as food shortage or crowding will "social feeding" result. Hence the conclusion that the phenomenon "may be a response to stressful conditions." The same signaling molecular process at work here is also on display with fruit flies and honeybees, though the molecules themselves differ. Oh yes, the NPR-1 protein is related to the neuropeptide Y receptor grouping found in mammals. It is important to ascertain whether the *for* gene (which works in different ways in the fruit fly and honeybee) is located in the same neuronal pathway as *npr*-1 in *C. elegans*. It is even more important to reach agreement on whether nematodes eating together are actually indulging in social behavior. Just what exactly *is* social behavior, or just what exactly *is* political behavior, anyway, even entomologists wonder.[132]

Now consider slime molds, the simplest of eukaryotes. When the threat of starvation looms, they not only feature sociality, but they actually feature altruism. Social scientists are forever trying to dope out why humans should ever be altruistic to non-kin; conventional wisdom among all scientists has been that one mechanism they could surely discount was "green beard" recognition. A "green beard" is a simple gene for altruism that is species adaptive in conditions where organisms display green beards, recognize one another, and respond preferentially. How, biologists asked, could "green beards" possibly trigger complex social behaviors? Well, green beards do exist, and the proof of principle is found in the slime mold. When in need of food, two cells, each bearing the *csA* allele, will adhere to one another at their joint gp80 protein sites and coalesce into a larger mass consisting of reproductive spores and nonreproductive stalk cells. The stalk cells will then sacrifice themselves to enhance the reproductive cells' food-gathering opportunities. In laboratory tests, molds in which this gene was knocked out emerge as "cheaters" when "playing the game"; they would have the capacity to latch onto the proliferating mass while giving nothing in return, so green bearded alleles repulse them.[133] There is evidence that the green beard syndrome is at work in mammalian maternal-fetal interactions as well as in certain game theoretic contexts where green beards take the form of "tags" or display markings of some sort. In the latter case, computer simulations demonstrate that cooperation through donation (classic altruism) will show itself a fit strategy over time as measured by rates of offspring proliferation. Note the lack of reciprocity in this exercise; the players had never before seen one another.[134]

Sociogenetics will continue to provide important results for the foreseeable future. Already, though, a faster gun has appeared. In the view I take, a viable research paradigm should contain at least three new constituent elements, the first two of which ought to be readily apparent as the stranger in town seeks to establish "his" credentials of superiority. First, the emergent paradigm must set out a grand theory, a never-before-articulated "worldview" of the relevant knowledge universe. Second, the germinating paradigm will require for implementation a new epistemology unique in cause and effect to the unique grand theory. Third, as scholars build a literature bearing witness to the efficacy and utility of the paradigm, the paradigm itself will commence to display a constitution. This constitution will comprise the value system of the paradigm, the attitudes and workways of its practitioners, and the coherent body of truths emerging from the trial and error testing of various hypotheses that, logic tells us, should flow intuitively from paradigmatic assumptions. Where does sociogenomics stand in these regards?

The underlying theory informing the sociogenomics paradigm flows directly from human genome investigations. We now have at our disposal virtually the entire *Homo sapiens* DNA sequence and, as critical aids to making sense of that information, the sequences of the yeast, the fruit fly, the nematode, the mouse, and many, many microorganisms. Sociogenomics seeks to pool these data for comparative purposes in order to home in on the genomic precursors of human social behavior. The underlying tools of the sociogenomics trade are new methodologies permitting scientists to perceive and measure the workings of constellations of genes as they, individually and collectively, contribute to behavioral propensities. Already we can see a significant literature taking form, as the following synopses show.

The zebra fish is a key organism for genomics inquiry; it reproduces rapidly and features a transparent embryo. Scientists shot a battery of genetically engineered retroviruses into the genomes of subject zebra fish to create mutations. Employing a PCR derivative, they located the alien intruders and sequenced the nucleotide arrangements on either side of each, looking for telltale signs of the malfunctioning genes. In this fashion they located 75 "culprits" responsible for zebra fish deformities. There are hominid homologies for each and every one of the 75.[135] Of equal importance to the zebra fish in the world of comparative genomics is the fugu or puffer fish. The fugu genome contains as many genes as the human genome, but it is much easier to study because it doesn't contain endless reams of junk DNA. About 75% of the creature's genes have human counterparts; investigators trolling for fugu genes in the human genome gold pot have found almost 1,000 previously unknown human genes.[136] Now let's return to the social insects. Researchers wanted to investigate the relationship between gene expression and winglessness in ants. They studied several wing development genes for level of expression in four ant species. For queens and males, the results conformed to expectation: the same genes for wingedness at work in other insects were at work here. For workers, which do not fly, a counterintuitive result emerged: each species displayed its own discrete pattern of gene expression. Wingless workers make up a separate, distinct social caste with a particular role to play in the ant reproductive cycle. The functional routines of the various genes among the ant species constitute an aesthetically pleasing study of evolutionary nuance befitting the complexity of this important social behavior.[137] For years, biologists have pondered the genetic proximities between mouse and man. Certain comparable chromosomes have now been investigated to the point of marshaling specific similarities. The mammals share about 200 homology segments, that is, chromosomal regions. Most mouse chromosome 16 genes (over 500) have human counterparts that show up in expected patterns; however 30% (over 200) don't.[138] And finally we have the age-old question of the chimp-man likeness. Invoking the most exotic of the new genomics technologies, called the microarray procedure (which I shall describe in detail below), researchers attacked the question of whether the two primates were as proximate genetically as the conventional 98.5% similarity estimate indicated. The answer now appears to be no. They found instead sequence additions and subtractions ranging from 200 base pairs to 10,000 base pairs. Some of these differences showed up only in introns, but others appeared in exons, the protein-coding regions. To make full sense of all this, a chimpanzee sequence project has now been launched.[139]

In January 1999, Gene Robinson, as principal investigator, and six co-principal investigators, including five biological scientists and one political scientist (myself), received a $200,000 grant from the University of Illinois to commence a serious study of sociogenomics using the honeybee as a model organism. The specific purpose of the grant proposal was to create a set of expressed sequence tags culled from the bee brain. To quote Robinson: "[W]e know more about honey bees than just about any other animal on earth."[140] That knowledge stems from the species' role as honey producer and plant pollinator. And because we know so much about bees, scientists can manipulate with precision their social environments, thus facilitating the investigation of genetic variation. At the time our application was submitted, there were no plans in the hopper to sequence the honeybee; yet, the need to move from a "candidate gene" strategy to a design emphasizing the simultaneous study of gene behavior was clear if our sociogenomic aims were ever to materialize. Waiting to be utilized was Craig Venter's EST technology, which involved sequencing coding region samples of genes. To do this, we proposed isolating mRNA molecules and experimentally converting them to cDNA, a well-standardized procedure. The complementary DNA would then be cloned, and a few hundred base pairs of each would be sequenced, thus producing an EST bank. We knew that if one had an EST in hand, one could ascertain if the gene itself were homologous to known genes in other species, and we would know exactly what those species were. The messenger RNA would come from the brain, the seat of social behavior. Because the project proposal presented no hypotheses for testing, we realized NIH and NSF assistance was unrealistic. The University of Illinois, our home institution, filled the breach by providing the requisite funding, and the bee brain bank housing 5,000 nonredundant ESTs opened for business in January 2002. If you do research on fruit fly genetics, if you do research on human genetics, you can make good use of the bank to check for homologies. And for honeybee specialists, it is a treasure trove.[141]

The methodological lynchpin of sociogenomics research is the microarray gene expression technology developed in Patrick Brown's Stanford laboratory. By 1998, there were more DNA sequences in GenBank, the largest data repository in the United States, than there were related publications in the literature. In an array experiment, cDNA from two sources of interest, for example, the brains of honeybee foragers and nurses, are laid on a glass substrate, and researchers measure the levels of expression occurring simultaneously in what can be a very large cluster. It is the ultimate (thus far) deployment of gene chips to assess genetic function.[142] In the chimp-human comparison research mentioned above, scientists placed 13 *billion* DNA pieces on the chip, each one about 25 bases in length. When contrasting the human chromosome 21 with the chimpanzee chromosome 22, the investigators were busy inspecting 27 million nucleotide sequences. In the honeybee, some workers display reproductive activity; most do not. What is the genetic difference? A sociogenomic study involving changes in hive conditions followed by microarray scrutiny to ferret out the precise genes being expressed and those not being expressed is underway.

Scientists specializing in the sociogenomics of a nonhuman population can make a powerful contribution toward understanding the dynamics of hominid behavior by employing the microarray technology to discover salient genes active in the display of various social behaviors and then conducting data bank searches for homologies down the phylogenetic tree and up the phylogenetic tree. An even greater challenge awaiting microarray application is ferreting out human single nucleotide polymorphisms (SNPs) and determining their sociogenomic role. Recall, a SNP refers to solitary letter changes in the genome. SNPs might be referred to as commonplace mutations in that, by convention, more than 1% of the population must share the letter substitution. Most SNPs have nothing to do with species behavior; in fact, most SNPs do not even occur in exons. The numbers are these: the human genome may contain 9 million SNPs; the coding sequence may contain 400,000 SNPs; SNPs responsible for amino acid composition shifts could number 200,000. Also to be accounted for are promoter region SNPs, which could have a marked influence in gene expression and, therefore, human differences. Each SNP variation yields a unique allele. "The Holy Grail in the SNP race is the identification of all functional SNPs."[143]

The not very visible SNP race is a microcosm of the human genome race. Abbott Laboratories and Genset started the ball rolling in 1997: Abbott pledged more than $40 million to support Genset research activities, and Genset, with Daniel Cohen in the leadership saddle, mobilized to develop human SNP sets. To no one's surprise, the sets would be patented and licenses negotiated. SNPs have excellent potential for use as disease gene markers, and pharmaceuticals figured to pay hefty prices to obtain access.[144] Alarmed at the prospect of biotech outfits patenting SNPs, NHGRI director Collins persuaded NIH to set up a SNP project.[145] The competition really heated up in 1998 when Celera, Incyte, and CuraGen all announced plans to find and hoard SNPs.[146] Now it was the big drug makers who were running scared, so much so that they formed an extraordinary alliance with an old, natural enemy, the Wellcome Trust, the group enlisting as SNP foragers heavy-hitting scientists at MIT, the Sanger Centre, and Washington University. The pharmaceutics would provide $45 million and Wellcome $14 million. Arthur Holden was chosen to direct the fledgling TSC (The SNP Consortium), and his group was given a directive: locate 300,000 SNPs in two years. Collins's SNP shop was aiming for between 60,000 and 160,000. As with the NHGRI program, TSC promised to put all its data in the public domain with no strings attached.[147] In what might be described as a proof-of-principle study, Whitehead Institute (MIT) scientists used a microarray system to hunt down SNPs. The experiment was highly successful. The team employed over 16,000 STSs, two-thirds taken from ESTs and one-third from random sequence. With many subject individuals contributing DNA data, broad chromosomal coverage was guaranteed. Demonstrating a remarkably low error rate for those early (1997) years of 10%, the MITers identified more than 3,000 SNPs, and then went on to locate more than 2000 of them. They were able to genotype about 550 SNPs on one chip, a number that today seems almost incredibly primitive. The fruits were sent to GenBank.[148] Francis Collins has joined Eric Lander's and many others' chorus of praise for the role of microarrays in tracking down the nature and function of SNPs.[149]

It would not be long before the link between sociogenomics and SNPs would make itself evident. The occasion was the important discovery that SNPs traveled in blocks; these blocks—haplotypes—were strikingly similar. In fact, researchers seemed to think human haplotypes came in only a handful of versions. Obviously, there were racial and ethnic SNP identifiers, and while scientists drooled over the prospect of "hapmapping," some ELSI observers were quick to remind them of the possibility that to formulate minigenomes highlighting such sensitive information might somehow convey the erroneous (to them) impression that these data amounted to something! One can read the collective mindset of these skeptics: if scientists find an allele for this and a QTL for that, pretty soon people will start thinking key social behaviors have genetic links. Clearly, haplotyping poses a clear and present danger to the culturalists.[150] Meanwhile, the pace of SNP detection was quickening. In 2001, GenBank held three million, and Celera bragged that it had even more under lock and key. SNP manipulation also received a boost when scientists used the technology to find mutations causing Crohn's syndrome and a form of diabetes.[151]

So what do we know about haplotypes? Two large studies demonstrate their potential significance. In the first, David Cox's team at Perlegan Sciences took on the daunting task of finding all the SNPs on human chromosome 21 and delineating their haplotype locations. Putting to splendid use the microarray strategy, they obtained a complete chromosomal sequence catalogue for 20 subjects. Their important finding: these individuals carried only three common haploytypes, and 80% of all SNPs could be found in these three clusters. Said one reviewer: "Now, one can aspire to analyze all of the unique DNA sequences in the genome simultaneously."[152] In the second, Eric Lander and associates at Whitehead/Harvard Medical School addressed the makeup of 51 autosomal regions evenly spaced throughout the genome with each containing one SNP authenticated by TSC. The subject universe was 275 persons of European, Asian, and African ancestry. Again employing microarrays, researchers genotyped 3,700 candidate SNPs, and the majority cropped up across the three population sets. All told, 928 haplotype blocks were observed in these samples, and although block size varied greatly, they were ultimately reducible to from three to five common architectures. The African group displayed a haplotypic average of 5.0; the European sample averaged out to 4.2; the Asian grouping yielded a mean figure of 3.5, all of this bearing witness to the greater genetic diversity of the older continental lineages. Summing up: if one knows the haplotype location for a SNP in a particular continental ancestral cluster, then that SNP will almost certainly reside in the same haplotype in another cluster; moreover, the vast majority of SNPs sit in designated haplotypes, and a small number of SNPs can identify any and all haplotypes.[153]

To Francis Collins, these findings meant one thing: the road to SNP discovery lay in haplotype tracking, and the founding and funding of an international haplotype consortium was the order of the day. He recruited five nations to join NIH to a total tune of $100 million. The U.S. government has allocated $40 million, the Sanger Centre $25 million, Canada $10 million, with Japan, China, and TSC also slated to contribute. In 2005, we could very well know the complete haplotype story.[154]

Recent literature reviews are exceedingly cautious if not skeptical about the long-term future of sociogenetics, and rightly so. They are modestly upbeat about the future of sociogenomics and rightly so given the fact that they are already behind on the rapidly changing learning curve. These theorists appreciate the role of SNPs in identifying QTLs, the stuff of all complex behavior whether sociogenomic or otherwise, and they understand the role of human genome exploration in facilitating the process.[155] They also appreciate the need for plugging in biopsychology, for example, in plotting causal paths from serotonin regulatory differences to amygdala function and then to phenotypic response rather than hypothesizing some old gene-to-behavior cause-and-effect model, and they hint vaguely at the role of microarrays in scanning simultaneous DNA activities.[156] They do not discuss SNP banks, haplotypes, or comparative species genomics. If they did and if they thought seriously about the sorts of experiments waiting to be done, they would not be modestly upbeat; they would be enthusiasts.

In 2001, entomologists, under Gene Robinson's intrepid leadership, tried to persuade the federal government to sequence the honeybee. They were rebuffed. "[U]nless it applies to human health, the NHGRI is not likely to get involved," said Francis Collins. Other agencies were also unpersuaded. The fact that mastering the honeybee genome would provide unique insights into species' evolutionary history was much too much the stuff of pure science for any of them, including the National Science Foundation. Nobody was talking about sociogenomics. Way too touchy![157] Robinson and his associates kept applying pressure. The big sequencing factories had available space; the human genome and the nematode had been wrapped up. Now was the time to dip into evolutionary theory and sociogenomics, they argued, raising the bar of discourse. The NHGRI relented. Honey bee sequencing, performed at Baylor, is being finished as I write (May 2003).[158]

The world of sociogenomics will explode in the next two decades. All key genomes will be sequenced; all human haplotypes will be mapped; all human SNPs will be known; microarray offshoots and fine-tunings will effortlessly yield homologous regions over many species.[159] Given that the brain and the pattern of group organization are simpler and more accessible in the social insects and the lower-order primates than in humans, research on these species will inform and energize in bottom-up fashion what we in academia call the social sciences. By that time, there may not be, and perhaps should not be, any discipline, no matter how broadly conceived, that is called the social sciences. The field of political science will have become a true science—joined cheek to jowl with many other sciences, social and otherwise—or it will cease to be anything except an intellectual museum piece. All of this will frighten the entrenched academic interests and will more than frighten those who have staked their careers in formulating on a pinhead the role of genetics in human behavior. They will call sociogenomics the new eugenics, and they will demand that academicians everywhere resist the "new political incorrectness." Can we devise an academic program—a macrointellectual discipline—to rebut their every negative presumption? That is the story of the final chapter.

6

Consilience

Is it not possible that the real relationship of students of politics is with biology or neurology rather than with psychology? Do we yet know what changes may be wrought in the individual through biological modifications? How far may attitudes and behavior be influenced or determined by biochemical processes . . . through . . . physiological functions, or neural mechanisms?[1]

So asked one of the great visionary political scientists of the 1920s, Charles Merriam. Note how much he sounded like the founders of his discipline and how little he sounds like the political science of today as I described it in chapter 1. The task of my concluding remarks is to provide a viable synthesis bringing political science into the mainstream of scientific discourse and investigation. Framing any sort of "grand unification" is heady stuff. However, several others have preceded me in this endeavor, and so my task is lightened immeasurably. These commentators have fought their own battles; blowing the trumpet of paradigm shift is not a vocation for the faint of heart, and many never survive the arrows of disciplinary injury. From my perspective, the leading trumpet blowers have been evolutionary biologists. I discuss their trials and tribulations, their insights and their inadequacies. From the tumult has emerged a new theoretical umbrella for political scientists to ponder and then, hopefully, adapt to their own needs and aspirations.

I then revisit my own academic home armed with the weaponry of sociogenomics fortified now by this larger, consilient frame of reference. In chapter 1, I talked of all our failings in the contemporary world of disciplinary discourse. And yet as Professor Merriam's questions illustrate, there has always been a minority sentiment among us dating back even before his time. That sentiment has spawned a literature deserving of greater recognition, some of which I have already cited. More will come. These writings serve as a modest stepping-stone; they show us the way toward a well-lighted place if we only knew where to find the lanterns. In fact, the lanterns are in place, ready for use. I conclude not simply by reciting an agenda for the future in general terms but by formulating a set of educational specifics, creating, if only on paper, a course of study for those who wish to pursue a graduate education in political science worthy of any comparable set of preparatory requirements.

A. Sociobiology and Beyond

If we are going to mold a "grand theory" of anything from the clay of evolutionary biology, we had best begin with Darwin.[2] The founding father's credo can be encapsulated in two principles. First we have the idea of natural selection. It directs the evolutionary process by creating fitter life forms through generational adaptation in an environment of random stimuli. Second we have the idea of reproductive success. It is the process through which selection operates; certain life forms simply do a better job of proliferating their numbers than competing life forms. Eventually, some survive while others perish, assuming critical environmental parameters remain in equilibrium.[3] In chapter 1, I noted the fusion of Darwinian evolutionary theory with Mendelian genetics to form what is called the "modern synthesis." Thirty-odd years ago, advocates of this school forged a path by zeroing in on the principal unanswered question held over from Darwin's time: "Survival of the fittest *what?*"[4] The "what," they argued, was neither a species nor a population subset. The unit of success was the individual or some inherent characteristic thereof.[5] If the individual, whose only game is fitness maximization through reproductive achievement, is the driving force of human survival, then what becomes of sociality? The answer, said Hamilton, was "kin selection" or "inclusive fitness": an organism will perform acts of altruism if these acts enhance the creature's *genetic* fitness, that is, the reproductive opportunities available to the entity's DNA as represented by relatives.[6] But there persisted the dilemma of good will toward friends and even strangers. The answer here, said Trivers, was "reciprocal altruism": individuals will assist others if they know they can bank on those "others" to help them down the road. So the underlying watchword for contracting parties who agree to help one another now or in the future is genetic fitness.[7] From this literature came sociobiology. Put far too succinctly, this "new synthesis" maintains that those social behaviors possessing the greatest survival capacity are precisely the behaviors best suited to carry forward the participating organism's genomic constitution. Taken to its logical extreme, human culture becomes the dependent variable of population genetics.

Sociobiology's guru was/is Edward O. Wilson. Let us do justice to what I consider a major conceptual breakthrough in thinking about human nature by reflecting on his central argument. "Sociobiology," says Wilson, "is defined as the systematic study of the biological basis of all social behavior." Its adversary in the proper study of humans is sociology; sociology is inherently flawed because of its "nongenetic approach." Human behavior can never be explicated by emphasizing an "empirical description of the outermost phenotypes" by which Wilson presumably means our elevated species ripped from the context of species generally. "[S]ociology and the other social sciences, as well as the humanities, are the last branches of biology waiting to be included in the Modern Synthesis."[8] "Evolution," he reminds us, "mean[s] changes in gene frequency";[9] it follows that "the organism is only DNA's way of making more DNA" and, hopefully for its survival, adaptive DNA.[10]

In chapter 5, I introduced James Q. Wilson's notion of *Homo*'s "moral sense." Edward O. Wilson would never invoke such a term. His problem would not be with the word "moral"; it would be with the word "sense." "Sense" is too vague, too normative, too unscientific. Instead E. O. W. talks about "the morality of the gene."[11] Dawkins would write a highly controversial book around this time called *The Selfish Gene,* essentially describing the full play of genetic survival mechanisms and capacities given the validity of Darwinian theory.[12] But Dawkins's model organism was the beetle, so critics could simply accuse him of hyperbole, nothing more damning. Wilson's model organism is *life,* all life. We can understand "the evolution of social behavior" of any species if we understand, he concludes, relevant demographic trends combined "with information on the behavioral constraints imposed by the *genetic constitution* of [that] species" (emphasis added).[13]

Over the next 25 chapters of his abridged edition covering 265 pages, Wilson expounds on social behavioral propensities throughout the kingdoms of life, including only the most general and therefore noncontroversial references to our species. Thus: "Human societies approach the insect societies in cooperativeness and far exceed them in powers of communication." "Absolute genetic identity [among lower invertebrates] makes possible the evolution of unlimited altruism," an obvious impossibility for humans.[14] (Query: Under what conditions, if ever, would the altruism arising from human cloning be an adaptive strategy? Is this a question for sociobiologists, political scientists, whom? Answer: everyone.) Wilson's concluding chapter discusses humans and only humans. He takes the role of a "zoologist from another planet completing a catalog of social species on Earth."[15] He is in search of the "human biogram," that is, "the behaviors and rules by which individual human beings increase their Darwinian fitness through the manipulation of society." Is this biogram, he wants to know, "an adaptation to modern cultural life" or "a phylogenetic vestige"?[16] (I, of course, would substitute the word "constitution" for the word "biogram.") His search for rules takes him through the contributions of Maslow, Homans, the primatologists, the anthropologists, and the linguists. Theories of ethics are advanced for the purpose of ascertaining their potential for biologization. Locke, Rousseau, and Rawls are found wanting, because they have no knowledge of how the human brain actually operates: they talk of "justice as fairness," but "the human genotype . . . [was] fashioned out of extreme unfairness." What we need, he argues, is a theory of the "*genetic evolution of ethics.*"[17] Finally, he provides a prognosis: in about 125 years, after man's ecological circumstance has achieved "steady state" and sociobiology has cannibalized all other paradigms of human understanding, we will be able to frame "a full, neuronal explanation of the human brain," and the "genetic basis of social behavior" will be sufficiently clear. Turning political scientist, he finds the "planned society" toward which we are pointed "inevitable," and he fears that "social control [will] rob man of his humanity" by depriving our species of its "Darwinian edge."[18]

Sociobiology was published in 1975. Richard Nixon had resigned the presidency; Watergate was fresh in everyone's mind. The Vietnam War had ended in travesty; the political "left" was riding high in its claims of responsibility. *Roe v. Wade* was a new addition to our law; the "culture wars" were assuming shape and form. Almost impervious to all this, E. O. Wilson assumed the role of Diogenes, in search not of honesty but of truth. In a very real sense he was an Archimedes, an innocent. As Trevor Howard said of Joseph Cotton in *The Third Man:* he was "born to be murdered."

In response to Wilson's mammoth commentary, a cluster of his Harvard colleagues declared war. Calling themselves the Sociobiology Study Group, they numbered among their members a who's who of what one impartial, notable observer called "the radically oriented" scientific professoriate. The list included Richard Lewontin, Stephen Jay Gould, Jonathan Beckwith, and Ruth Hubbard.[19] Denouncing the belief that *any* attribute of human social behavior was genetically orchestrated, the group called Wilson's argument a defense of the prevailing political order, which, for them, was steeped in elitism, racism, and sexism. He was even accused of sounding the same eugenic themes as the Nazis.[20] This melodrama in higher education McCarthyism descended into pure burlesque when members of a group called the International Committee Against Racism ran onto the stage at a AAAS symposium on sociobiology in 1978 and poured water on panelist Wilson's head as he was preparing to speak.[21]

Wilson, ever the gentleman, took partial blame for the brouhaha. He had written, he later said, two books in one. The first section, the entire essay minus the last chapter, was a dissertation on lower-order social behavior wrapped in the conventional trappings of evolutionary theory; the second section, devoted to humans, was presented in the form of social science data reassessed from a biological sciences perspective. This latter part lent itself, he now had decided, to the criticism of genetic determinism, a claim one clearly could not make about the sociobiological paradigm taken as a whole if one had bothered to digest the book's first 95%.[22] I wish I could agree with his retrospective, but I cannot. As I tried to show above, *Sociobiology* is quite unidimensional in character; it may contain some references here and there to the fact that genes are less significant in human behavior than in other species' behavior, but there can be little doubt that as of 1975 DNA was the name of the new game. That said, Wilson's sociobiology ushered in a fresh attempt to rethink the mainsprings of human behavior. No longer would the Freuds, the Skinners, and the Meads (I shall not even deign to include the Marxes) hold sway as the grand architects of the human psyche, its passions, its rationalities, its very moral calling. Biology, for the foreseeable future, was now a force to be reckoned with in the idea marketplace of liberal arts and sciences inquiry.[23]

Smarting under the criticisms from his peers and also appreciating that much more work was needed to absorb humankind into the sociobiological mainstream, Wilson, assisted now by Charles J. Lumsden, a physicist, embarked on the formulation of what I would call a new model of the earlier-posited paradigm.[24] They called their contribution the "gene-culture coevolutionary cycle." It works as follows: genes provide a context for neurological structure and function; the mind perceives and discriminates among cultural phenomena, expanding in power accordingly; human behavior will, in the long run, adapt to those cultural conditions that favor species survival; genotypic probabilities, therefore, are subject to evolution in the light of these conditions just as the conditions themselves are selected for by the relevant limbic-cortical processes. So genes affect minds that affect cultures that affect genes; each agent possesses a degree of independence but is constrained by one another's "leash." At each stage, the nexus is not mere convenience; it is a matter of necessity. The human brain needs culture in order to perform accustomed routines because there is insufficient DNA in the germline to instruct the requisite behavioral orderings. To sum up, the study of genetics coupled with the study of cultures relegates sociobiology's shining stars, inclusive fitness and reciprocal altruism, to subordinate status as explanatory variables in the macrodrama of human relations.

The salient terms in the Lumsden-Wilson conception are "epigenetic rule" and "culturgen." Epigenetic rules are biological processes that orchestrate cranial activity.[25] These processes represent halfway houses between their ultimate source, DNA, and the cultural data they spawn; they commence their work with RNA transcription and include all mental responses to environmental stimuli. Although the authors admit their model has never been satisfactorily operationalized, one cannot doubt that humankind possesses innate tendencies to accept certain cultural modes while rejecting others.[26] Moreover, the human brain does observe various predictable regularities—I would call them algorithms—such as chunking, satisficing, model matching, and reification, as well as the deeper dynamic flows of cooperation, defection, and reciprocity. The term "culturgen" combines two Latin words: *cultur(a)* (culture) and *gen(o)* (produce).[27] Culturgens are sets of artifacts, behaviors, or mentifacts sharing significant functional features.[28] The brain accepts culturgenic sets, and in essence, it learns by constructing networks of linked nodes or concepts around them that are stored in the long-term memory banks.[29] Culturgenic nodes "colonize" the mind, and some of these "hot" linkages implicate the limbic area because of their emotional content.[30] In contradistinction to the static notion of cultural trait, the authors see the term "culturgen" as conveying a dynamic image of human actions and ideas. Evidently, once a culturgenic unit or set such as "constitutionally acceptable conduct" or "belief in X rules of the game" is absorbed by the individual's epigenetic apparatus, a field of node links develops therefrom, and the idea of constitutional value as a socially defined thought-action phenomenon is merely an expression of "the totality of shared culturgens."[31]

Lumsden and Wilson use the incest taboo as their model phenomenon to demonstrate how the system works. Why is it, they inquire, that virtually every human culture rejects sibling sex? We possess, they conclude, an epigenetic rule strongly disfavoring incest in conditions where boy and girl were raised together in close domestic proximity during their earliest years. The rule has evolved from our genetic constitutions; these ensembles seek to maximize fitness, and brother-sister incest will increase the likelihood of flawed genetic progeny. So we build culturgenic rule and norm structures treating incest as a form of forbidden knowledge.[32] Over time, these structures will favor the genes that generate the incest avoidance mental predisposition, and the authors further argue that it takes about 50 generations or 1,000 years for the alleles orchestrating epigenetic rules producing adaptive culturgens to unseat the less competitive genes.[33]

If there is a key to unlocking *genetically* the coevolutionary model, then that key lies in our species' use of language. Language, as humans employ it, clearly and neatly separates us from other primates; through language, Lumsden and Wilson would say, individuals share culturgens and build communities, even nation-states, based on common understandings. If the authors are correct, there should be epigenetic rules fostering certain universal grammars, and as Chomsky has shown, indeed there are.[34] Children inherit faculties enabling them to master quickly their parents' tongue, and they have an innate sense of how words should be strung together to create meaning. There is no rationality vehicle by which individuals pick and choose from all competing syntactical formulations to arrive at optimal choices; instead, the brain features a cast of heuristic devices that permit them to home in on "sensible" patterns.[35] These should be genetic in nature; hence, scientists ought to be able to find them.

The major breakthrough came when Anthony Monaco and his colleagues at Oxford traced an inherited speech disorder to the FOXP2 gene on chromosome 7. A mutation occurs when an A turns up in place of a G, on which occasions the gene shuts down. When this happens, subjects are unable to articulate certain sounds necessary to produce various words and sentences. What we have here is not the lack of the proper epigenetic rule or process; rather, it is the lack of a motor mechanism necessary to mobilize that process. The mutation causes dysfunction in the basal ganglia, severely impairing sound production. Hominid language skills could not have evolved without a hospitable genetic code supplying the requisite physiological launching pad for the universal grammar to make itself manifest.[36] A Leipzig team then set about sketching the gene's evolutionary history. After sequencing the FOXP2 in the chimpanzee, the gorilla, the orangutan, the rhesus macaque, and the mouse, the researchers were able to show that the human version came to be formed by natural selection well after we parted company with the chimpanzee. "This is the best candidate for a gene that enabled us to become human," said renowned geneticist Mary-Claire King. We still do not know where to find the genes coding for a universal grammar, nor do we know if these genes became "fixed" in human populations before or after the FOXP2 in its current amino acid form.[37] With the rapidly maturing theory of sociogenomics and the methodology of microarray comparison, the prognosis for tracking down this and many other human epigenetic rules is bright indeed. In fact, the journey has already commenced. The Leipzig group pooled mRNA from humans, chimpanzees, and rhesus macaques and ran the samples over a microarray tray on which sat roughly 15,000 human cDNAs to ascertain gene expression levels. Previous research had shown human brains to be much bigger than chimp brains, yet the protein sets in these two species were strikingly similar. As it turns out, enhanced gene expression in the human brain since our species went its own way is on the order of three to four times the degree of performance found in our closest primate relatives. The next step is to compare these genes one by one.[38]

Adding credence to the Lumsden-Wilson formulation is primatological research on what scientists once considered that most human of characteristics—culture. Well, chimpanzees have plenty of culture, too. So do macaques. They develop norm systems regarding food sharing; they punish norm violators; they reciprocate; and they engage in "consolation behavior" for purposes of making the community whole after a fight. Just as humans have developed elaborate cultural norms for checking and balancing aggression, so have these other primates. And now we know orangutans are culture producers as well. Because these motifs are under fairly severe genetic control in these animals, it would be quite a leap to argue that human culture is constrained by no genetic leash whatever. Children are clearly superior at "culturizing" than apes, but who is daring enough to think their immature creative powers are unbound by physiological terms and conditions? And the only "evidence" we have that adults possess an infinite capacity to culturize whereas children do not is strictly theological. We appear to reside on a gene-culture sliding scale linked now and forever to the incredible richness and diversity of all other life forms.[39]

No matter how you slice it, *Genes, Mind, and Culture* was a more subtle and sophisticated account than *Sociobiology*. Wilson had acceded to the notion that hominids were different, yet in the end, the difference was really one of degree, not of kind. The sociogenomics paradigm sharpens our awareness of this degree: we test for genetic differences among all species, and we track down the cause and effect of each and every difference and each and every combination of differences at the level of social behavior. If the human mind does special things such as housing unique language faculties, then we get the genes; if cultures possess, to greater or lesser extent depending on the species, physiological embodiments in brains, then we get the genes. Perhaps the absence of a sociogenomic perspective helps explain the odd reception Wilson's coevolutionary conception has received over the years. When the book first appeared, it was the subject of intense scrutiny after which everybody more or less forgot about it.[40] Another reason why the argument failed to achieve long-term visibility (a prerequisite to long-term fashion) may be that the political climate had changed: the Vietnam War and Watergate were behind us, Ronald Reagan was president, and even Marxists get tired of calling names. Yet a third reason likely was that the book is very hard going even for a sympathetic scholarly audience, while *Promethean Fire,* its lay counterpart, was seen as too "soft." When Wilson, cohosted by Gene Robinson and this writer, visited the University of Illinois campus in 1999 to kick off our sociogenomics initiative, we prevailed on him to engage in an autograph-signing session. Students were invited to present their favorite piece of Wilsonia for his signature. I watched as copies of *Understanding Human Nature, Naturalist,* and *Consilience* were passed before him. When he finished I remarked, "Notice that nobody gave you a copy of *Genes, Mind, and Culture* to sign." "No," he sighed in resignation; "it's too formidable, too inaccessible."

In 1998, E. O. W. went back to the drawing board he had constructed 20 years earlier and presented the intellectual community with *Consilience.*[41] This crowning achievement constitutes for my purposes a whole new ballgame, not because it is any more inventive or even any more brilliant than what had previously flowed from his pen, but because it clearly and pointedly throws down the gauntlet of challenge to the social sciences, and certainly to political science, holding up to obloquy the very fabric of their idea constitutions. As such, it is of significantly larger compass than the model that is *Genes, Mind, and Culture;* more, it is of larger compass than the paradigm that is *Sociobiology. Consilience* stakes out what I would call a worldview of human knowledge, an estate precious few modern belief systems have achieved.

A key mission of this chapter is to give the consilience worldview a full and complete airing. The discussion will proceed in three phases. I will begin by providing a statement of the argument as I understand it, not merely its essence but its several significant permutations. Next, I will present the major counterarguments critics have lodged against the thesis, most of which I think are wide of the mark. Finally, I will outline my own reservations and amendments. These, I hope, will prove serviceable as instruments of clarification and assist in building a sturdier bridge between Wilson's professional world (entomology and evolutionary biology) and my professional world (political science and sociogenomics).

The notion of consilience owes an intellectual debt to the Ionian Enchantment: "a belief in the unity of the sciences . . . [holding] that the world is orderly and can be explained by a small number of natural laws." The Enchantment dates from Thales, much admired by Aristotle, whose theory of water as primal unit was the first explicit understanding of "the material basis of the world and the unity of nature."[42] Thales sensed unity, but he did not know about consilience, which "is the key to unification." Consilience, a term coined by Whewell in 1840, means the "jumping together" of various levels of human understanding. Thus, "[t]he Consilience of Inductions takes place when an Induction, obtained from one class of facts, coincides with an Induction, obtained from another different class. This Consilience is a test of the truth of the Theory in which it occurs."[43] These inductions arise from the application at various levels of analysis of the strategies and tactics of natural science investigation. Such tools will provide "the deliberate, systematic linkages of cause and effect across the disciplines."[44]

Fundamental to the consilient research temperament is reductionism, says Wilson, by which he means "the breaking apart of nature into its natural constituents." Ultimately, consilience seeks mastery of the complex, but scientific statements of complex phenomena are impossible without penetrating to the deepest, simplest categories and then negotiating the many rungs of the bottom-up ladder. In the end, synthesis, not analysis, will be the greatest challenge; then and only then can we envision "true consilience, which holds that nature is organized by simple universal laws of physics to which all other laws and principles can eventually be reduced."[45]

Of course, it is the human mind that is doing the reducing and the synthesizing, so critical to the successful journey toward consilience is "the master unsolved problem of biology: how the hundred billion nerve cells of the brain work together to create consciousness."[46] The brain is the most complex system known to man, and "the greatest challenge today . . . in all of science, is the accurate and complex description of complex systems."[47] If scientists can understand genetic diseases, he believes, they can "link the genes and their products into functional pathways, circuits, and networks."[48]

At this point in his account, Wilson recapitulates his gene-culture coevolutionary cycle. He says: "[T]o genetic evolution the human lineage has added the *parallel track* of cultural evolution, and . . . the two forms of evolution are linked" (emphasis added).[49] As certain genes and epigenetic rules are favored by natural selection, so "*certain cultural norms also survive and reproduce better than competing norms causing culture to evolve in a track parallel to and usually much faster than genetic evolution. . . .* Gene-culture coevolution is a special extension of the more general process of evolution by natural selection."[50] To penetrate this dynamic and recursive pattern, we must "search for the basic unit of culture." Earlier referred to as a culturgen—now called a meme out of deference to Dawkins's nomenclature—this basic unit is "the node of semantic memory and its correlates in brain activity." At the level of consilience, then, the task is to forge "causal connections between semiotics and biology." Meanwhile, as we await the inevitable neuroscientific conquest of our cranial constitution, we must admit that "particular features of culture have sometimes emerged that reduce Darwinian fitness at least for a time."[51] He does not explain what brings this phenomenon about.

The coevolutionary cycle is best understood, he emphasizes, as an exercise in fitness. The generalization emerging from the play of genes on culture and the play of culture on genes "sounds like a tautology—the fit survive and those who survive are fit—yet it expresses a powerful generative process well documented in nature. . . . Human nature is adaptive, or at least was at the time of its genetic origin." We again are left hanging: How are we to tell whether a set of memes is adaptive or maladaptive? Once more seeking reduction, he reposits the central message of sociobiology— now called by its practitioners "evolutionary psychology" in order to provide Political Correctness cover: "[F]rom the first principles of population genetics and reproductive biology [scientists] predict the forms of social behavior that confer the greatest Darwinian fitness. The predictions are then tested [in the field]." Sociobiology has produced a slew of useful categories and concepts, he argues: kin selection, parental investment, mating strategy, status, territorial expansion and defense (particularly "when some vital resource serves as a 'density-dependent factor'"), contractual agreement, and, of course, incest avoidance, the last of these continuing to be the classic example of how the coevolutionary parameters work in perfect harmony and also in a fashion consistent with the best of rational choice theory.[52] After all, isn't it rational *ceteris paribus* to do that which is consistent with already-proved-adaptive DNA/mental sets?

So what does the new worldview have to say about the social sciences, the humanities, ethics and religion? Social scientists, Wilson begins, "are expected to tell us how to moderate ethnic conflict, convert developing countries into prosperous democracies, and optimize world trade." They fail fairly miserably. He notes the great strides in the medical sciences, whose practitioners work closely with various kinds of biologists. "The crucial difference between the two domains is consilience: The medical sciences have it and the social sciences do not." Social science is riven by "bitter ideological disputes": from its *sancta sanctora* have come "Marxist-Leninism," "Social Darwinism," "laissez faire capitalism," "radical socialism," and "post-modernist relativism." "But never . . . have social scientists been able to imbed their narratives in the physical realities of human biology and psychology."[53] He reviews the research paradigms of cultural anthropology, sociology, and economics, finding them all held hostage to the Standard Social Science Model of culture as the product of "environment and historical antecedents." (Nothing is said whatever of political science theory, literature, methods, and understandings, i.e., its constitutional norms, practices, and accrued wisdoms.) Put bluntly, "they lack . . . a true scientific theory."[54] We know "the translation from individual to aggregate behavior is the key analytic problem. Yet in these disciplines the exact nature and sources of individual behavior are rarely considered." Wilson lauds Herbert Simon's "satisficing" mode of reasoning (Simon is called an economist; his ties to political science go unmentioned) and also Kahneman and Tversky's cognitive heuristics approach. Both of these comport much better with the human brain as he understands its workings than does the rational choice paradigm that currently reigns supreme.[55]

He saves some of his heaviest salvos for ethicists and preachers. The core difference of opinion is between "transcendentalists" and "empiricists." Whichever, "the true answer will eventually be reached by the accumulation of objective evidence." "Moral reasoning," if grounded in such evidence, "is at every level intrinsically consilient with the natural sciences"; "a law-giving God in the traditional Judaeo-Christian sense" is not. "Ethics . . . is driven by hereditary predispositions in mental development—the 'moral sentiments' of the Enlightenment philosophers." He is candid about his own faith or lack thereof: "On religion I lean toward deism but consider its proof largely a problem in astrophysics. . . . [but] the existence of a biological God, one who . . . intervenes in human affairs . . . is [dubious]." Science, unlike either ethics or religion, is "the first medium devised able to unite people everywhere in common understanding. . . . It is the base of a truly democratic and global culture." The "moral sentiments" to which he refers have as their "primary origin . . . the dynamic relation between cooperation and defection"; "game theory, particularly the solutions to the famous Prisoner's Dilemma" is a good way to appreciate how cooperative genes and therewith these sentiments spread throughout populations.[56]

As for today's ethics, they are "a mess." For one thing, we are driven by epigenetic rules "such as quick hostility to strangers and competing groups" [that] have become generally ill-adapted and persistently dangerous," given our advanced, high-tech culture. For another thing, we are much misdirected by "novel," "volatile," relativistic standards of right and wrong. This "chimera" makes of our ethics a bear pit of nonscientific, hopelessly philosophic controversy. The chimera also explains the tragic failure of political science, which he defines as "applied ethics." The struggle against transcendentalism is an uphill battle because "the human mind evolved to believe in the gods. It did not evolve to believe in [the science of] biology"; still, "science has always defeated religious dogma point by point when the two have conflicted,"[57] and he predicts more of the same.

The reception accorded *Consilience* in the academic world was decidedly chilly: "reviews . . . tended to be united in their hostility."[58] *Science* magazine assigned the task of assessment to a philosopher, the kiss of death. The resulting review was a worthless fulmination: "the central thesis of the book is vague, the argument presented generally difficult to discern, and many of the opinions expressed are quite eccentric."[59] Geneticists were hardly more helpful. My Illinois colleague Jerry Hirsch, who has been nipping at Wilson's heels ever since the appearance of *Sociobiology,* at least admitted that "it would be presumptuous for me to claim the expertise required to judge such an ambitious treatment of so many branches of learning." This nod toward modesty did not keep him from asserting that the author had flunked Genetics 101 and had furthermore committed the cardinal sin of taking seriously heritability quotients derived from twin studies, which Hirsch called "the mainstay of arguments for the legitimacy of racism." All in all, said Hirsch, without so much as a glance at the gene-culture coevolutionary model, "the purported links between the laws of physics and the cultural behaviour of human beings are fatally weak."[60] Wilson's worldview vision also received short shrift from scholars who study "the rhetoric of science." *Consilience* reeked of conquest rather than negotiation, cutting social scientists and humanists no slack whatever.[61] Richard Rorty found this damning account "imaginative and plausible" in his favorable *Science* review.[62] Finally, there were the backhanded slaps administered by advocates of "wholistic biology": "[O]ld-fashioned science 'tried to explain observable phenomena by reducing them to an interplay of elementary units. . . .' Contemporary science, on the other hand, recognize[s] the importance of 'wholeness,' defined as 'problems of organization, phenomena not resolvable into local events.' "[63] Wilson was saying: First reduction, then complexity, then synthesis. These critics were saying: First complexity, then synthesis, who needs reduction?

Some critics were kinder but ultimately negative. There was, for example, the claim that *Consilience* was sexist. Surely men and women were behaviorally distinguishable because they were biologically distinguishable as Wilson was arguing, said one observer, but still "I get rather tense" when encountering such "spots." Eventually, this "tension" got the better of him. As a result, he chided Wilson for disparaging the existence and the utility of free will. This criticism is quite odd and quite wrong on two counts. Wilson

clearly believes we have free will, because he talks about how we have sometimes chosen to depart from our epigenetic rule posterity. Also, how can we
overcome features of our no-longer-adaptive phylogenetic heritage except
via free will? The "tension" also became too much to bear when *Consilience*
discusses ethics, for Wilson was ignorant, we are told, of metaethics. Wilson admits he finds metaethics "obsolete." In his view, ethics arises from
brains organized by genes, inhabiting certain environments, and therefore
creating competing meme understandings that are subject to standards of
adaptation. Ethics is not the stuff of disembodied spiritualism.[64]

Another balanced, though in the end critical, assessment comes from
sociology. Wilson is right, we are advised, to upbraid social scientists who
refuse to take biology seriously. However, "total consilience" is neither
"possible [n]or desirable." The essential failure lies in Wilson's "prescrip-
tive" determination to "bring us back into line" because modern man indulges too much in culturgenic aimlessness. I find this a misreading. In a
rebuttal, Wilson calls *Consilience* "the outline of a research agenda," and he
is surely correct. Again and again, he states that our epigenetic rule structure is not well adapted in certain of its dynamics to a civilized environment it was never meant to address. This same commentator next takes on
Wilson's conception of cultural evolution as running on a "parallel track"
to genetic evolution. I must confess to having some difficulty with this
characterization myself. The idea of parallel trackism does not appear in
Genes, Mind, and Culture, or, at least, I can't find it. In its earlier incarna-
tion, the coevolutionary cycle features the brain as culture creator, and it
seemed to me that the prevailing culture might or might not take on an existence in sync with postulated epigenetic preferences. The idea of "parallel
tracking" appears deterministic, unless Wilson is simply using it as short-
hand for the cycle itself. I absolutely do not agree with my colleague sociologist that the track works to make of cultural evolution nothing but the
enforcer of already-established epigenetic trends and tendencies. To repeat, just as epigenetic phenomena may or may not be adaptive, so meme
networks may or may not be adaptive.[65] Usually, of course, they are, and
maybe this is what, "parallelism" means. I think some specifics would help
here. Chronic complainers such as Jesse Jackson notwithstanding, the
American constitutional polity looks pretty adaptive to most, no matter
what standards of adaptation one cares to apply. On the other hand, I can't
imagine criteria that would support the argument that Hitler's "constitutional polity" was adaptive even before he marched on Moscow.

Unfortunately, the balance of the negative commentary a respectable
scholar is obliged to report centers around matters of theology. And so

when Wilson talks about using consilience as a tool for social control (he probably has in mind his biodiversity policy agenda), he was immediately accused by a professor of divinity studies of perhaps advocating the destruction of Down's syndrome or other "different" fetuses. Disparaging the entire notion of consilience, this same commentator argued that "the point of the *revelation* and coming together of all things awaits the endtime" (emphasis added).[66] In the same vein, though far more frenetic, are essayists who consider Wilson the prophet of a new religion called science, an intellectual bent on uprooting and then destroying the wonders of all things beautiful, things inherently the province of the spiritual.[67] Then we have Wilson's "religious friends" who know him as an ethical and good man (so they say). *Consilience,* they have decided rather condescendingly, is not a "work of scholarship" at all! It is an "alternative to Christianity," which explains why the book sits on the bestseller list. So Wilson is a sort of Norman Vincent Peale in Darwinian clothing. Just as true Christians cannot accept Wilson, the argument runs, so those who cannot accept *any* vision will and should reject him. The God of Evolution deserves to fail, as do all other gods peddling "ultimate meaning."[68]

Such is the intellectuals' assessment of *Consilience*. Let me try my hand. I begin by stating my value premises, which, by now, should be all too obvious: I think Wilson's argument is fundamentally sound. I also see precious little need to carp over many of his grace notes and footnotes. Writ large, consilience is the only avenue that can maximize fully the impact and understanding of sociogenomics. However, for my ulterior purposes, the worldview needs some work, else I cannot peddle it to my political science peers.

First, it will do all of us a lot of good—including Wilson—if we can remove religion from the debate agenda without scuttling the guts of his thesis. I think we can meet the challenge. Wilson himself must take some responsibility for permitting matters theological to enter the dialogue fray. I do *not* fault him for honestly informing his readership that Darwin's theories "released [him] from the confinement of fundamentalist religion." I do not even fault him for saying that "science is religion liberated and writ large."[69] What really counts are the ways he defines and then circumscribes the terms "science" and "religion." In these regards, I think he comes up short, and two sentences cited earlier are my exhibits A and B. In exhibit A, Wilson states: "On religion I lean toward deism but consider its proof largely a problem in astrophysics. . . . [but] the existence of a biological God, one who . . . intervenes in human affairs . . . is [dubious]."[70] Even if we knew to a certainty the total system of simultaneous nonlinear equations governing "our" universe, such astrophysical knowledge would place us absolutely no closer to God as defined by the planet's major religions than we are today. Deism presupposes an impersonal first mover whose existence can no more be proven empirically than can the existence of a personal first mover. Which brings us to exhibit B, where Wilson states: "Moral reasoning," if grounded in such evidence, "is at every level intrinsically consilient with the natural sciences"; "a law-giving God in the traditional Judaeo-Christian sense" is not.[71] Wilson argues that consilience precludes a god who responds to prayer. If a god can ordain and establish the theory of natural selection, then why can't a god grant favors to those "he" loves?[72] The scientist who relies solely on the forces of nature for explanation forever runs into the theological riposte: "God did it. God created the rules, and God can break any or all of those rules at 'his' whim." How can the scientist disprove the counterclaim? How can the scientist ever prove there is no hereafter? Impossible, so let's forget it.

Now I turn to something vastly more interesting, at least to me: politics. By calling political science "applied ethics," by considering a medical school education and a political science education kindred analytical pursuits, Wilson badly misconceives the nature of the political science undertaking. I hope the reader will recall here my chapter 1 essay and its treatment of the discipline's founders, principally Burgess and Lowell. Our business is fleshing out empirically the "laws of power relationships." In other words, political *scientists*, if they were acutely aware of their intellectual roots, would dedicate themselves to addressing human behavior in precisely the same way Wilson addresses ant behavior. Sure, we must perforce employ different tools on occasion; honeybee researchers don't necessarily copy social amoeba researchers in conducting experiments. And sure, some of the knowledge we have acquired will help us "solve" a problem or two, but that is not the core of the mission. Here is a sample of how a political scientist thinks: Wilson cites at one point Murdock's "universals of culture," of which there are 67; the first thing I asked myself was "how many of these are political in nature according to any acceptable definition, and how can we study them"; I counted 13.[73] Wilson, on the other hand, treats political "problems" as though they were sick patients, lambasting us because we can't tell George W. Bush what to do about nuclear proliferation.

Consider now the difference between Wilson and Carmen on what the former calls "the most profound intellectual and ethical choices humanity has ever faced," namely, "[t]he prospect of . . . 'volitional evolution'—a species [having the power to] decid[e] what to do about its own heredity." What he has in mind, of course, is that molecular biology will shortly provide us with the tools to override natural selection in determining our species' future.[74] Turning a kind eye toward the philosophers and ethicists whose biological ignorance he had earlier derided, Wilson argues they should leave the pastures of grand theory to the geneticists and neuroscientists and swell the ranks of "public philosophy" where they can usefully explore "intuitive analysis narrowly focused on contemporary problems in medical ethics, law, and *political science*" (emphasis added).[75] I can't think of much of anything political science could profit from less than a debate between Leon Kass and Arthur Caplan on how many stem cell lines should be permitted to dance on a human cloned pinhead. A close second would be a debate between Gregory Stock and Francis Fukuyama on whether the Congress should criminalize human germline gene enhancement.[76] The reason why I wrote chapters 2, 3, and 4 in considerable measure was to show the inherently political nature of these research agendas, a nature meet for political science description and analysis. As one of Wilson's critics puts it, disparaging his apolitical naïveté: "politics is and always will be."[77] And central to politics is leadership, a concept Wilson never addresses. How else to explain the roles, attitudes, and priorities of Watson, Collins, Venter, and Wilson himself in the great unfolding drama that is the dawn of consilience. "Who dares suggest that Mozart can be reduced to atoms? Well, I confess, I suppose I do," said Wilson.[78] Fine and dandy, and while we're about it, let's reduce Venter and Wilson to atoms, too.

The business of political science, the business of scholarship and leadership, provides the keys unlocking the mysteries of Wilson's missing causal linkages. Why does humankind stray from its epigenetic heritage? What is maladaptive about cultures totally severed from that heritage? Under what conditions should cultures rein in maladaptive epigenetic rules? What does it mean to "rein them in" anyway? All of this requires political science knowledge about "who gets what, when, how, and why," the guts of influence and power. Wilson predicts that our species will be "genetically conservative." We will want to "save the emotions and epigenetic rules of mental development, because these elements compose the physical soul of [who we are]. . . . Why should a species give up the defining core of its existence?"[79] Well, we both know the temptations to surrender that posterity in the name of new visions of our "proper posterity" will be great, indeed. Here is where leadership comes in. It will be up to political scientists to define the parameters of adaptive leadership in selected environments, to be able to assess the qualities of adaptive statecraft among contending parties and interests. In this, political science will have at its disposal everything a mature consilience has to offer in the form of a serviceable knowledge base when moments of truth arise. Such will be the new constitution of political science.

B. Biopolitics

Throughout my discussion I have had occasion to introduce important research findings and observations by political scientists who have contributed to what some of us call the subdiscipline of biopolitics. For example, I have cited Douglas Madsen on power seeking and serotonin uptake, Andrea Bonnicksen on IVF and cloning policy, James Q. Wilson on humankind's moral sense, Gary Johnson on sociality, and Glendon Schubert on sociobiology. In this section, I will trace the surge of interest in biopolitics from its beginnings 30 years ago to the present, giving credit where credit is due to the full range of relevant research inquiry. If I omit some names and some lines of investigation, it is only because of page constraints and my desire to emphasize the nexus between biopolitics and sociogenomics. I will also document the explicit antipathies of the political science establishment toward biopolitics. In chapter 1, I recounted the ways in which modern political science is so out of touch not only with the teachings of our founders but also with the evolutionary psychology and molecular biology agendas. Unfortunately, that was only half the story. The other half involves direct and indirect disparagements of our subfield by various American Political Science Association insiders. I find such documentation distasteful but necessary in order to provide therapeutic cleansing in the form of cards-on-the-table dialogue, a marketplace forum meant to maximize intellectual honesty and minimize the savage stillness of silence.

The term "biopolitics" was initially coined more than a half-century ago, its conceptual utility serving to highlight the seeming plausible analogy of the governmental system to a living organism.[80] David Easton, a leading political scientist during and beyond the "behavioral revolution," picked up on the comparison and tried to construct a disciplinary paradigm around it, but the work received a jaundiced eye from my graduate school mentors, being regarded then as premature given the paucity of valid and reliable political science data.[81] A mere eight years later, Albert Somit took the initiative in pointing out the considerable extent to which biological science can instruct political science. He argued that two particular lines of research were especially suggestive: ethology and psychopharmacology.[82] For the first time in the pages of a mainline political science journal emerged the message that laboratory practitioners could alter human behavior and confirm the central ethological tenet that significant elements of human thought and action were rooted in our species' genetic constitution.

The Association for Politics and the Life Sciences, which currently numbers 200-odd members, was formed about 25 years ago, and the association's journal, *Politics and the Life Sciences,* made its initial appearance in 1982. The first executive director of the association and the first editor of the journal was Thomas Wiegele of Northern Illinois University, an organizational spirit of uncommon commitment. Wiegele himself named the following political scientists as subdisciplinary cofounders, though he must be given major credit for bringing them and others who would shortly follow under the same roof: Lynton K. Caldwell, Peter Corning, James C. Davies, Roger Masters, Glendon Schubert, Albert Somit, Thomas Thorson, and Fred Willhoite. I was a johnny-come-lately, arriving on the scene in 1985. How have these political scientists and others built a biopolitical research program?

Communication is a key attribute of political behavior. Standard political science investigation, arguing from the premise that among humans language is the major avenue of communication, has analyzed many a candidate's campaign speech for verbal clues and symbols. Lower-order primates, as we know, are quite adept in getting across the information needed for the politics of survival, but they must rely to a far greater degree on nonverbal gestures. In particular, they employ facial displays. The question arises: To what extent do nonverbal communications form a significant medium of political expression for humans, especially in a day and age when television, with its emphasis on the nonverbal, has become the dominant carrier of such expression?

Roger Masters and Denis Sullivan have addressed this and related inquiries in the context of contemporary American and French politics. They argued that facial displays are "expressed and decoded in similar ways across different human cultures" but that "members of different societies exhibit somewhat different expressive behavior, especially in public settings." They also contended that dominance, or perhaps more accurately, that generally accepted *legitimate* dominance is a salient independent variable in determining the efficacy of agonic (threatening or competitive) and hedonic (reassuring) facial displays.[83] With these considerations in mind, the authors presented to selected audiences soundless videotapes of presidential candidates delivering campaign speeches. Three kinds of displays and their differential impacts on groups of watchers seemed particularly worthy of study: happiness/reassurance, anger/threat, and fear/evasion, where the term preceding each slash mark delineates an emotion transmitted by the display and the term succeeding each slash mark delineates its purpose and significance.

Their first finding of note was that panels of French viewers responded to the three kinds of facial displays expressed by a set of French presidential candidates in the same fashion as panels of American viewers responded to the three kinds of facial displays expressed by a set of American presidential candidates. When a French or American public figure shows happiness, or anger, or fear, that emotion—that medium of nonverbal communication—conveys the same meaning and the same *correct* meaning in both political environments.

A second important finding assessed the intervening variables of partisanship and ideology. One might hypothesize that personal leadership characteristics would be less important to French than to American respondents, because the French multiparty system emphasizes issues and policy orientation in contradistinction to the American two-party system where, oftentimes, it is hard to distinguish differences among candidates except in individual cases. That hypothesis proves tenable. Whereas American viewers found Ronald Reagan's displays reassuring—no matter what they thought of his politics—French viewers' perceptions of Laurent Fabius and Jacques Chirac varied appreciably with whether they shared these candidates' political point of view. The same phenomenon is at work when the candidate's facial display is one of anger or fear.

These intervening variables also "wash out" variations in the ways respondents perceive different facial displays. In the United States, viewers react with much greater affinity to happiness/reassurance expressions than to anger/threat expressions, whether the candidate is a liberal Democrat or a conservative Republican. Not so for French respondents whatever their politics. They react with the same level of positive emotion to happiness/reassurance displays as to anger/threat displays and with the same level of negative emotion to anger/threat displays as to fear/evasion displays.

Nonverbal communicative displays between leaders and followers are crucial parameters of political cohesion among primates. What is required now is to address the nature of gene-culture interaction among humans. We have learned that verbal language cannot be considered solely cultural; it would be a mistake to assume that nonverbal language is solely genetic. Moreover, the influence of television defies neat distinctions between the two. How we perceive the stimuli generated by high technology and how we react to those stimuli are in part a function of the camera itself, guided, of course, by human hands. If we do not strip this process down to its gene-culture elementals, our constitutional republic cannot truly be an artifact of our own understanding.

This line of classical behavioral investigation shows that students of biopolitics are as adept in empirical research as any of their political science peers. Does the same hold true in the time-honored subfield of political theory? Indeed, yes. Students of politics are forever debating the nature of man: is humankind inherently a voice of reason, a voice of passion, or some mix of the two? Plato is squared off against Aristotle; Hobbes against Locke; Rousseau against Burke; the Athenians vs. Hobbes vs. Hegel. In contemporary political science debates, it is the students of biopolitics vs. rational choice theory, the only difference being that the latter barely acknowledges the existence, much less the legitimacy, of the former.

Foremost among political theorists of biopolitical bent is Herbert Simon, one of the great American social scientists of the past century. Simon's principal concern was capturing the lineaments of "human nature" by constructing models of decision making.[84] He particularly touted the modern field of cognitive psychology, the roots of which he traced back through Graham Wallas to William James. This is a much neglected intellectual tradition among political scientists, unappreciated even by many scholars of biopolitics. James has been called the father of pragmatism. We need to consider a quite different sobriquet for him: the father of decision theory. In his day, conventional understanding had it that people ran from bears because they were afraid. He dissented. People, James argued, were afraid because they ran. In other words, we have here a sequence of events commencing with some "arousing influence" and culminating in a "conscious emotional experience." If this be so, what psychologists need to study are the natures of these physiological upheavals from stimulus to bodily response to brain to their conversion in the brain to such emotional constructs as fear, anger, trust, and affection. This is an "evolutionarily old system" best tested for in political contexts not through the verbal symbols of conscious processing but the nonverbal measures of unconscious processing.[85] His disciples offered a refinement: the brain must somehow appraise stimuli in order to induce bodily reaction; the process was not automatic. Perhaps, though surely the brain at this level of function is not involved in ratiocination. And just where would all of this go on in the cranium? James suggested the cortex, but when scientists removed the cerebral cortex from cats, emotional processes were unimpaired. What was impaired was emotional *control,* that is, the times, places, and manner of emotional display. Enter the thalamic "relay station" regions that send messages to the hypothalamus before they ever reach the cortex and various bodily response systems. Later would come Paul MacLean's triune brain theory—now generally considered valid—with its emphasis on the limbic system as the emotional response center.[86] Later still, empirical studies fleshed out the relationship between the amygdala and the nucleus basalis, and they also showed that the autonomic nervous system acts selectively, perhaps triggering different combinations of hormonal response to different kinds of stimuli.[87] Very recent research on this subject is highly edifying. Inhibited infants grow up generally to be "avoidance" adults, while uninhibited infants grow up generally to be novelty-seeking adults as measured by functional MRI amygdalar screenings. Moreover, fine-tuned functional MRI tests have also detected the differential roles of the left amygdala and the right amygdala in reacting to "anger faces" and "fear faces," with the former showing a high degree of sensitivity as to whether the stimulus gaze is frontal or averted whereas the latter is unresponsive to such nuances.[88]

It is important to distinguish James's insights from Freud's insights. When political scientists of Lasswellian persuasion talked of the unconscious, they were primarily concerned with psychopathology. The political scientists of Jamesian persuasion have been concerned with the decisional apparatus of the typical human, attempting to characterize the impact of these pathways of the unconscious.

Graham Wallas was a student of James's, who turned essentially to political theory. In 1908, he published his seminal essay, which attempted to apply the world of Jamesian psychology to the world of political science:[89] According to Simon, Wallas's fundamental concern was "the ubiquitous workings of instinct, ignorance, and emotion in . . . politics [particularly] the interplay of . . . rational and nonrational components."[90] Note the publication date of 1908, the very same year Arthur Bentley published *The Process of Government*. As political scientists then fixated on the study of formal governmental arrangements would ignore Bentley, so they would also ignore Wallas.

For Simon, building explicitly on the James-Wallas foundation, the new field of cognitive psychology, which studies what "[happens] inside the head" of political actors, is the light and the way of today's political science. Cognitive psychology, for him, goes hand in hand with his own formulation, bounded rationality. People are decidedly limited in their abstract reasoning capacities, and they search very heuristically, incompletely, and sometimes inadequately through partial sets of information, their algorithms of choice often veiled in ignorance and yielding, therefore, hopefully satisfactory, rarely optimal, solutions. Simon likens "bounded rationality" to procedural due process of law, where the "choosing *organism* . . . uses methods of choice that are as effective as its decision-making and problem-solving means permit. . . . [This is] behavior that is *adaptive* within the constraints imposed *both* by the external situation and by the capacities of the decision maker" (emphasis added in first two instances). With procedural due process and bounded rationality the process of decision is rational (constitutional?) if reasonable. Simon then goes on to lock horns with rational choice theory, which he likens to substantive due process. With these models, rationality or fairness lies in the result itself. Of little consequence to rational actor theory is the nature of the organism in any empirical sense; one need only know the organism's utility function and "the objective characteristics of the situation." I note that substantive due process was the guiding paradigm for constitutionalizing laissez-faire economics during the 1890–1936 Supreme Court era and for constitutionalizing the woman's absolute right to obtain an abortion in the first trimester and other rights of "personhood" during Justice Brennan's heyday in the 1970s. For Simon, as with E. O. Wilson years later, rational choice analysis will not do because the human brain does not work that way. What does work is the "flesh[ing] out by a myriad of facts, . . . harvested by laborious empirical research . . . our aspirations for . . . [general laws using as a model] *the complexities of molecular biology*" (emphasis added).[91]

Simon was no student of behavioral genetics or even physiological psychology. Yet, his bounded rationality thesis comports well with the literatures of these disciplines, and immersing himself in his later years in the quandaries of sociobiology, he set out to demonstrate the inadequacies of the reciprocal altruism argument. Rational choicers, of course, cotton to reciprocal altruism; for them, mutual back-scratching in the name of money, power, or genes is the apotheosis of contracts freely and rationally entered into. They even think reciprocal altruism in the form of political economy is the most adaptive strategy for solving tragedies of the commons.[92] Human fitness, Simon explained, is much enhanced by social learning or "docility," the ability to "accept well the instruction society provides." Because human rationality is bounded, docility carries big payoffs; people simply cannot optimize their capacity to select what they need to learn. Over time, docility becomes favored by natural selection, as it at least is reasonable to learn what society seems to think we should learn. If the social arbiters teach altruism, then altruism ought to eventually spread throughout the population, thus enhancing the survival and genetic success of the altruists. The increments of fitness lost on account of doing things for others is more than compensated for by the knowledge and skills accruing from docility. Docility is bestowed, then, by the genes and is not the product of rational choice: "By virtue of bounded rationality, the docile person cannot acquire the personally advantageous learning that provides [a gross increase in offspring] without acquiring also the altruistic behaviors that cost the decrement [of lost offspring arising thereby]." Correctly, he chides Lumsden and Wilson for refusing to face up to the inadequacies of reciprocal altruism. However, in the end he fudges the question of whether altruism is a cultural or genetic construct.[93] I should think that the substance is likely the former, while the process is likely the latter, a conviction again attesting to the robustness of the due process dichotomy earlier cited.

James Q. Wilson's "moral sense" notion jibes nicely with Simon's bounded rationality, and, I think, takes the major thrust of the biobehavioral paradigm a step further. Recall that Wilson rejected social compact explanations with their reliance on rational self-interest, arguing that "people everywhere have a natural moral sense" in the form of "sentiments" that "constitute the fundamental glue of society." The preeminent or "founding sentiment" is the "parent-child relationship," which clearly transcends the individual actor's utility-optimization calculus.[94]

Wilson supported this assertion by citing evidence from anthropology and social psychology, thus folding in empirical evidence earlier that was not sufficiently investigated by biopolitics specialists. Children are every bit as indulged by the precivilized on the Kalahari as they are by the suburbanites in the megalopolis; infanticide is practiced only under extreme environmental stresses that could occur anywhere; youngsters develop by assuming eventually the mantle of responsibility for the well-being of their aging forebears, though they gain little in tangible reward for so doing; the brain of the newborn is not a blank sheet of paper, for children prefer human sounds to other sounds, prefer their mothers' sounds above all other human sounds, and express the broad array of nonverbal facial displays mentioned in the above-cited literature before anyone can teach them anything. Wilson summed up these data as follows: "bonding is driven by powerful biological forces."[95]

Aristotle's guiding theoretical light for our species was "reason and nature," and Wilson is, if nothing else, an Athenian with empirical shoes on. So when Aristotle talked of a "natural striving to leave behind another that is like oneself" and said that a "parent would seem to have a natural friendship for a child, and a child for a parent," and that "in the household" one sees exhibited the rudiments of "political organization," which by nature tends toward (that is, evolves over time into) the city state, Wilson found justification for each link in this chain of reasoning in "modern science." But here Wilson improves upon Aristotle's theories. If the fundamentals of political order are subsumed in nature, then perhaps the fundamentals of political *rights* are also subsumed in nature, he inquired. For him, the evidence is overriding. Social psychologists have shown that "fair play"—that is, procedural due process—"is a necessary condition for the child to satisfy its natural sociability." More controversial is his conclusion that primary school youngsters have a natural tendency to spin from the yarn of "follow-ing the rules" the cloth of "equal shares," that is, substantive due process. But children soon learn, we are told, that "equality in results" is manifestly unfair when their peers don't "deserve" what they receive.[96] In my view, what social psychology had actually shown is that prepubescents naturally accept a rather different constitutional principle: equal protection of the law. This tenet speaks to the unfairness of arbitrary and irrational classifi-cations in the prevailing rule structure (e.g., all Jews go to concentration camps, no blacks can vote) and ought not be confused with more general-ized notions of equalitarianism. There certainly is no evidence in the litera-ture of "modern science" to support the idea that youngsters (or oldsters) have a greater disposition toward equality than to liberty. And for political scientists, equality and liberty are in "natural" tension.

Like Simon, Wilson was leery of reciprocal altruism. His proofs, how-ever, do not arise from abstract model building; they arise from what he sees as a deficit in mainline sociobiological theory. The argument is that whatever strengths are provided by the paradigm cannot readily be trans-lated into viable predictions for individual behavior. That is because "evo-lutionary biologists ordinarily do not specify the psychological mechanism by which a trait that has been selected for governs behavior in particular cases."[97] Simon, as well as Lumsden and Wilson, falls short on this count. If only they all—including J. Q. Wilson—had sociogenomics staring them in the face.

The last biopolitical theorist I think should be included in this summary is Francis Fukuyama, J. Q. Wilson's colleague on President Bush's blue-ribbon bioethics panel. In a sense, he is an odd choice. His major contri-bution, *Our Posthuman Future,* contains no original research and, quite frankly, no terribly original ideas. As one critic has reported: "There is much not to like about [this] splashy foray."[98] Yet, Fukuyama is important for my purposes, the first being that he builds a bridge between biobehav-ioralism and public policy. He begins in lockstep with the two Wilsons: "[H]uman nature . . . has provided a stable continuity to our experience as a species. . . . Human nature shapes and constrains the possible kinds of po-litical regimes. . . . [W]e share a common humanity that allows every human being to potentially communicate with and enter into a moral rela-tionship with every other human being on the planet." Political science must take the natural sciences very seriously because "modern biology is fi-nally giving some meaningful empirical content to the concept of human nature." The power of evolution is such that man cannot be endlessly ma-nipulated by social engineers with any degree of success. Modern biology

has now assumed for political scientists a second dimension of vital importance: "the possibility that it will alter human nature and thereby move us
into a 'posthuman' stage of history."[99] Human engineering predicated upon
a sophisticated knowledge of the "neurological structure or biochemical
basis of the brain," that is, "the genetic sources of behavior" is so powerful
an instrument that *we should use the power of the state to regulate it.*"[100] He
goes so far as to posit an algebraic equation-like operational definition of
human nature so we will know exactly what form our endangered essence
takes: "human nature is the sum of the behaviors and characteristics that
are typical of the human species, arising from genetic rather than environmental factors."[101] My corrected formulation would read this way: human
nature is the sociogenomic constitution of the human species. The peril is
so great that even if "human genetic engineering" "never materializes," the
three earlier stages of biotechnological innovation—"greater knowledge
about genetic causation, neuropharmacology, and the prolongation of
life—will all have important consequences for the politics of the twenty-
first century. [They] . . . will challenge dearly held notions of human equality . . . , will change our understanding of human personality . . . , upend
existing social hierarchies . . . [and] give societies new techniques for controlling the behavior of their citizens."[102] "[Political scientists and others]
need to start thinking concretely now about how to build institutions that
can discriminate—[that can find] middle ground—between good and bad
uses of biotechnology."[103]

The policy guidelines and recommendations Fukuyama endorses are, in
my opinion, quite oversimplified and, if his voting record as a Bush committee member on human cloning be a benchmark, quite wrong. All this is
beside the point. It is a major step forward when someone who is not a
working member of the biopolitics establishment such as it is and who possesses a considerable measure of political science cachet touts biopolicy as a
fundamental area of scholarly research, and who appreciates that the study
of *content,* of *substance,* is critical if political scientists are to come to grips
with the full range of our species' adaptive strategies. The term "political
science cachet" deserves explication for the uninitiated. Some have argued
that biopolitics—or, if one must, "politics and the life sciences"—has never
gotten off the ground as a viable political science subfield because it has
never had the concerted backing of anyone with substantial political science
cachet.[104] Herbert Simon did not even work in a political science department. James Q. Wilson went into the job market before the term "biopolitics" existed in political science, and he built his reputation by writing about
all kinds of important things having nothing to do with biology. Cachet

they have in abundance; biopolitical standard bearers they never were nor have been. I do not think Fukuyama is in their class, but he meets today's minimal cachet standard because he could quite conceivably get a job in one of the top ten departments in the country, and (this is what makes him unique) he could do so while advertising himself as a committed scholar of biopolitics. By the top ten departments I mean, moving geographically from east to west, Harvard, Yale, Cornell, Columbia, Princeton, Michigan, Chicago, Berkeley, Stanford, and the reader may supply the tenth. An avowed student of biopolitics has never landed a tenured professorship and today could not land a tenured professorship in any of these political science departments unless special circumstances obtained. Fukuyama passes the test: his major books carry bestseller gloss, and the *New York Times* made sure to interview him at some length upon the publication of *Our Posthuman Future.*[105] In other words, Fukuyama is an academic "star," whatever that term may mean. A highly competent, even brilliant, biopolitics "nonstar" could not even get a job interview at Ohio State, much less at those citadels of scientific research, MIT and Cal Tech.

There is one final biopolitical research thrust requiring attention here; in my opinion, it is the most innovative and insightful program in our literature. "Men think in terms of models."[106] Models are a vehicle for converting chaos into order, for converting political indecision into strategic pathways and tactical sequences consonant with our epigenetic rule structure. Which model of the human mind—which model of human decision theory—best comports with the Darwinian perspective?

The models most favored in the political science literature are drawn from the theory of games. In "game theory," players develop logics or blueprints of action for the purpose of securing victory in naturally competitive situations. Examples of the fruits of victory would include money, power, and enhanced reproductive fitness. Of these models, many students of politics consider Prisoner's Dilemma (PD) as providing the deepest insights. That is because it yields an uncommonly parsimonious solution to James Q. Wilson's "central problem": Should humans cooperate or should they defect?

For those unfamiliar with PD, consider these facts. Two unrelated thieves are arrested and locked in separate cells. The question is whether they should say nothing to the police (cooperate with one another to their long-term benefit) or betray an ally (defect). Unable to consult and fearful that each will implicate the other, they both avert the "loss" of being played for a sucker, thereby sealing their respective fates. The lesson seems to be that, at least in the short run, that is, a single "play of the game," the best strategy is always to defect even though the results could not be more disastrous. From a political science point of view, then, asociality wins and governance loses, unless somehow it becomes adaptive for the player to focus on more altruistic considerations. The PD format seems particularly relevant to the technologically sophisticated, impersonal politics of today, where people see one another relatively infrequently and where cheaters are hard to recognize. Note also that if the two culprits are brothers, then inclusive fitness theory will dictate cooperative responses from each no matter what the other does. That is because their coefficient of relatedness is .50; therefore Brother A will get one-half of Brother B's payoff bonus plus what he would ordinarily derive in the non-kin PD scenario.

The leading figure studying the biopolitical implications of PD is Robert Axelrod, a political scientist at the University of Michigan. I regard Axelrod as one of the ten leading political scientists on the current scene, and he has political science cachet aplenty. Needless to elaborate, the two are not related. Axelrod has no affiliation with the Association for Politics and the Life Sciences and seemingly no interest in identifying himself as a student of biopolitics. For example, in the American Political Science Association 1988 Biographical Directory, he lists his areas of expertise as "Formal/Positive Theory; International Relations/World Politics; Public Policy [International Security Affairs]; in the 1994 –96 American Political Science Association Directory of Members, he gives his fields as "Public Policy, International Politics, Political Thought: History, International Security." Not a hint of his interest in Darwin, not a hint of his interest in sociobiology, not a hint of his interest in genetics. From my biased perspective, it is disheartening when we biopolitics types are unable to welcome into our midst some of the best and the brightest either because we have failed to convey properly our message or because our audience finds the prospect of membership professionally problematic (or both). So where does Axelrod's cachet come from? For one thing, he is a master computer simulator, and today's political science loves people who can manage high-speed information-processing technologies. Second, he is as proficient a formal model-builder as we have in our profession, and rational choicers revel in formal models. In other words, I contend that Axelrod's well-deserved cachet results from all the wrong reasons. I am absolutely convinced that the overwhelming majority of political scientists who think him superior don't have a clue as to the larger significance of his work. In fact, much of this research has been published in *Science* and *Nature*, journals read by biologists, not political scientists.

In an attempt to provide a solution to PD, Axelrod joined forces with W. D. Hamilton, one of the most brilliant geneticists of the past

half-century. E. O. Wilson has spoken in glowing terms of Hamilton's contribution to the mathematics of kinship, to the spread of altruism among
social insects in particular using quantitative assignments based on genetic
relationships. In so doing, he provided "the tools for real, empirical advances in sociobiology."[107] I regard the Axelrod-Hamilton alliance as consilience made in heaven.

They began by drawing a correspondence between "single-play" PD
among non-kin and "single-play" interactions of unrelated organisms in
nature: defection is always the preferred strategy for maximizing survival
and fecundity. Even when the number of interactions is multiple but
known, defection maximizes dividends as no other decision rule can.[108] But
where the game is an iterated PD and the number of plays is unknown
but multiple, other strategies become competitive. To determine the relative
robustness of possible strategies, Axelrod conducted a computer tourna-
ment for PD. The winner was a format called TIT FOR TAT. Its algorithm
was simple enough: cooperate on the first move and then ape the other
party's reply. In other words, it does pay to cooperate, provided defectors
are punished immediately but never excessively. How does TIT FOR TAT
establish a toehold in a jungle of defectors? The answer lies, of course, in
genetic kinship theory and its altruistic incentives. And if *Homo sapiens* can
structure their politics so as to enhance interpersonal contact, free riderism
can be deterred and reciprocity will flourish. TIT FOR TAT works in theory
for humans and even bacteria; it is not a rational choice artifact. This has
been shown empirically. TIT FOR TAT proves an evolutionarily stable
strategy among vampire bats and vervet monkeys, and experimental simulations show its robustness for tree swallows and stickleback fish.[109]

Axelrod has also tested the fitness value of TIT FOR TAT by using the
genetic algorithmic format.[110] He converted various PD strategies into
chromosomal form featuring a 2-bit DNA code: C for cooperate, D for defect. Whereas it would require an expert system eons to discriminate among
winners and losers, genetic algorithms make the competition manageable.
After one all-play-all round, the most robust chromosomes were permitted
to mate with one another; the less robust were dropped. Critical to the play of
this game is that two well-documented processes of genetic adaptation,
crossover and mutation, are here employed to order the DNA sequences of
chromosomal progeny. Eventually, the high-scoring routines came to resemble in large degree TIT FOR TAT. Axelrod summed up this work as follows:

> [T]he main advantage of simulations can already be glimpsed. . . . Instead
> of having to rely only on our observations of real biological systems . . . we

are able to approach genetics and evolution as a theoretical design problem. . . . We can begin investigating alternative ways genetics might have evolved. . . . Today microbiologists are developing the techniques to alter our genetic heritage. Perhaps now is also the time to think about doing some "as if" experiments to better appreciate the fundamental properties of the genetic system.[111]

More conservatively, but not less significantly, one may observe that we have here a second proof predicated on genetic theory and methods to show the evolutionary potential of cooperation-reciprocity choice making as the most viable political strategy.

In providing operational definitions of humans' bioconstitutional politics, I have often emphasized the importance of rules and norms in evolutionary perspective. Axelrod has employed a computer simulation exercise to flesh out the enduring quality of norms.[112] His "norms game" is also based on evolutionary theory. Players can either cooperate (live within the established norm structure) or defect (violate the norm structure). But unlike PD, a defector can be punished by an observant onlooker. And so competing strategies will consist of sundry combinations of boldness and vengefulness (i.e., the probability of defection and the probability of inflicting punishment on others). Initial strategies are selected at random and pitted against one another, with points awarded according to the mix of successful defections, punishments meted out, and unsuccessful defections. Each of these strategies turns out to be a sort of genetic algorithm, with Axelrod instructing big winners to mate and eliminating big losers. He also once again includes a mutation variable. This sequence is repeated over 100 generations to provide the evolutionary component. The upshot: defection rules the day. But in a "metanorms game," where players can punish not only defectors but those who observe defection yet decline to mete out punishment, norms become stable and resist violation. The important lesson to be drawn is that an enlightened politics requires constitutional arrangements, however these arrangements lack adaptive capacity unless metaconstitutional mechanisms are at work.

Although biologists have found versions of game theory useful in their studies of lower-order animal life as I reported earlier,[113] Axelrod himself has pointed to the inherent limitations in these formats:

> [Genetic algorithms] do not capture the idea that players may have a great enough understanding of the situation to do some forward-looking calculations as well as backward-looking comparisons with others.[114]

Recall as well that TIT FOR TAT cannot benefit from consultations with allies. Making it even less useful in "game of life" trials is the fact that it remembers only one move back; this has the effect of rendering impossible any attempt to arm it with a sense of history, even a sense as to the relevance of some literature containing empirical and theoretical decision rules. I shall suggest improvements momentarily.

Before I take my leave of Axelrod, I want to point out the far superior "fitness" of his evolutionarily stable strategy motif (originated, incidentally, by a biologist, Maynard Smith) when squared off against its rational choice analogue, the Nash equilibrium. The Nash equilibrium notion assumes that players behave perfectly rationally based on their perfect understanding of their rivals' intentions, both of which parameters, as Herbert Simon demonstrated, are nonsense.[115] That is why it is difficult to predict when a Nash equilibrium will take hold, and we all know how economists covet prediction. I also want to point out the influence Axelrod's work has had outside of political science.[116] In simulations more nearly mirroring the typical biological context in which stochasticities abound in the form of a heterogeneity of player population, TIT FOR TAT is less robust in the long run than a more forgiving generosity strategy wherein Player A spots Player B two uncooperative moves.[117] And a recent simulation study shows that altruism can proliferate even in larger populations when a measure of indirect reciprocity is introduced: social status. Altrusitic behavior begets altruistic behavior by others if the others see that such behavior elevates one's status among watchful peers. So here we have a formal proof showing how social status can make for cooperation and community, therefore taking its place among other evolutionarily stable strategies such as docility, forms of generosity, and "green beard" commonalities.[118] Of late, a new field of study has emerged—neuroeconomics—providing grist for my mill. The idea is to demonstrate empirically the cranial mainsprings of cost-benefit calculation. In a recent experiment, functional MRIs were performed on responders playing the "take-it-or-leave-it" game, demonstrating that unfair offers triggered heightened responses in the anterior insula of the brain centers, a seat of emotional response, and that these activities prompted rejections to a statistically significant degree. Another setback to the dogma that utility maximization would in these circumstances yield positive responses because getting something is better than nothing.[119] And the underlying leitmotif of all these is bounded rationality.[120] Finally, groundbreaking empirical tests show respondents employ well-specialized cognitive processes in order to engage successfully in "social exchange," this term meaning cooperation for mutual benefit. So the human brain is not an all-purpose calculator but a biological system containing discrete adaptational coping devices.[121]

If human beings are species of bounded rationality, then perhaps we ought to start labeling lower-order species creatures of bounded irrationality. Take our specimen of choice, the honeybee. We have just now found out that these social insects are able to trade their usual rate of foraging speed for greater accuracy in selection when faced with tough choices.[122] Too bad we can't use functional MRI on bumblebees, though Gene Robinson and colleagues are sure to get us started by ferreting out the relevant decision-making DNA. Sophisticated technological tools can be applied to monkeys to the extent that when those animals make certain food choices scientists are able to distinguish between the respective roles of neurons in the lateral and medial frontal lobes.[123] The DNA precursors of these decision processes will undoubtedly be ascertained in timely order. On to humans! A consilience of sociogenomic constitutionalism awaits.

A critical question is whether "expert systems," what I call the "Great White Hope" of rational choice theory, can defeat biologically (read: genetically) grounded algorithms relying as they must on "satisficing" and "muddling through." The most advanced "thinking machines" extant are chess computers, so they provide a good test of the possibilities. Chock full of information gleaned from the great matches and tournaments of history as well as the vast array of investigations into opening and end game theory, the best of these softwares can calculate tactics and variations with a speed no human living or dead could ever match. Moreover, machines neither sweat nor tire. Still, in recent matches World Champion Vladimir Kramnik and ex-World Champion Gary Kasparov held their own against Deep Fritz and Deep Junior respectively. With all of their formidable "talents," these programs cannot *plan*. Brute force "number-crunching power" is no substitute for the strategic rationalities and artistic epiphanies that are the hallmarks not just of human genius but the armamentarium of the average human brain.

What does this discussion have to do with biopolitics? Like mathematics and music, chess is an abstract form of cranial calculation, a talent that appears well divorced from cultural influence. Mathematical, musical, and chess precocity among selected preadolescents is well documented; however, unlike the first two, chess playing is a competitive exercise, rather like a sport. Chess is also much like politics in that the struggle to win is a struggle for power: the pieces on the board facing one another, engaging one another, are "fields of force."[124] More specifically, chess involves the application of general principles of correct play—well-established rules and norms—to the resolution of concrete problems that arise during the course of a game. In this sense, one could readily analogize the problems confronting the chess expert to the problems confronting the constitutional expert who, sitting on the United States Supreme Court, must apply the rules and norms of the law to the resolution of cases coming before the high bench. The latter task is also inherently political, because social scientists, even lawyers, now understand that judges are policymakers. They exercise discretion and often permit their values—their sense of the "rightness" of some rules (but not other rules) when applied in particular cases—to dictate the choices they must make. And yet it is one thing to say that liberals and conservatives pick and choose differently from an array of constitutional strategies; it is quite another thing to say that judges think differently in a fashion that resembles the way chess players think.

I have attempted to enhance an understanding of the "mind" component in the Lumsden-Wilson genes-mind-culture coevolutionary cycle by investigating the robustness of chess models in decision making.[125] First, I reviewed the literature of the game, finding that great players with pronounced stylistic tendencies fall into three clusters: structuralists, heuristicians, and functionalists. Structuralists develop plans according to the application of rationally formulated rules, conventions, and understandings of proper play. Heuristicians are artists specializing in intuition and "hunch" in problem solving. Often, they defy convention. Functionalists are highly political in their approach, seeing the pieces on the board, including the stylistic preferences of their opponents, as a gravitational field of dynamic relationships where the "totality of data" is generally greater than the sum of its parts. Functionalism does not lie at the confluence of structuralism and heuristicism but, rather, arises *from* the confluence of these two. So functionalism is science, plus art, plus politics. Second, I reviewed the literature of Supreme Court opinion writing, finding that "players" of constitutional jurisprudence with pronounced stylistic tendencies fall into the same three clusters: some were structuralists, some were heuristicians, some were functionalists. A judge's place in a particular cluster was totally independent of whether that judge was a liberal or a conservative, no matter how one defines those culturally loaded terms of value preference. I like to think that this research exemplifies consilience: "a 'jumping together' of knowledge by the linking of facts and fact-based theory across disciplines to create a common groundwork of explanation."[126]

From the biopolitical perspective, these findings seem exceedingly probative. The chess algorithms described here are presumptively neurologic in their underlying features. That is, to take one example, the chess rationalism of José Raúl Capablanca and Anatoly Karpov is roughly comparable to the constitutional law rationalism of Warren Burger and William J. Brennan. There is no evidence that chess masters learn or are socialized to be "scientists" (structuralists) or "artists" (heuristicians). Under the official sponsorship of the Communist Party, the Soviets made chess a state religion, but the formidable U.S.S.R. grand-master population of the 1935–1985 period featured many protagonists in both schools. There is also no evidence that judges learn or are socialized in their law school apprenticeships or anyplace else to be structuralists, heuristicians, or functionalists. And so in their hands as well, these algorithms are presumptively neurologic in their underlying features. That is, to take another example, William O. Douglas's constitutional romanticism, shared by his ideological archenemy James McReynolds but *not* shared by his ideological comrade-in-arms Hugo Black, is roughly comparable to the chess romanticism of Alexander Alekhine and Kasparov. Of course, players can observe and learn, but the argument is that the central problem-solving tendencies set out in this research are at root genetic. What political scientists need to do is to subject chess player brains to the same empirical tests as one might apply to participants in various other "power games."

From the Darwinian perspective, the examples drawn from the chess literature show beyond doubt that functionalism defeats both structuralism and heuristicism. In the particular environment of chess, functionalism is adaptive and should, in the long run, achieve evolutionary stability. A question of profound theoretical interest is whether functionalism outperforms these and other game theoretic algorithms in similar political contexts bounded by well-understood rules of the game. One of these contexts is the American political system. I believe that at least on the Supreme Court the play of the functionalist format enhances adaptational decision making and therefore political legitimacy for the Constitution of the United States as well as for the Court itself. To name names: the King of the Functionalists in chess was Emanuel Lasker, the deepest "player of the game"; the King of the Functionalists on the Supreme Court was Felix Frankfurter, the deepest "player of that game." The research trail to be blazed entails administering to political decision makers the same tests for cranial physiology as have and will be developed for other subject sets.

Writing in the early 1980s, one student of the history of our discipline wrote: "[The new field of biopolitics] by now [has become] well established as a viable development within the profession."[127] Much closer to empirical reality then and now was John Wahlke's APSA presidential address in which he chided his colleague political behavioralists for publishing almost exclusively "pre-behavioral" research reports, studies that he contended eschewed "fundamental [political] questions" about the "functioning individual human organism" in "biobehavioral perspective," that is, "from the standpoint of the entire human species, in the context of its evolutionary history."[128] Wahlke did not even have the heart to mention the utter dearth of biopolicy investigations, the sorts of "laborious empirical research" Herbert Simon talked about but never did. The APSA establishment provided explicit testimony to Wahlke's bleak assessment when it failed to mention biopolitics in its literature review of disciplinary activity.[129] In 1991, the American Political Science Association, in concert with other social science professional organizations, successfully petitioned the NSF to move their research-funding program out from under the Directorate for Biological, Behavioral, and Social Sciences.[130] Testifying for the APSA at a hearing on the petition, Warren Miller, a leading member of the discipline, stated: "Biology is largely irrelevant to the center of gravity of political science."[131] Miller had directed my MA thesis more than 30 years earlier; I knew well his thinking. He had coauthored *The American Voter*, at that time on the cutting edge of behavioral political science research. But he never permitted himself to grow beyond that and related achievements. Here we were in 1991, and he was oblivious to the genetics revolution and its ELSPI agenda exploding all around him. So much for the possibility under the guidance and subsidy of NSF for what has become all the rage: multidisciplinary research involving natural and social scientists. And without multidisciplinary research, what are the hopes for consilience? In 1993, thirteen years after its first installment, the APSA revisited its survey of the leading subfields of interest in political science. Nineteen chapters were included, and once again biopolitics by whatever name failed to receive even a nodding glance.[132]

I conclude not with an example of negligent neglect but of malignant neglect. Evidently fearing that new genetic revelations will entice political scientists to desert their time-honored preoccupations, Robert Dahl, perhaps the most lionized member of our discipline over the past half-century, recently warned his colleagues not to waste their time and energies.[133] According to Dahl, the causal linkages from genes to neurology to Darwinian survivalism all the way up the ladder to the traditional stuff of politics is so long, winding, and tenuous that their impact has evaporated many rungs earlier. Of particular concern for Dahl is the play of "human institutions," which, he says, are cultural in origin and which our species has learned to manipulate in sundry ways to cope with survival pressures and to keep our genetic fallibilities in check. Sounding like the president of an Ivy League university extolling the racial and ethnic diversity of the campus student and faculty makeup, Dahl trumpets the "diversity of *political* cultures and institutions . . . [and their] powerful *impact on human action*."[134] Without belaboring details, the key difference between Dahl and students of biopolitics is that he is a "top-down" thinker and we are "bottom-up" thinkers. For us, the commonalities among humans—the universals we share—as exemplified by our 99-plus-percent genetic congruence are far more important for the study of our political nature and our political propensities than are the cultural disparities we exhibit. The question whether a parliamentary system is to be preferred over a presidential system pales in comparison to explaining precisely and scientifically how, why, and when the human brain decides to allocate various attributes of power. Ordinarily I would say no more except to remark that "reasonable people can differ." Not so here. By far the most alarming feature of Dahl's analysis is that in framing his indictment, he fails to cite a single example of the biopolitics literature: no Axelrod, no Simon, no Fukuyama, no Masters, no J. Q. Wilson, *no E. O. Wilson*. Dahl cannot be bothered to indulge in even the rudiments of a literature review. His essay stands as an epitaph to paradigms dead and dying, a Canute-like last stand by the APSA's oldest guard.

C. A Teaching Constitution for Human Genomics

I desire to end my commentary on a highly positive note. Human genomics is more than a research agenda; it is an instructional agenda. Consilience is more than a scholarly call to action; it is a pedagogical format for imparting and disseminating knowledge. Today's classrooms beg for new paradigms of teaching, educational models transmitting to students the ELSPI of human genomics synthesized fully into a consilient format. These paradigms, like all paradigms, feature agreed-upon understandings about the scope and methods of the new agenda, the new way of looking at this emerging world of human purpose. These understandings are yet another instance of a working, living constitution.

My notion of a genomics/consilience teaching program differs considerably from proposals—fairly vague suggestions—others have offered. As an example, I note the generous sums of money pouring into public university coffers largely as a result of the agreement entered into between many of the states and the tobacco industry. Several campuses are using these assets to fund genomics research facilities, and school administrators envision spin-offs enhancing the social sciences and humanities. The University of Michigan has created an Institute for the Study of Biological

Complexity and Human Values, and in order to overcome the "two cultures" problem, various academic units have organized "lectures and seminars" addressing the ethical, social, and legal implications of genomics discovery. However, no new departments are envisaged, and the disciplines, as defined by the various tenured faculty establishments, will remain undisturbed. Other such initiatives are cropping up around the country.[135] To me, this sounds like another version of "same-old, same-old." As these institutions gamely seek practical pedagogical accommodations to a fundamental research program fairly bursting with new insights, so philosophers of education look for broader, more sweeping changes in curricular arrangement. As an example of these recommendations, I cite E. O. Wilson himself. He advocates a total reform of the "core curriculum in colleges and universities" that would emphasize "the cause-and-effect connections among the great branches of learning."[136] In this way, "the crumbling structure of the liberal arts" can be revived. It will not suffice, he says "[to] force-feed . . . students with some-of-this and some-of-that."[137] Unfortunately, Wilson does not have the time to descend into details. If he did, I think he would encounter severe logistical problems. Wilson would probably argue: the pathology is rampart so the cure must be equally traumatic in arresting the "business as usual" ethic. Robert Maynard Hutchins probably thought much the same. But this isn't 1925. The economic facts of life are such that a spanking new St. John's College of Consilient Education is out of the question. A third possible approach would entail an overhaul of graduate education. Perhaps we can't reconfigure the arts and sciences BA degree programs at Yale and Stanford, but just maybe we can recraft their PhD programs. If you asked any of my political science colleagues around the country about the prospect of synthesizing biological science and social science doctoral programs (forget the humanities) into a viable consilient format, they would simply chuckle ear to ear.

What we need is a practical halfway house, a plausible, realistic starting point, which would bridge the gap between undergraduate and graduate study and therefore would have the potential to extend eventually to both. This program proposal must be sufficiently elaborate to cover the genomics/consilience basics and yet be budgetarily feasible once good times return, as ultimately they will, to higher education. The course of study I propose is entitled the "MA/MS Program in Natural/Social Science Consilience" (see Figure 2). It would be a two-year degree-granting enterprise and would serve as an add-on to the standard undergraduate baccalaureate or as a launching pad for the more specialized doctorate. It would exclude, except tangentially, the literatures of the humanities. They will have to wait a while.

Figure 2. MA/MS Program in Natural/Social Science Consilience (2-year program)

Fall Semester (I)

1) *Evolutionary Theory:* Darwin; Dawkins; Hamilton; reciprocal altruism; sociobiology and its critics; genes-mind-culture coevolution; *Consilience;* reductionism and its critics; cooperation and conflict strategies over time; the development of social institutions; "cultural evolution"
2) *Decision Theory:* genes and cranial capacities; the brain as problem solver; the brain as culture producer and vice-versa; personality and behavioral genetics; the neurotransmitters; bounded rationality; rational choice vs. natural selection models; "evolving brains"
3) *The Human Genome:* what is it; how was it discovered; comparisons with other genomes; functional genomics

Spring Semester (I)

1) *Ecological Genetics; The Science of Fitness:* models of adaptation: biological and cultural; what does it mean to say that social behaviors, institutions, and processes are "fit"?
2) *Social Communication:* languages verbal and nonverbal; universal grammars of social expression; exchange theory; the role of media technology and its evolutionary consequences
3) *Genetic Interventions:* the ELSPI of human gene therapy and enhancement; cloning; germline alterations

Fall Semester (II)

1) *Laws of Nature; Laws of Man:* comparing and contrasting the structure and function of biological rules and society's rules
2) *Human Nature:* theories of biology, philosophy, and psychology
3) *Eugenics v. Euculture:* Francis Galton, Sigmund Freud, Margaret Mead, and their disciples

Spring Semester (II)

1) *Biological and Genetic Diversity:* competing models of demographic change; population genetics; man as the evolutionary prime mover; "artificial evolution"; *The Future of Life*
2) *The Social Scientist in the Laboratory:* sociogenomics; MA/MS research paper
3) *Biopolicy:* public and private regulatory mechanisms; who decides? how do they decide? what influence do they have? what is the content of their missions and accomplishments?

The rationales for and the orderings of the various specific courses described in Figure 2 deserve some detail. The two years are subdivided into four semesters, and each semester presents material in an escalating level of particularity. The student would take three courses per term, not a very heavy load in itself; the subject matter, though, would be more than sufficient to challenge any of today's budding PhD hopefuls. There are no electives and no waivers! It perhaps is not self-evident that the courses labeled with a number 1 build on each other, as do the courses labeled number 2 and number 3. I would call the number 1 rubric "Darwin's Legacy"; number 2, "Our Epigenetic Posterity"; and number 3, "The New World of Genomics." Let's take each in turn.

With number 1, we begin with "Evolutionary Theory," in which students wrestle with the writings of the "masters" from Darwin to E. O. Wilson, ultimately facing the challenge of operationally defining the term "cultural evolution."[138] They move on to the spring term, where they engage the concepts "fitness" and "adaptation"; this exercise leads directly to the second year's fall semester study of the general laws governing the behavior of all things organic. In the spring capstone course, emphasizing biodiversity issues, they confront the fact that humans are now redirecting evolutionary paths all the time,[139] that "artificial evolution" is developing into a field of study,[140] and that leading scholars are postulating grand designs for salvaging ecosystems and perhaps even our own survival.[141]

Sequence number 2 commences with a macroscopic investigation of the human brain entitled "Decision Theory," moves forward with a discussion of language and communication broadly construed, grapples with competing conceptions of human nature in the third semester, and culminates in a research practicum in which students and faculty of many disciplinary foci meet in the laboratory to engage in serious empirical research of a genomics ilk. As Wilson has said: "[The traditional process of investigation in which] the genes affecting epigenetic rules and the rules themselves are . . . searched out independently by different teams of researchers [should end]."[142]

Sequence number 3 is the most straightforward: the course arrangement begins with the human genome, flows naturally into the many policy debates I have addressed in this book and more, tracks down the genetic and culturalist advocates of perfectionism, and ends with an investigation of the strengths and weaknesses of regulatory control, a "fitting" bow to the inherently political dynamic informing this degree program.

I have little hope that any such curriculum will be adopted today or tomorrow. A big hurdle will be finding a teaching faculty. Even now they are out there, but they will need inducements. A less substantial hurdle will be finding students. If the program carries elitist cachet, they will enroll, especially if they can skip their last (or first) two years of undergraduate boredom. I don't underrate the many challenges. Nonetheless, we are badly in need of a model—not a model filled with banalities and generalities but a model delineating specific plans forged in specific overviews and conceptions. Perhaps 10 or 20 years down the road, somebody, somewhere, with the gumption to take the kinds of steps I envision here, will get things going. That person is in for rough sledding. The apostles of forbidden knowledge will rise up and attempt to slay the enemy: consilience in practice. In their paradigms, the natural sciences and the social sciences should no more commingle than should the domains of theology and secular knowledge. On the latter score, they are dead right; on they former score, they are dead wrong.

The world of human genomics features many constitutions. We have charted the rules and norms of genomic discovery. We have surveyed the rules and norms of genomic usages. We have outlined the rules and norms of genomic purposes. We have set forward the rules and norms of our own genomic "mindsets," by which I mean the ways in which we think genomics sheds light on who we are. And we have once again seen that a constitution, these constitutions, are at root not legal instrumentalities but political instrumentalities. Only the world of politics, Aristotle's world of public discourse and accountability, can prudently make of these constitutions adaptive instrumentalities.

This understanding places an enormous burden on political science, the discipline of politics. To make of our academic calling a major player in the human genomics debate, in the human genomics power structure, we must readdress in a fresh light the great intellectual dilemmas that have dogged us since the founders' time more than 100 years ago. The dilemmas, in my view, are two in number. The first is whether political science should stand alone or join forces with kindred specialties. I actually believe we can have our cake and eat it, too. In the world of consilience, we simply must consider ourselves social scientists first and political scientists second. And yet we need not sacrifice our identities on some alter of intellectual homogeneity. Always there will be special questions of power, leadership, and statecraft. To these we can bring our own unique, indispensable expertise. The second dilemma is whether political science—whether social science— should be in the business of providing moral insight and guidance. We have had great disciplinary spokespeople making the case for and the case against. Even our founders were of two minds: Burgess saw no place for a moral mission; Lowell considered a moral perspective indispensable.[143] Arthur Bentley and William Riker, one a sociologist/psychologist and the other a frustrated physicist/economist who became the guru of political science rat. choice, followed Burgess. Harold Lasswell, reaching his prime during the days when totalitarianism achieved the status of genuine menace to Western Civilization, argued that the "policy sciences" were fundamental to political science.[144] I believe, optimistically, that even here we can have our cake and eat it, too. Larry Arnhart, a theoretician of strong biopolitical conviction, has cited the ethical considerations of constitutional standing that suffuse Aristotle's writings.[145] There are compelling empirical data demonstrating the adaptive consequences arising from our species' display and implementation of these ethical preferences. If by "morality" we mean his and other such adaptive strategies—the essence of our sociogenomic constitution—we can continue to articulate to the world a viable place and context for ethics in the affairs of *Homo politicus*. If, however, we mean by morality a system of inalienable truths, revealed or otherwise, based not on who we really are but on what some shaman says we are, then we will depart the world of political science and commit our energies to the world of political humanism. Some among us think that is the way to go, and ultimately some will go that way. I will not be among them.

Note the one basic question with which we have wrestled since our founding that I have not mentioned: What is the constitution of political science? It has yet to achieve a robustness worthy of elucidation. But this I know: its mainsprings, on that day of maturity and elegance, will be clear enough. They will be the mainsprings of a bioconstitutional politics paradigm.

Notes
Bibliography
Index

Notes

1. Political Science, Constitutional Politics, Genomics

1. The ensuing commentary is drawn from Albert Somit and Joseph Tanenhaus, *The Development of Political Science* (Boston: Allyn and Bacon, 1967).

2. Ibid., pp. 19, 24, 33.

3. Ibid., p. 71; emphasis added.

4. Ibid., pp. 72, 75.

5. I know that some of my colleagues will say I have shortchanged the role of their heroes in this section, and that I have not said enough about various aspects of political science. For example, I do not specifically allude to the fields of international relations and comparative politics. But, to repeat, we are dealing in large measure with American political science, so why should we be surprised when the paradigms of American politics as a field of study become the paradigms of the profession? Besides, there is a generous overlap. What has become of area studies? To say one's research arena is German or French politics is a formula for not getting a job interview. The data of area studies—often idiosynchratically cultural—are hard to quantify.

6. Truman's classic work is *The Governmental Process* (New York: Knopf, 1951), and Bentley's now legendary work is *The Process of Government* (Chicago: University of Chicago Press, 1908).

7. The story of the recombinant DNA controversy from a constitutional politics perspective was recounted a decade later in Ira H. Carmen, *Cloning and the Constitution* (Madison: University of Wisconsin Press, 1986), hereinafter cited as *Cloning*.

8. Edward O. Wilson, *Consilience* (New York: Vintage Books, 1999), pp. 224–226.

9. Margaret Masterman, "The Nature of a Paradigm," in *Criticism and the Growth of Knowledge,* ed. Imray Lakatos and Alan Musgrave (Cambridge: Harvard University Press, 1970), pp. 59–89.

10. The G-ers would break ranks here, as many of them find it intriguing to explore the possible biological precursors of homosexual orientation.

11. An earlier variation on the theme of bioconstitutional politics appeared in Ira H. Carmen, "Bioconstitutional Politics: Toward an Interdisciplinary Paradigm," *Politics and the Life Sciences* 5 (1987): 193–207.

12. Walter F. Murphy, James Fleming, and Sotirios A. Barber, *American Constitutional Interpretation*, 2nd ed. (Westbury, N.Y.: Foundation Press, 1995).

13. William F. Harris II, "Bonding Word and Polity: The Logic of American Constitutionalism," *American Political Science Review* 76 (1982): 34–45.

14. Thomas H. Hammond and Gary J. Miller, "The Core of the Constitution," *American Political Science Review* 81 (1987): 1155–1174.

15. James G. March and Johan P. Olsen, *Rediscovering Institutions* (New York: Free Press, 1989).

16. Ibid., p. 111.

17. Ibid., pp. 22–23.

18. Ira H. Carmen, *Power and Balance* (New York: Harcourt Brace Jovanovich, 1978), pp. 25–26. See also two essays that I continue to find rewarding: Karl N. Llewellyn, "The Constitution as an Institution," *Columbia Law Review* 34 (1934): 1–40, and Glendon Schubert, "The Rhetoric of Constitutional Change," *Journal of Public Law* 16 (1967): 16–50.

19. Lynne R. Baker, *Persons and Bodies: A Constitutional View* (Cambridge: Cambridge University Press, 2000), p. 32.

20. Marya Schechtman, *The Constitution of Selves* (Ithaca: Cornell University Press, 1996), pp. 93–96.

21. Matt Ridley, *Genome* (New York: Perennial, 2000), pp. 12–13.

22. Gregor Mendel, "Versuche über Pflanzen-Hybriden," *Verhandlungen Des Naturforschenden Vereines, Abnahndlungen, Brunn*, 4 (1866): 3–47.

23. Charles Darwin, *The Origin of Species by Means of Natural Selection* (New York: John B. Alden, 1859).

24. G. C. Williams, *Adaptation and Natural Selection* (Princeton: Princeton University Press, 1966).

25. O. T. Avery, C. M. MacLeod, and M. McCarty, "Studies on the Chemical Nature of the Substance Inducing Transformation of Pneumococcal Types: Induction of Transformation by a Deoxyribonucleic Acid Fraction Isolated from Pneumococcus Type III," *Journal of Experimental Medicine* 79 (1944): 137–158.

26. J. D. Watson and F. H. C. Crick, "A Structure of Deoxyribose Nucleic Acid," *Nature* 171 (1953): 737–738.

27. Thomas S. Kuhn, *The Structure of Scientific Revolutions*, 2nd ed. (Chicago: University of Chicago Press, 1970), p. viii.

28. Philip Kitcher, "1953 and All That: A Tale of Two Sciences," *Philosophical Review* 93 (1984): 335–373 at 336, 340.

29. Jerry E. Bishop and Michael Waldholz, *Genome* (New York: Simon and Schuster, 1990), pp. 204–208.

30. Ibid., ch. 4. Virginia Morell, "Huntington's Gene Finally Found," *Science* 260 (1993): 28–30.

31. Kevin Davies, *Cracking the Genome* (New York: Free Press, 2001), p. 54.

32. See the discussion in text at note 28.

33. The complete story is told in Carmen, *Cloning,* ch. 3.

34. See the discussion in text following note 16.

35. Ira H. Carmen, "Debates, Divisions, and Decisions: Recombinant DNA Advisory Committee (RAC) Authorization of the First Human Gene Transfer Experiments," *American Journal of Human Genetics* 50 (1992): 245–260.

36. Theodore J. Lowi, *The End of Liberalism* (Chicago: Norton, 1969).

37. Robert F. Merton, *Social Theory and Social Structure* (New York: Free Press, 1968).

38. Bishop and Waldholz, *Genome,* ch. 2; Robert Cook-Deegan, *The Gene Wars* (New York: Norton, 1994), p. 29.

39. Daniel J. Kevles, "Out of Eugenics: The Historical Politics of the Human Genome," in *The Code of Codes,* ed. Daniel J. Kevles and Leroy Hood (Cambridge: Harvard University Press, 1993), pp. 3–36 at p.18.

40. Quoted in Davies, *Cracking the Genome,* p. 12.

41. Davies, *Cracking the Genome,* pp. 15–16.

42. Bishop and Waldholz, *Genome,* pp. 217–218.

43. Renato Dulbecco, "A Turning Point in Cancer Research: Sequencing the Human Genome," *Science* 231 (1986): 1055–1056.

44. Ridley, *Genome,* p. 8.

45. Davies, *Cracking the Genome,* pp. 13–14.

46. Roger Lewin, "Proposal to Sequence the Human Genome Stirs Debate," *Science* 232 (1986): 1598–1600.

47. Roger Lewin, "Politics of the Genome," *Science* 235 (1987): 1453.

48. Leslie Roberts, "Academy Backs Genome Project," *Science* 239 (1988): 725–726; Cook-Deegan, *Gene Wars,* pp. 132–133.

49. Bishop and Waldholz, *Genome,* pp. 219–221.

50. Wojciech Makalowski, "Not Junk After All," *Science* 300 (2003): 1246–1247.

51. Quoted in Davies, *Cracking the Genome,* p. 28.

52. Roger Lewin, "Genome Projects Ready to Go," *Science* 240 (1988): 602–604.

53. Marjorie Sun, "NIH and DOE Draft Genome Pact," *Science* 241 (1988): 1596.

54. Kevles, "Out of Eugenics," pp. 23–26; Bishop and Waldholz, *Genome,* p. 223; Cook-Deegan, *Gene Wars,* p. 161. Through all this, the DOE kept a stiff upper lip. Charles R. Cantor, "Orchestrating the Human Genome Project," *Science* 248 (1990): 49–51.

55. Bernard D. Davis and colleagues, "The Human Genome and Other Initiatives," *Science* 249 (1990): 342–343.

56. Daniel J. Kevles and Leroy Hood, "Reflections" in Kevles and Hood, *Code of Codes,* pp. 300–328 at pp. 303–304.

57. Ibid., pp. 306–308.

58. James D. Watson, "A Personal View of the Project," in Kevles and Hood, *Code of Codes,* pp. 164–173 at pp. 168–169.

59. Quoted in Davies, *Cracking the Genome,* p. 30.

60. Daniel E. Koshland Jr., "Sequences and Consequences of the Human Genome," *Science* 246 (1989): 189.

61. James D. Watson, "The Human Genome Project: Past, Present, and Future," *Science* 248 (1990): 44–49 at 46.

62. Richard Saltus, "Crash Effort to Map Human Genes Urged," *Boston Globe,* June 9, 1986, p. 53.

63. Jon Beckwith, "Forward: The Human Genome Initiative: Genetics' Lightning Rod," *American Journal of Law and Medicine* 17 (1–2) (1991): 1–13 at 5–6.

64. Ibid., p. 6.

65. Edward O. Wilson, *Sociobiology: The New Synthesis* (Cambridge: Harvard University Press, 1975).

66. Jonathan Beckwith and fourteen cosigners, "Against Sociobiology," *New York Review of Books* 22 (18) (November 13, 1975).

67. Charles J. Lumsden and Edward O. Wilson, *Promethean Fire* (Cambridge: Harvard University Press, 1983), p. 43.

68. Readers can judge for themselves by perusing Jon Beckwith, *Making Genes, Making Waves* (Cambridge: Harvard University Press, 2002).

69. R. C. Lewontin, "The Dream of the Human Genome," *New York Review of Books* 39 (10) (May 28, 1992): 31–40 at 31.

2. Conducting Genome Research

1. Edward O. Wilson, *Naturalist* (New York: Warner Books, 1995), pp. 218–219.

2. Ibid., p. 224.

3. Cook-Deegan, *Gene Wars,* p. 359.

4. Ibid.

5. Joseph Palca, "Genome Projects Are Growing Like Weeds," *Science* 245 (1989): 131.

6. Cook-Deegan, *Gene Wars,* pp. 176, 184.

7. Quoted in Evelyn Fox Keller, "Nature, Nurture, and the Human Genome Project," in Kevles and Hood, *Code of Codes,* pp. 281–299 at p. 294.

8. Watson, "Personal View of the Project," p. 165.

9. Leslie Roberts, "New Game Plan for Genome Mapping," *Science* 245 (1989): 1438–1440.

10. Leslie Roberts, "The Genetic Map Is Back on Track after Delays," *Science* 248 (1990): 805.

11. Leslie Roberts, "Genome Project: An Experiment in Sharing," *Science* 248 (1990): 953.

12. Leslie Roberts, "Plan for Genome Centers Sparks a Controversy," *Science* 246 (1989): 204–205; Leslie Roberts, "Genome Center Grants Chosen," *Science* 249 (1990): 1497.

13. Leslie Roberts, "Report Card on the Genome Project," *Science* 253 (1991): 376.

14. Kevles, "Out of Eugenics," p. 28.

15. David Dickson, "A Soviet Human Genome Program?" *Science* 240 (1988): 140.

16. David Dickson, "Go-Ahead for Gene Sequencing Venture," *Science* 240 (1988): 1728.

17. Quoted in Kevles, "Out of Eugenics," p. 28.

18. Quoted in ibid., p. 29.

19. David Dickson, "Watson Floats a Plan to Carve Up the Genome," *Science* 244 (1989): 521–522.

20. Leslie Roberts, "Genome Project Under Way, at Last," *Science* 243 (1989): 167–168; Roger Lewin, "Genome Planners Fear Avalanche of Red Tape," *Science* 244 (1989): 1543.

21. Cook-Deegan, *Gene Wars*, p. 209.

22. Watson, "Personal View of the Project," p. 172.

23. Leslie Roberts, "Watson versus Japan," *Science* 246 (1989): 576, 578.

24. Mark H. Crawford, "HUGO: Genome Data Open to Scientists," *Science* 246 (1989): 1565.

25. Leslie Roberts, "HUGO Takes on Role as Marriage Broker," *Science* 254 (1991): 932.

26. Constance Holden, "Where HUGOing?" *Science* 255 (1992): 27.

27. Lori B. Andrews, *The Clone Age* (New York: Henry Holt, 1999), pp. 185, 194.

28. An excellent example of Nancy Wexler's writing in this area is "Clairvoyance and Caution: Repercussions from the Human Genome Project," in Kevles and Hood, *Code of Codes*, pp. 211–243.

29. Kevles, "Out of Eugenics," p. 35.

30. Cook-Deegan, *Gene Wars*, pp. 237–238.

31. Andrews, *Clone Age*, p. 195.

32. Watson, "Personal View of the Project," p. 173; emphasis in text.

33. Cook-Deegan, *Gene Wars*, p. 275.

34. "Ethical, Legal, and Social Implications Program, NCHGR Activities Report," May 1993, p. 3.

35. Cook-Deegan, *Gene Wars*, pp. 245–246, 268–270, 273–274.

36. Ibid., pp. 263, 278, 279–280.

37. Quoted in Andrews, *Clone Age*, p. 206.

38. Leslie Roberts, "Genome Patent Fight Erupts," *Science* 254 (1991): 184–186 at 184.

39. Leslie Roberts, "Who Owns the Human Genome?" *Science* 237 (1987): 358–361.

40. Ira H. Carmen, "Ownership of the Human Genome," *Science* 237 (1987): 1555; see Carmen, *Cloning*, pp. 44–46.

41. U.S. Constitution, art. 1, sec. 8, cl. 8.

42. Carmen, *Cloning*, p. 5.

43. Quoted in ibid., p. 7.

44. Rebecca S. Eisenberg, "Patenting the Human Genome," *Emory Law Journal* 39 (1990): 721–745 at 726–728.

45. Genes that do not express for proteins would continue to resist detection, as would some genetic "start" and "stop" signals that are located in the junk. Hence

Watson's argument, noted earlier, that we need to sequence everything to know everything important.

46. For useful discussion and a palatable policy alternative, see Linda J. Demaine and Aaron X. Fellmeth, "Natural Substances and Patentable Inventions," *Science* 300 (2003): 1375–1376.

47. Mark D. Adams et al., "Complementary DNA Sequencing: Expressed Sequence Tags and Human Genome Project," *Science* 252 (1991): 1651–1656 at 1651.

48. Cook-Deegan, *Gene Wars*, p. 328.

49. Kevles and Hood, "Reflections," p. 314; David Dickson, "Europe Says No to Animal Patents," *Science* 245 (1989): 25.

50. Richard Stone, "Brits and EC at Odds over Gene Patenting," *Science* 256 (1992): 727.

51. Leslie Roberts, "Scientists Voice Their Opposition," *Science* 256 (1992): 1273–1274.

52. Leslie Roberts, "Genome Patent Fight Erupts," p. 184.

53. Leslie Roberts, "NIH Gene Patents, Round Two," *Science* 255 (1992): 912–913.

54. Cook-Deegan, *Gene Wars*, pp. 313–315.

55. Larry Thompson, "Healy Approves an Unproven Treatment," *Science* 259 (1993): 172.

56. Cook-Deegan, *Gene Wars*, p. 328.

57. Ibid., pp. 317, 329.

58. Ibid., p. 336.

59. Leslie Roberts, "Why Watson Quit as Project Head," *Science* 256 (1992): 301–302.

60. Watson, " Human Genome Project," p. 46.

61. Cook-Deegan, *Gene Wars*, pp. 327, 337, 339.

62. Leslie Roberts, "Rumors Fly over Rejection of NIH Claim," *Science* 257 (1992): 1855.

63. Francis S. Collins, "Positional Cloning: Let's Not Call It Reverse Anymore," *Nature Genetics* 1 (1992): 3–6.

64. Davies, *Cracking the Genome*, pp. 4, 72.

65. Quoted at ibid., p. 69.

66. The root causes of this phenomenon and its considerable implications for American constitutional politics are explored in Ira H. Carmen, "God and Man in and around the White House" (unpublished manuscript, 2000). The paper cannot be published in a political science journal for the reasons cited in ch. 1.

67. Quoted in Andrews, *Clone Age*, p. 188.

68. Quoted in ibid., pp. 188–189.

69. Jean L. Marx, "Genome Project Plans Described," *Science* 260 (1993): 152–153.

70. Leslie Roberts, "Taking Stock of the Genome Project," *Science* 262 (1993): 20–22.

71. Davies, *Cracking the Genome*, pp. 76, 77.

72. Leslie Roberts, "NIH Takes New Tack on Gene Mapping," *Science* 258 (1992): 1573.

73. Christopher Anderson, "Genome Project Goes Commercial," *Science* 259 (1993): 300–302.

74. Richard Preston, "The Genome Warrior," *New Yorker*, June 12, 2000, pp. 66–83 at p. 66; Davies, *Cracking the Genome*, p. 92.

75. Davies, *Cracking the Genome*, pp. 64–65, 95; Preston, "Genome Warrior," pp. 71–72; Eliot Marshall, "The Company That Genome Researchers Love to Hate," *Science* 266 (1994): 1800–1802.

76. Davies, *Cracking the Genome*, pp. 97–100.

77. Preston, "Genome Warrior," p. 72; Rachel Nowak, "Bacterial Genome Sequence Bagged," *Science* 269 (1995): 468–470.

78. Davies, *Cracking the Genome*, pp.109–110; Ridley, *Genome*, pp. 8–9.

79. Mildred K. Cho et al., "Ethical Considerations in Synthesizing a Minimal Genome," *Science* 286 (1999): 2087–2090.

80. Carl Zimmer, "Tinker, Tailor: Can Venter Stitch Together a Genome from Scratch?" *Science* 299 (2003): 1006–1007.

81. Carol J. Bult et al., "Complete Genome Sequence of the Methanogenic Archaeon, Methanococcus jannaschii," *Science* 273 (1996): 1058–1073.

82. Daniel E. Koshland Jr., "Ahead of Schedule and on Budget," *Science* 266 (1994): 199.

83. Leslie Roberts, "Controversial from the Start," *Science* 291 (2001): 1182–1188 at 1186.

84. A. Goffeau et al., "Life with 6000 Genes," *Science* 274 (1996): 546–567.

85. Eliot Marshall, "A Strategy for Sequencing the Genome 5 Years Early," *Science* 267 (1995): 783–784.

86. Jocelyn Kaiser, "British Genome Boost," *Science* 270 (1995): 903.

87. Eliot Marshall and Elizabeth Pennisi, "NIH Launches the Final Push to Sequence the Genome," *Science* 272 (1996): 188–189.

88. Patricia Kahn, "Sequencers Split over Data Release," *Science* 271 (1996): 1798–1799.

89. Eliot Marshall, "Genome Researchers Take the Pledge," *Science* 272 (1996): 477–478; Michael Balter, "Generous Funding Wins a Seat at the Genome Top Table," *Science* 274 (1996): 1293; Gary Zweiger, *Transducing the Genome* (New York: McGraw Hill, 2001), p. 64.

90. Eliot Marshall, "Fraud Strikes Top Genome Lab," *Science* 274 (1996): 908–910.

91. Richard Stone, "Nonprofit to Launch Gene-Mapping Effort," *Science* 267 (1995): 443; Eliot Marshall, "Genomic's Odd Couple," *Science* 275 (1997): 778; J. Craig Venter, "Human Genome Agreements," *Science* 275 (1997): 601–602; Jocelyn Kaiser, "Commercial Gene Kingdom Splits Up," *Science* 276 (1997): 1959.

92. Preston, "Genome Warrior," pp. 72–73.

93. Davies, *Cracking the Genome*, pp. 141–143.

94. The last three paragraphs are based on Eliot Marshall and Elizabeth Pennisi, "Hubris and the Human Genome," *Science* 280 (1998): 994–995; Davies, *Cracking the Genome*, pp. 145–150; Preston, "Genome Warrior," p. 75.

95. Elizabeth Pennisi, "Funders Reassure Genome Sequencers," *Science* 280 (1998): 1185.

96. A note on acronyms: as of January 1997, Collins's center had been elevated to a full-fledged NIH institute, that is, the National Human Genome Research Institute.

97. Eliot Marshall, "NIH to Produce a 'Working Draft' of the Genome by 2001," *Science* 281 (1998): 1774–1775.

98. Quoted in Davies, *Cracking the Genome,* p. 152.

99. Elizabeth Pennisi, "Worming Secrets from the *C. elegans* Genome," *Science* 282 (1998): 1972–1974; Davies, *Cracking the Genome,* pp. 90–91.

100. Elizabeth Pennisi, "Fruit Fly Researchers Sign Pact with Celera," *Science* 283 (1999): 767; Davies, *Cracking the Genome,* pp. 154–159.

101. Davies, *Cracking the Genome,* pp. 165–167.

102. Elizabeth Pennisi, "Fruit Fly Genome Yields Data and a Validation," *Science* 287 (2000): 1374; Davies, *Cracking the Genome,* pp. 201–203.

103. Eliot Marshall, "Commercial Firms Win U.S. Sequencing Funds," *Science* 285 (1999): 310.

104. Elizabeth Pennisi, "Academic Sequencers Challenge Celera in a Sprint to the Finish," *Science* 283 (1999): 1822–1823; Davies, *Cracking the Genome,* pp. 162–164.

105. Dennis Normile and Elizabeth Pennisi, "Team Wrapping Up Sequence of First Human Chromosome," *Science* 285 (1999): 2038–2039; Davies, *Cracking the Genome,* pp. 194–196.

106. Leslie Roberts, "Controversial from the Start," p. 1188.

107. Elizabeth Pennisi, "Mouse Sequencers Take Up the Shotgun," *Science* 287 (2000): 1179–1181.

108. Eliot Marshall, "Talks of Public-Private Deal End in Acrimony," *Science* 287 (2000): 1723–1724.

109. Quoted in Davies, *Cracking the Genome,* p. 205. See also Eliot Marshall, "Clinton and Blair Back Rapid Release of Data," *Science* 287 (2000): 1903.

110. Quoted in Davies, *Cracking the Genome,* p. 205.

111. Eliot Marshall, "How a Bland Statement Sent Stocks Sprawling," *Science* 287 (2000): 2127.

112. Elizabeth Pennisi, "Chromosome 21 Done, Phase Two Begun," *Science* 288 (2000): 939.

113. Davies, *Cracking the Genome,* pp. 236–237, 238.

114. Quoted at ibid., p. 242.

115. Eörs Szathmáry et al., "Can Genes Explain Biological Complexity?" *Science* 292 (2001): 1315. But at this writing, who knows the correct ballpark figure? Application of the SAGE methodology yields 70,000 genes. Ben Shouse, "Revisiting the Numbers: Human Genes and Whales," *Science* 295 (2002): 1457.

116. Eliot Marshall, "Celera and *Science* Spell Out Data Access Provisions," *Science* 291 (2001): 1191.

117. Ben Shouse, "Less Can Be More, U.K. Study Finds," *Science* 293 (2001): 1238–1239.

118. Jennifer Couzin, "Taking Aim at Celera's Shotgun," *Science* 295 (2002): 1817; Eliot Marshall, "Genome Teams Adjust to Shotgun Marriage," *Science* 292 (2001): 1982.

119. Eliot Marshall, "Public Group Completes Draft of the Mouse," *Science* 296 (2002): 1005; Jocelyn Kaiser et al., "Celera, NIH Make a Deal," *Science* 293 (2001): 191.

120. Denise Casey, "Human Genome Working Draft: First-Edition Travel Guides," *Human Genome News* 11 (3–4) (2001): 3–4; S. Blair Hedges and Sudhir Kumar, "Vertebrate Genomes Compared," *Science* 297 (2002): 1283–1285.

121. ScienceScope, "De-Celeration," *Science* 296 (2002): 635.

122. Bartha Maria Knoppers and Ruth Chadwick, "The Human Genome Project: Under an International Ethical Microscope," *Science* 265 (1994): 2035–2036. For one of many analogous reports from an American perspective, see Henry T. Greely, "Legal, Ethical, and Social Issues in Human Genome Research," *Annual Review of Anthropology* 27 (1998): 473–502.

123. Agreeing with me is Andrea L. Bonnicksen, "National and International Approaches to Human Germ-Line Gene Therapy," *Politics and the Life Sciences* 13 (1) (1994): 39–49.

124. Bartha Maria Knoppers, "Genetic Benefit Sharing." *Science* 290 (2000): 49.

125. Leslie Roberts, "Taking Stock of the Genome Project." The quotation is at p. 22.

126. Kathy L. Hudson et al., "Genetic Discrimination and Health Insurance: An Urgent Need for Reform," *Science* 270 (1995): 391–393; Karen Rothenberg et al., "Genetic Information and the Workplace: Legislative Approaches and Policy Challenges," *Science* 275 (1997): 1755–1757.

127. Andrews, *Clone Age*, pp. 194–201.

128. Lori B. Andrews and Dorothy Nelkin, "The Bell Curve: A Statement," *Science* 271 (1996): 13.

129. Eliot Marshall, "Panel Urges Cloning Ethics Boards," *Science* 275 (1997): 22.

130. For the latest NHGRI five-year plan including a new enumeration of ELSI goals such as addressing "ways in which socioeconomic factors and concepts of race and ethnicity influence . . . understanding . . . of genetic information," see Francis S. Collins et al., "New Goals for the U.S. Human Genome Project: 1998–2003," *Science* 282 (1998): 682–689.

131. Elizabeth Pennisi, "Finally, the Book of Life and Instructions for Navigating It," *Science* 288 (2000): 2304–2307.

132. Eliot Marshall, "Whose Genome Is It, Anyway?" *Science* 273 (1996): 1788–1789.

133. Officials at Celera declined to discuss with me the substance of the protocol beyond the information they have previously released for public consumption citing the anonymity principle. How this principle controls the salient questions I raise in text is a mystery.

134. J. Craig Venter, "A Part of the Human Genome Sequence," *Science* 299 (2003): 1183–1184.

135. Nicholas Wade, "Scientist Reveals Genome Secret: It's Him," *New York Times,* April 27, 2002, p. 1; Arthur L. Caplan, "His Genes, Our Genome," *New York Times,* May 3, 2002, p. A27.

136. Zweiger, *Transducing the Genome,* pp. 99–103.

137. Theodore J. Lowi, *The Politics of Disorder* (New York: Norton, 1971).

3. Cloning

1. Daniel J. Kevles, *In the Name of Eugenics* (New York: Knopf, 1985); see especially pp. 186–192.

2. Quoted in ibid., p. 190; emphasis in text.

3. Ibid., pp. 260–263.

4. Quoted in ibid., p. 97.

5. Quoted in ibid., p. 98.

6. Ibid., pp. 100, 106.

7. 274 U.S. 200 (1927).

8. 316 U.S. 535 (1942).

9. Ira H. Carmen, "Chess Algorithms of Supreme Court Decision Making: A Bioconstitutional Politics Analysis," *Political Behavior* 11 (1989): 99–121 at 112.

10. *Griswold v. Connecticut,* 381 U.S. 479 (1965).

11. *Eisenstadt v. Baird,* 405 U.S. 438 (1972); emphasis in text.

12. 410 U.S. 113 (1973).

13. *Planned Parenthood v. Casey,* 505 U.S. 833 (1992).

14. The European "constitutional outlook" is unclear. An EC group of advisers has called procreative choice fundamental and strongly endorses prenatal diagnosis. If PND yields evidence of genetic infirmity, the pregnant woman's decisions are controlling. Abortions based on "non-medical reasons" e.g., sex selection, are "ethically unacceptable" and should not be permitted. This ELSI declaration and *Roe* are at loggerheads. Group of Advisors on the Ethical Implications of Biotechnology, "Ethical Aspects of Prenatal Diagnosis," *Politics and the Life Sciences* 15 (2) (1996): 329–334. As one might expect, the United Kingdom's regulations and France's regulations are quite different.

15. Andrea L. Bonnicksen, *In Vitro Fertilization* (New York: Columbia University Press, 1989), p. 5.

16. Leon R. Kass, "Making Babies—The New Biology and the 'Old' Morality," *Public Interest* 26 (1972): 18–56; Paul Ramsey, "Shall We Reproduce?" *Journal of the American Medical Association* 220 (1972): 1346–1350, 1480–1485.

17. James D. Watson, "Moving toward the Clonal Man—Is This What We Want?" U.S. Senate, 92nd Cong., 1st sess., *Congressional Record* 117, no. 61 (April 29, 1971): 12,751–12,752.

18. The following factual discussion tracks Bonnicksen, *In Vitro Fertilization,* pp. 77–82. Policy assessments are my own and differ somewhat from hers.

19. *Harris v. McRae,* 448 U.S. 297 (1980).

20. Robert M. L. Winston and Alan H. Handyside, "New Challenges in Human in Vitro Fertilization," *Science* 260 (1993): 932–936.

21. Leslie Roberts, "Ethical Questions Haunt New Genetic Technologies," *Science* 243 (1989): 1134–1136.

22. The following discussion is taken from Ira H. Carmen, "Washington Politics and Genetic Engineering Research: When Worlds Collide," *Human Gene Therapy* 7 (1996): 97–108 at 100–103.

23. Eliot Marshall, "Varmus Puts His Stamp on NIH," *Science* 270 (1995): 1290.

24. Jocelyn Kaiser, "Congress Bans Embryo Study Funds," *Science* 271 (1996): 585.

25. Eliot Marshall, "Varmus Grilled over Breach of Embryo Research Ban," *Science* 276 (1997): 1963.

26. J. A. Thomson et al., "Embryonic Stem Cell Lines Derived from Human Blastocysts," *Science* 282 (1998): 1145–1147.

27. Eliot Marshall, "A Versatile Cell Line Raises Scientific Hopes, Legal Questions," *Science* 282 (1998): 1014–1015.

28. Nicholas Wade, "Clinton Asks Study of Bid to Form Part-Human, Part-Cow Cells," *New York Times,* November 15, 1998, p. 25; Eliot Marshall, "Claim of Human-Cow Embryo Greeted with Skepticism," *Science* 282 (1998): 1390–1391.

29. Eliot Marshall, "Ruling May Free NIH to Fund Stem Cell Studies," *Science* 283 (1999): 465–466.

30. Mark P. Lanza et al., "Science over Politics," *Science* 283 (1999): 1849–1850.

31. I discuss in the next section the committee's assessment of the cow-human SCNT hybrid.

32. Cf. Louis M. Guenin, "Morals and Primordials," *Science* 292 (2001): 1659–1660.

33. Eliot Marshall, "Ethicists Back Stem Cell Research, White House Treads Cautiously," *Science* 285 (1999): 502; Andrea L. Bonnicksen, *Crafting a Cloning Policy* (Washington, D.C.: Georgetown University Press, 2002), pp. 82–83.

34. Gretchen Vogel, "Researchers Get Green Light for Work on Stem Cells," *Science* 289 (2000): 1442–1443.

35. Elizabeth Finkel et al., "ScienceScope," *Science* 291 (2001): 807; Gretchen Vogel, "NIH Pulls Plug on Ethics Review," *Science* 292 (2001): 415–416.

36. Gretchen Vogel, "Rumors and Trial Balloons Precede Bush's Funding Decision," *Science* 293 (2001): 186–187.

37. Gretchen Vogel, "Bush Squeezes between the Lines on Stem Cells," *Science* 293 (2001): 1242–1245; Constance Holden, "NIH's List of 64 Leaves Questions," *Science* 293 (2001): 1567.

38. Constance Holden and Gretchen Vogel, " 'Show Us the Cells,' U.S. Researchers Say," *Science* 297 (2002): 923–925.

39. Constance Holden, "HHS Inks Cell Deal; NAS Calls for More Lines," *Science* 293 (2001): 1966–1967; Gretchen Vogel, "Stem Cells Not So Stealthy After All," *Science* 297 (2002): 175–177; Marcus Grompe, "Adult versus Embryonic Stem Cells: It's Still a Tie," *Molecular Therapy* 6 (2002): 303–305.

40. Holden, "HHS Inks Cell Deal."

41. James McGrath and Davor Solter, "Inability of Mouse Blastomere Nuclei Transferred to Enucleated Zygotes to Support Development in Vitro," *Science* 226 (1984): 1317–1319.

42. Ann Cook, "Cloning Humans Next? Ex-UI Student Doing It with Rabbits," *Champaign-Urbana (Ill.) News-Gazette,* June 11, 1987, p. B-9; Jean L. Marx, "Cloning Sheep and Cattle Embryos," *Science* 239 (1988): 463–464.

43. Release by Donald J. Quigg, assistant secretary of the Department of Commerce and commissioner of Patents and Trademarks, April 21, 1987.

44. Rebecca Kolberg, "Human Embryo Cloning Reported," *Science* 262 (1993): 652–653. For the media hysteria, see Center for Biotechnology Policy and Ethics, "Baby Boom: Human Embryo Cloning Hits the Mass Media," *Center for Biotechnology Policy and Ethics Newsletter* 3, no. 4, January 1, 1994.

45. I. M. Wilmut et al., "Viable Offspring Derived from Fetal and Adult Mammalian Cells," *Nature* 385 (1997): 810–813.

46. Elizabeth Pennisi and Nigel Williams, "Will Dolly Send in the Clones?" *Science* 275 (1997): 1415–1416.

47. "Memorandum on the Prohibition on Federal Funding for Cloning of Human Beings," *Weekly Compilation of Presidential Documents* 33 (March 4, 1997): 281; Eliot Marshall, "Panel Weighs a Law against Cloning," *Science* 276 (1997): 1185.

48. Letter from Bill Clinton to Harold Shapiro, chair, National Bioethics Advisory Commission, February 24, 1997, reprinted in "Cloning Human Beings," Report and Recommendations of the NBAC, June 1997; hereinafter cited as NBAC Report.

49. Eliot Marshall, "Mammalian Cloning Debate Heats Up," *Science* 275 (1997): 1733; Dan Eramian, "The Real Message from Dolly," *Bio News,* April/May 1997, pp. 3–5; Constance Holden, "Scholars' Group Defends Cloning," *Science* 276 (1997): 1341.

50. NBAC Report, pp. 107–110.

51. Eliot Marshall, "Clinton Urges Outlawing Human Cloning," *Science* 276 (1997): 1640.

52. The following discussion is culled from Ira H. Carmen, "Should Human Cloning Be Criminalized?" *Journal of Law and Politics* 13 (1997): 745–758. Among other critical assessments were Susan M. Wolf, "Ban Cloning? Why NBAC Is Wrong," *Hastings Center Report* 27 (1997): 12–15, and John A. Robertson, "Liberty, Identity, and Human Cloning," *Texas Law Review* 76 (1998): 1371–1456.

53. Harold T. Shapiro, "Ethical and Policy Issues of Human Cloning," *Science* 277 (1997): 195.

54. NBAC Report, p. 95.

55. Ibid., pp. 93–94.

56. Ibid., p. 100.

57. Human Genome Organization, "General Assembly of HUGO Members," *Genome Digest* 4 (2) (April 1997): 4–5 at 5.

58. "HUGO Ethics Committee—Statement on Cloning," *Eubios Journal of Asian and International Bioethics* 9 (1999): 70.

59. The European Parliament and the Council of the European Union, Common Position, April 8, 1998. This document has all the legal force of a treaty when ratified by the requisite number of member states. Ratification was finalized in March 2001. Bonnicksen, *Crafting*, p. 173.

60. UNESCO's statement on the human genome specifically calls for a ban on human reproductive cloning, though the term is never defined. The United States, not being at that time a member of UNESCO, did not sign the document. In December 1998, the UN General Assembly approved it. Constance Holden, "UN Weighs In on Cloning," *Science* 278 (1997): 1407. Bonnicksen, *Crafting*, p. 171. As of this writing, the UN is paralyzed: some countries want to ban therapeutic cloning, while others do not. The deadlock prevents even a stop order on reproductive cloning. In all this, the UN mirrors the Congress of the United States. Gretchen Vogel, "U.N. Split over Full or Partial Cloning Ban," *Science* 298 (2002): 1316–1317.

61. Jocelyn Kaiser, "Cloning Ban Waits in the Wings," *Science* 277 (1997): 1191.

62. Elizabeth Pennisi, "Transgenic Lambs from Cloning Lab," *Science* 277 (1997): 631; Angelika E. Schnieke et al., "Human Factor IX Transgenic Sheep Produced by Transfer of Nuclei from Transfected Fetal Fibroblasts," *Science* 278 (1997): 2130–2133.

63. Constance Holden, "Calf Cloned from Bovine Cell Line," *Science* 277 (1997): 903.

64. Oliver Morton, "First Dolly, Now Headless Tadpoles," *Science* 278 (1997): 798.

65. Arthur Caplan and J. Craig Venter, "Using One's Head," *Science* 278 (1997): 1547–1548.

66. David Kestenbaum, "Cloning Plan Spawns Ethics Debate," *Science* 279 (1998): 315.

67. "Physicist Says He'll Clone Himself with Wife's Help," *Champaign-Urbana (Ill.) News-Gazette*, September 7, 1998, p. A-5.

68. Richard G. Seed, "Inevitable Human Cloning . . . On the Road to Immortality" (unpublished outline in the author's possession).

69. Quoted in Bonnicksen, *Crafting*, p. 56.

70. Eliot Marshall, "Biomedical Groups Derail Fast-Track Anticloning Bill," *Science* 279 (1998): 1123–1124.

71. Nancy Bradish, "Cloning Forces Congress to Do a Double-Take," *Bio News*, February/March 1998, pp. 1, 3.

72. Dennis Normile, "Bid for Better Beef Gives Japan a Leg Up on Cattle," *Science* 282 (1998): 1975–1976; Elizabeth Pennisi, "Cloned Mice Provide Company for Dolly," *Science* 281 (1998): 495, 497.

73. Elizabeth Pennisi and Dennis Normile, "Perseverance Leads to Cloned Pig in Japan," *Science* 289 (2000): 1118–1119.

74. A. W. S. Chan et al., "Clonal Propagation of Primate Offspring by Embryo Splitting," *Science* 287 (2000): 317–319.

75. Gretchen Vogel, "In Contrast to Dolly, Cloning Resets Telomere Clock in Cattle," *Science* 288 (2000): 586–587.

76. Elizabeth Pennisi and Gretchen Vogel, "Clones: A Hard Act to Follow," *Science*

288 (2000): 1722–1727; David Humpherys et al., "Epigenetic Instability in ES Cells and Cloned Mice," *Science* 293 (2001): 95–97. For an accessible overview of the development and future of "cloning science," I recommend Anne McLaren, "Cloning: Pathways to a Pluripotent Future," *Science* 288 (2000): 1775–1780.

77. Dennis Normile, "Human Cloning Ban Allows Some Research," *Science* 290 (2000): 1872.

78. Michael Baker, "Korean Report Sparks Anger and Inquiry," *Science* 283 (1999): 16–17; Michael Baker, "Report Casts Doubt on Korean Experiment," *Science* 283 (1999): 617, 619.

79. Constance Holden, "Clone Rangers," *Science* 284 (1999): 2083; Constance Holden, "Company Gets Funds to Clone Baby," *Science* 289 (2000): 2271.

80. Gretchen Vogel, "Cloning: Could Humans Be Next?" *Science* 291 (2001): 808–809.

81. John Pickrell, "Experts Assail Plan to Help Childless Couples," *Science* 291 (2001): 2061, 2063.

82. Rudolph Jaenisch and Ian Wilmut, "Don't Clone Humans!" *Science* 291 (2001): 2552.

83. Gretchen Vogel, "Human Cloning Plans Spark Talk of U.S. Ban," *Science* 292 (2001): 31; Gretchen Vogel, "Cloning Bills Proliferate in U.S. Congress," *Science* 292 (2001): 1037.

84. Gretchen Vogel, "Bush Grapples with Stem Cells, Cloning," *Science* 292 (2001): 2409, 2411; Constance Holden, "Would Cloning Ban Affect Stem Cells?" *Science* 293 (2001): 1025.

85. President George W. Bush, "Address to the Nation," reprinted in *The Future Is Now,* ed. William Kristol and Eric Cohen (Lanham, Md.: Rowman and Littlefield, 2002), pp. 306–310.

86. See, e.g., Kathi E. Hanna, Robert M. Cook-Deegan, and Robyn Y. Nishimi, "Finding a Forum for Bioethics in U.S. Public Policy," *Politics and the Life Sciences* 12 (1993): 205–219.

87. The following analysis draws from Ira H. Carmen, "Bioethics, Public Policy, and Political Science," *Politics and the Life Sciences* 13 (1994): 79–81.

88. Susan Biggin, "Italian Committee Loses Its Balance," *Science* 267 (1995): 326.

89. Executive Order no. 12,975, *Federal Register* 60 (October 5, 1995): 52,063–52,065.

90. Constance Holden, "All Quiet on the Bioethics Front," *Science* 281 (1998): 169.

91. Eliot Marshall, "Bioethics Panel Urges Broader Oversight," *Science* 292 (2001): 1466–1467.

92. See the discussion in text at note 28 above.

93. Bonnicksen, *Crafting,* pp. 80–81.

94. Eliot Marshall, "New Chair of Bioethics Panel Wants National Debate on Issues," *Science* 293 (2001): 1243. For the PCB's de jure mission statement, see Executive Order no. 13,237, *Code of Federal Regulations,* title 3, sec. 821 (2001).

95. Leon R. Kass, "Genetic Tampering," reprinted in Kristol and Cohen, *Future Is Now,* pp. 31–32.

96. Leon R. Kass, "Making Babies—The New Biology and the 'Old' Morality," reprinted in Kristol and Cohen, *Future Is Now*, 54–60; emphasis in text.

97. Leon R. Kass, "'Making Babies' Revisited," reprinted in Kristol and Cohen, *Future Is Now*, pp. 61–69.

98. Leon R. Kass, "The Wisdom of Repugnance," reprinted in Kristol and Cohen, *Future Is Now*, pp. 116–118.

99. Leon R. Kass, "Why Not Immortality?" reprinted in Kristol and Cohen, *Future Is Now*, pp. 321–332.

100. Leon R. Kass, "Preventing a Brave New World," reprinted in Kristol and Cohen, *Future Is Now*, pp. 219–241.

101. Wilson's thesis is well summarized in his essay "The Moral Sense," *American Political Science Review* 87 (1993): 1–11.

102. James Q. Wilson, "The Paradox of Cloning," reprinted in Leon R. Kass and James Q. Wilson, *The Ethics of Human Cloning* (Washington, D.C.: AEI Press, 1998), pp. 61–74.

103. James Q. Wilson, "Sex and Family," in Kass and Wilson, *Ethics of Human Cloning*, pp. 89–100.

104. Francis Fukuyama, "A Milestone in the Conquest of Nature," reprinted in Kristol and Cohen, *Future Is Now*, pp. 77–80.

105. Francis Fukuyama, "Separating Good Biotech from Bad," reprinted in Kristol and Cohen, *Future Is Now*, pp. 242–245.

106. Michael J. Sandel, *Liberalism and the Limits of Justice*, 2nd ed. (Cambridge: Cambridge University Press, 1998).

107. Robert P. George, "Stem-Cell Research: Don't Destroy Human Life," reprinted in Kristol and Cohen, *Future Is Now*, pp. 289–291.

108. William Safire, "The Crimson Birthmark," *New York Times*, January 21, 2002, p. A19.

109. Sheryl Gay Stolberg, "Bush's Advisers on Ethics Discuss Human Cloning," *New York Times*, January 18, 2002, p. A19.

110. Sheryl Gay Stolberg, "Some for Abortion Rights Lean Right in Cloning Fight," *New York Times*, January 24, 2002, p. A23.

111. Eliot Marshall and Gretchen Vogel, "Cloning Announcement Sparks Debate and Scientific Skepticism," *Science* 294 (2001): 1802–1803.

112. Robert P. Lanza et al., "Cloned Cattle Can Be Healthy and Normal," *Science* 294 (2001): 1893–1894.

113. Jocelyn Kaiser, "Cloned Pigs May Help Overcome Rejection," *Science* 295 (2002): 25, 27. In August 2002, Ian Wilmut rejoined the research dialogue, publishing a report showing that he had successfully knocked out both copies. By that time, the human cloning debate had withered on the vine, only to be resurrected briefly following the Raëlians' announcement of human cloning reproductive success. "Cloned Pigs Lack Gene That Hinders Transplants," *Champaign-Urbana (Ill.) News-Gazette*, August 23, 2002, p. A-1.

114. Constance Holden, "Primate Parthenotes Yield Stem Cells," *Science* 295 (2002): 779–780.

115. Kenneth Chang, "Company Says It Used Cloning to Create New Kidneys for Cow," *New York Times,* January 31, 2002, p. A22.

116. Constance Holden, "Carbon-Copy Clone Is the Real Thing," *Science* 295 (2002): 1443–1444.

117. Constance Holden and Jocelyn Kaiser, "Report Backs Ban; Ethics Panel Debuts," *Science* 295 (2002): 601–602.

118. Stolberg, "Bush's Advisers"; Stolberg, "Some for Abortion Rights"; Constance Holden, "Battle Heats Up over Cloning," *Science* 295 (2002): 2009; Bill McKibben, "Unlikely Allies against Cloning," *New York Times,* March 27, 2002, p. A23; Carl Feldbaum, "Some History Should Be Repeated," *Science* 295 (2002): 975; Bert Vogelstein et al., "Please Don't Call It Cloning!" *Science* 295 (2002): 1237.

119. Sheryl Gay Stolberg, "Bush Rallies Opponents of Cloning," *New York Times,* April 10, 2002, p. A19; Sheryl Gay Stolberg, "Bush Makes Fervent Bid to Get Senate to Ban Cloning Research," *New York Times,* April 11, 2002, p. A23.

120. Sheryl Gay Stolberg, "Key Republican Backs Cloning in Research," *New York Times,* May 1, 2002, p. A20.

121. President's Council on Bioethics, "Human Cloning and Human Dignity: An Ethical Inquiry" (Washington, D.C., July 2002), http://www.bioethics.gov/topics/cloning_index.html (accessed May 4, 2004); hereinafter cited as PCB Report, pp. 3, 4.

122. Ibid., p. 8; emphasis in text.

123. Arguing along the same line is Walter E. Nance and John Fletcher, "Human Reproductive Cloning," *Science* 297 (2002): 1477.

124. The shifts and countershifts in various members' positions are discussed in Stephen S. Hall, "President's Bioethics Council Delivers," *Science* 297 (2002): 322–324.

125. The quotations can be found in the PCB Report, p. 10.

126. *United States v. Morrison,* 529 U.S. 598 (2000).

127. Jack M. Balkin, "The Cloning Conundrum," *New York Times,* January 30, 2002, p. A27. Both Balkin and the PCB are in error when they assert (without citation) that Congress can regulate human cloning because it has "nationwide implications" (PCB Report, p. 95). This argument was rejected long ago by the Supreme Court (*Kansas v. Colorado,* 206 U.S. 46, 1907), and I know of no justice today who supports it.

128. Ronald M. Green et al., "Overseeing Research on Therapeutic Cloning," *Hastings Center Report* 32 (2002): 27–32.

129. PCB Report, pp. 116–119.

130. Robertson, "Liberty, Identity, and Human Cloning."

131. *Planned Parenthood v. Casey,* p. 864.

132. For constitutional law aficionados: when I say "*Roe* should be overruled," I in no way mean to associate myself with Justice Scalia's *Planned Parenthood* dictum in dissent in which he says, "We should get out of this area," 505 U.S. 833 at p. 1002 (1992). There remains the time-honored version of substantive due process dating from Justice Holmes and approved in various forms by Justices Frankfurter,

Harlan II, and Souter among others. They would strike down laws "infring[ing] fundamental principles as they have been understood by the traditions of our people and our law" or which, over time, would now be seen as "arbitrary impositions" or "purposeless restraints." I believe statutes proscribing abortion in cases of rape (a felony), incest (a felony that also could produce genetically flawed progeny), and where the woman's *life* is at risk fall within the Holmes paradigm understood in the context of today's constitutional value system.

133. Sheryl Gay Stolberg, "Total Ban on Cloning Research Appears Dead," *New York Times,* June 14, 2002, p. A31.

134. Donald G. McNeill Jr., "Religious Sect Says It Will Announce the First Cloned Baby," *New York Times,* December 27, 2002, p. A13.

135. Calvin Simerly et al., "Molecular Correlates of Primate Nuclear Transfer Failures," *Science* 300 (2003): 297.

136. For exposition, see Roger Shattuck, *Forbidden Knowledge: From Prometheus to Pornography* (New York: St. Martin's, 1996). I thank Lawrence Lowell Putnam for calling this book to my attention.

137. Ira H. Carmen, *Movies, Censorship, and the Law* (Ann Arbor: University of Michigan Press, 1966); Richard S. Randall, *Freedom and Taboo* (Berkeley: University of California Press, 1989).

4. Germline

1. These paragraphs receive extensive elaboration in Carmen, *Cloning,* ch. 2.

2. *Home Building and Loan Association v. Blaisdell,* 290 U.S. 398 at p. 426 (1934).

3. Ronald Munson, *Intervention and Reflection,* 3rd ed. (Belmont, Calif.: Wadsworth, 1988), p. 278.

4. Arno G. Motulsky, "Impact of Genetic Manipulation on Society and Medicine," *Science* 219 (1983): 135–136.

5. The following discussion recapitulates Carmen, *Cloning,* ch. 3.

6. James D. Watson, "In Defense of DNA," *New Republic,* June 25, 1977, pp. 13–14.

7. This section amplifies Carmen, *Cloning,* ch. 5, pp. 162–170.

8. Quoted in Carmen, *Cloning,* p. 167; emphasis added.

9. This discussion draws on Carmen, "Washington Politics," pp. 98–110.

10. Motulsky, "Impact of Genetic Manipulation," p. 138.

11. President's Commission for the Study of Ethical Problems in Medicine and Biomedical and Behavioral Research, *Splicing Life: A Report on the Social and Ethical Issues of Genetic Engineering with Human Beings* (Washington, D.C.: Government Printing Office, 1982).

12. Carmen, *Cloning,* pp. 183–192.

13. President's Commission for the Study of Ethical Problems, *Splicing Life* (prepublication copy of the commission's "Summary of Conclusions and Recommendations"), pp. 4–5.

14. Barbara J. Culliton, "Gene Therapy Guidelines Revised," *Science* 228 (1985): 561–562; Harold M. Schmeck Jr., "U.S. Sets Guidelines in Using Gene Transplants in Humans," *New York Times,* September 24, 1985, pp. 1, 18.

15. The following paragraphs are taken from Carmen, "Debates, Divisions, and Decisions," pp. 247–252.

16. Apologia: I mention Dr. Murray's race only for the purpose of dispelling in a small way the canard that African Americans aren't involved in these scientific decisions at the highest level.

17. To this, the editors of two leading journals responded that when researchers present original data at conferences and government oversight meetings, they are in no way jeopardizing their publication rights. It is incredible that it took the ABR episode to settle what should much earlier have been obvious to all.

18. Quoted in Carmen, "Debates, Divisions, and Decisions," p. 251.

19. In the name of full disclosure, so also is the author.

20. Natalie Angier, "Gene Implant Therapy Is Backed for Children with Rare Disease," *New York Times,* March 8, 1990, pp. 1, B9; Barbara J. Culliton, "Gene Therapy Clears First Hurdle," *Science* 247 (1990): 1287.

21. National Institutes of Health, Meeting of the Human Gene Therapy Subcommittee of the Recombinant DNA Advisory Committee (RAC), March 30, 1990 (transcript of proceedings; Gaithersburg, Md.: StenoTech, 1990); Jeff Lyon and Peter Gorner, *Altered Fates* (New York: Norton, 1995), ch. 9.

22. Lyon and Gorner, *Altered Fates,* ch. 10.

23. The numbers come from "Human Gene Marker/Therapy Clinical Protocols," *Human Gene Therapy* 9 (1998): 2805–2852.

24. See Ira H. Carmen, "A Death in the Laboratory: The Politics of the Gelsinger Aftermath," *Molecular Therapy* 3 (2001): 425–428; Eliot Marshall, "Gene Therapy on Trial," *Science* 288 (2000): 951–957.

25. Marina Cavazzana-Calvo et al., "Gene Therapy of Human Severe Combined Immunodeficiency (SCID)-X1 Disease," *Science* 288 (2000): 669–672.

26. W. French Anderson, "The Best of Times, the Worst of Times," *Science* 288 (2000): 627, 629.

27. Eliot Marshall, "Gene Therapy a Suspect in Leukemia-Like Disease," *Science* 298 (2002): 34–35.

28. Eliot Marshall, "What to Do When Clear Success Comes with an Unclear Risk?" *Science* 298 (2002): 510–511.

29. Eliot Marshall, "Second Child in French Trial Is Found to Have Leukemia," *Science* 299 (2003): 320; Jocelyn Kaiser, "Seeking the Cause of Induced Leukemias in X-SCID Trial," *Science* 299 (2003): 495.

30. Allesandro Aiuti, "Correction of ADA-SCID by Stem Cell Gene Therapy Combined with Nonmyeloablative Conditioning," *Science* 296 (2002): 2410–2413.

31. This is exactly what R. D. Hotchkiss had in mind when he coined the phrase human genetic engineering. "Portents for a Genetic Engineering," *Journal of Heredity* 56 (1965): 197–202. For a contemporary example, see John H. Evans, *Playing God?* (Chicago: University of Chicago Press, 2002).

32. *Milwaukee Pub. Co. v. Burleson,* 255 U.S. 407 at p. 437 (1921).

33. John C. Fletcher and Gerd Richter, "Human Fetal Gene Therapy: Moral and Ethical Questions," *Human Gene Therapy* 7 (1996): 1605 –1614 at p. 1606. Evans denies that bioethicists today partake of these "thick" questions. He needs to read Leon Kass a little more carefully. I, myself, like William Powell in the 1930s movies, am a congenital "thin man." After all, I stupidly believe, contrary to Evans, in modern social science. As I write in ch. 6, I merely want to make it *more* modern.

34. Ibid., pp. 1606, 1609.

35. Esmail I. Zanjani and W. French Anderson, "Prospects for in Utero Human Gene Therapy," *Science* 285 (1999): 2084 –2088; Jennifer Couzin, "RAC Confronts in Utero Gene Therapy Proposals," *Science* 282 (1998): 27.

36. W. French Anderson, "The First Signs of Danger," *Human Gene Therapy* 3 (1992): 359.

37. Sheldon Krimsky, "Human Gene Therapy: Must We Know Where to Stop Before We Start?" *Human Gene Therapy* 1 (1990): 171–173.

38. Charles C. Culver and Bernard Gert, *Philosophy in Medicine* (New York: Oxford University Press, 1982).

39. Vincent Kiernan, "Cosmetic Uses of Genetic Engineering May Soon Be a Reality," *Chronicle of Higher Education,* October 3, 1997, pp. A17 –A18.

40. Cf. Judith Areen and Patricia King, "Legal Regulation of Human Gene Therapy," *Human Gene Therapy* 1 (1990): 151–161 at 156.

41. Richard Stone, "NIH to Size Up Growth Hormone Trials," *Science* 257 (1992): 739; Eliot Marshall, "Panel Approves Gene Trial for 'Normals,'" *Science* 275 (1997): 1561.

42. Theodore Friedman and Johann Olav Koss, "Gene Transfer and Athletics— An Impending Problem," *Molecular Therapy* 3 (2001): 819 –820.

43. Kiernan, "Cosmetic Uses of Genetic Engineering," p. A18.

44. John W. Gordon, "Genetic Enhancement in Humans," *Science* 283 (1999): 2023 –2024 at 2024.

45. John B. Attanasio, "The Constitutionality of Regulating Human Genetic Engineering: Where Procreative Liberty and Equal Opportunity Collide," *University of Chicago Law Review* 53 (1986): 1274 –1342.

46. Motulsky, "Impact of Genetic Manipulation," pp. 139 –140.

47. S. L. Byrand, "Evolutionary Wisdom or Genetic Roulette?" *Science* 278 (1997): 882.

48. Leon Kass, "New Beginnings in Life," in *The New Genetics and the Future of Man,* ed. Michael P. Hamilton (Grand Rapids, Mich.: Eerdmans, 1972), pp. 15 –63 at p. 61; emphasis in text.

49. Joseph Fletcher, *The Ethics of Genetic Control* (Garden City, N.Y.: Doubleday Anchor, 1974), pp. 14 –15.

50. NIH Recombinant DNA Advisory Committee, "The Revised 'Points to Consider' Document," reprinted in *Human Gene Therapy* 1 (1990): 93 –103 at p. 95.

51. NBAC Report, p. 97.

52. Barbara J. Culliton, "NIH Asked to Tighten Gene Therapy Rules," *Science* 233 (1986): 1378–1379.

53. *Eisenstadt v. Baird,* 405 U.S. 438 (1972).

54. Eliot Marshall, "Panel Reviews Risks of Germ Line Changes," *Science* 294 (2001): 2268–2269; Nicholas Wade, "After Scare, a Gene Therapy Trial Proceeds," *New York Times,* January 8, 2002, p. F6.

55. Trisha Gura, "Gene Therapy and the Germ Line," *Molecular Therapy* 6 (2002): 2–4.

56. Eliot Marshall, "Moratorium Urged on Germ Line Gene Therapy," *Science* 289 (2000): 2023; Erik Parens and Eric Juengst, "Inadvertently Crossing the Germ Line," *Science* 292 (2001): 397; Mark S. Frankel and Audrey R. Chapman, "Facing Inheritable Genetic Modifications," *Science* 292 (2001): 1303.

57. Eric T. Juengst, "Germ-Line Gene Therapy: Back to Basics," *Journal of Medical Philosophy* 16 (1991): 587–592.

58. LeRoy Walters, "Human Gene Therapy: Ethics and Public Policy," *Human Gene Therapy* 2 (1991): 115–122.

59. Quoted in ibid., p. 118.

60. Ibid., p. 118.

61. Eric T. Juengst, "The NIH 'Points to Consider' and the Limits of Human Gene Therapy," *Human Gene Therapy* 1 (1990): 425–433; cf. Andrea L. Bonnicksen, "Human Embryos and Genetic Testing: A Private Policy Model," *Politics and the Life Sciences* 11 (1) (1992): 53–62, arguing that biopolicy is inevitably the stuff of public law codification, an unwieldy mechanism for "governing" preimplantation diagnosis in preembryos. She thinks biopolicy is too political for such arcane tasks. I say the RAC has always been political, and that this is a *plus.* There are few "neutral principles" in these matters.

62. Ira H. Carmen, "Human Gene Therapy: A Biopolitical Overview and Analysis," *Human Gene Therapy* 4 (1993): 187–193.

63. Ira H. Carmen, *Movies, Censorship, and the Law.*

64. Richard Stone, "NIH to Study 'Germ-line' Therapy," *Science* 266 (1994): 1631; cf. Richard Stone, "Germ Cell Gene Panel," *Science* 255 (1991): 841. See generally Nelson A. Wivel and LeRoy Walters, "Germ-line Gene Modification and Disease Prevention: Some Medical and Ethical Perspectives," *Science* 262 (1993): 533–538.

65. W. French Anderson, "Human Gene Therapy," *Science* 256 (1992): 812.

66. In 1989, when the European Community initially endorsed the Human Genome Project, it also voted to proscribe "germ line *research*" (emphasis added). Kevles, in Kevles and Hood, *Code of Codes,* p. 33.

67. Bartha M. Knoppers and Sonia LeBris, "Recent Advances in Medically Assisted Conception: Legal, Ethical, and Social Issues," *American Journal of Law and Medicine* 27 (4) (1991): 329–361 at 361.

68. Bonnicksen, "National and International Approaches."

69. Robert M. L. Winston, "Germ-line Gene Therapy: An Exaggerated Threat," *Politics and the Life Sciences* 13 (2) (1994): 237–238; David Shapiro, "Germ-line Gene

Therapy: Do We Need an International Approach?" *Politics and the Life Sciences* 13 (2) (1994): 233–234; Ruth Chadwick, "Germ-line Gene Therapy, Autonomy, and Community," *Politics and the Life Sciences* 13 (2) (1994): 223–225; Robert Cook-Deegan, "Germ-line Gene Therapy: Keep the Window Open a Crack," *Politics and the Life Sciences* 13 (2) (1994): 217–220.

70. U.S. Department of HHS, PHS, NIH, *Recombinant DNA Research* 16 (January 1994): 239–243.

71. Eric S. Lander, "In Wake of Genetic Revolution, Questions about Its Meaning," *New York Times*, September 12, 2000, p. D5; Leslie Roberts, "Carving Up the Human Genome," *Science* 242 (1988): 1244–1246.

72. Constance Holden, "Sperm Factories," *Science* 266 (1994): 1482.

73. "Testicle Transplant Successful," *Champaign-Urbana (Ill.) News-Gazette*, August 15, 2002, p. A-2.

74. Li-Xin Feng et al., "Generation and in Vitro Differentiation of a Spermatogonial Cell Line," *Science* 297 (2002): 392–395.

75. "Researchers Make Strain of Pigs That Carry Human Genes," *Champaign-Urbana (Ill.) News-Gazette*, October 22, 2002, p. A-2.

76. See, for example, Nicholas D. Kristof, "Interview with a Humanoid," *New York Times*, July 23, 2002, p. A19, in which a cloned cow carrying the gene for the C-1 Esterase Inhibitor protein is again and again referred to as "part human."

77. Bernard D. Davis, "Germ-line Therapy: Evolutionary and Moral Considerations," *Human Gene Therapy* 3 (1992): 361–363.

78. Gordon, "Genetic Enhancement in Humans," p. 2024.

79. Bartha Maria Knoppers and Sonia LeBris, "Genetic Choices: A Paradigm for Prospective International Ethics?" *Politics and the Life Sciences* 13 (2) (1994): 228–230 at 229.

80. John C. Fletcher and W. French Anderson, "Germ-line Gene Therapy: A New Stage of Debate," *Law, Medicine and Health Care* 20 (1992): 26–39 at 29.

81. Wivel and Walters, "Germ-line Gene Modification," p. 536.

82. Carmen, "Ownership of the Human Genome."

83. Bonnicksen, "National and International Approaches," p. 45.

84. Cook-Deegan, "Germ-line Gene Therapy," p. 219.

85. Erik Parens, "Is Better Always Good? The Enhancement Project," in *Enhancing Human Traits*, ed. Erik Parens (Washington, D.C.: Georgetown University Press, 1998), p. 5.

86. The Parens collection cited in note 85 makes no reference that I can find to germline enhancement.

87. Quoted in Fletcher and Anderson, "Germ-line Gene Therapy," p. 36.

88. Constance Holden, "Didactics of *Gattaca*," *Science* 278 (1997): 1019.

89. Bill Clinton, "Science in the 21st Century," *Science* 276 (1997): 1951; Maxwell Mehlman and Jeffrey Botkin, *Access to the Genome: The Challenge to Equality* (Washington, D.C.: Georgetown University Press, 1997).

90. For normative noise in the context of genetic enhancement, see Attanasio,

"Constitutionality of Regulating Human Genetic Engineering," in which the armies of Robert Nozick's libertarianism, Bruce Ackerman's equalitarianism, and John Rawls's rationalism clash by night.

5. Sociogenomics

1. For a recent and refreshing exception, see Robert E. Gilbert, *Mortal Presidency* (New York: Fordham University Press, 1998).

2. H. D. Lasswell and Abraham Kaplan, *Power and Society* (New Haven: Yale University Press, 1950), ch. 4.

3. For suggestive data, see Steven A. Peterson, "The Health of the Polity: Results from a National Sample," *Politics and the Life Sciences* 10 (1991): 45–56.

4. Robert S. Robins and Henry Rothschild, "Ethical Dilemmas of the President's Physician," *Politics and the Life Sciences* 7 (1988): 3–11.

5. R. B. Handberg, "Talking about the Unspeakable in a Secretive Institution: Health and Disability among Supreme Court Justices," *Politics and the Life Sciences* 8 (1989): 70–73.

6. Robert Plomin, Michael J. Owen, and Peter McGuffin, "The Genetic Basis of Complex Human Behaviors," *Science* 264 (1994): 1733–1739.

7. Morell, "Huntington's Gene Finally Found," 28–30.

8. Guido Calabresi and Philip Bobbitt, *Tragic Choices* (New York: Norton, 1978); Rachel Nowak, "Genetic Testing Set for Takeoff," *Science* 265 (1994): 464–467.

9. Rachel Nowak, "Breast Cancer Gene Offers Surprises," *Science* 265 (1994): 1796–1799.

10. Robert Plomin and Stephen A. Petrill, "Genetics and Intelligence: What's New?" *Intelligence* 24 (1997): 53–77 at 55–56.

11. Nowak, "Breast Cancer Gene."

12. Richard Stone, "Taking Honest Abe's DNA Fingerprint," *Science* 256 (1992): 446.

13. Constance Holden, "Smoking Genes," *Science* 289 (2000): 2271.

14. Ridley, *Genome*, p. 236.

15. The above commentary is provided by Ridley, *Genome*, p. 191.

16. Gina Kolata, "Using Genetic Tests, Ashkenazi Jews Vanquish a Disease," *New York Times*, February 18, 2003, pp. D1, D6.

17. Martin Enserink, "Physicians Wary of Scheme to Pool Icelanders' Genetic Data," *Science* 281 (1998): 890–891; Martin Enserink, "Opponents Criticize Iceland's Data Base," *Science* 282 (1998): 859; Henry T. Greely, "Genomics Research and Human Subjects," *Science* 282 (1998): 625; Martin Enserink, "Iceland OKs Private Health Data Bank," *Science* 283 (1999): 13; Nicholas Wade, "A Genomic Treasure Hunt May Be Striking Gold," *New York Times*, June 18, 2002, pp. D1, D4.

18. Lone Frank, "Estonia Prepares for National DNA Database," *Science* 290 (2002): 31; Jocelyn Kaiser, "Population Databases Boom, from Iceland to the U.S.," *Science* 298 (2002): 1158–1161; Gwen Kinkead, "To Study Disease, Britain Plans a Genetic Census," *New York Times*, December 31, 2002, pp. D5, D8.

19. This paragraph relies on Ridley, *Genome,* ch. 5, and Wade, "Genomic Treasure Hunt."

20. Plomin, Owen, and McGuffin, "Genetic Basis."

21. John Travis, "New Piece in Alzheimer's Puzzle," *Science* 261 (1993): 828–829; Eliot Marshall, "A Clash over Testing for Alzheimer's Disease," *Science* 280 (1998): 1004.

22. Constance Holden, "Boxer Genes," *Science* 277 (1997): 321.

23. Jean L. Marx, "New Gene Tied to Common Form of Alzheimer's," *Science* 281 (1998): 507, 509.

24. Lars Bertram et al., "Evidence for Genetic Linkage of Alzheimer's Disease to Chromosome 10q," *Science* 290 (2000): 2302–2303.

25. Nilufer Ertekin-Taner et al., "Linkage of Plasma Aβ42 to a Quantitative Locus on Chromosome 10 in Late-Onset Alzheimer's Disease Pedigrees," *Science* 290 (2000): 2303–2304.

26. Amanda Myers et al., "Susceptibility Locus for Alzheimer's Disease on Chromosome 10," *Science* 290 (2000): 2304–2305.

27. Marshall, "Clash over Testing."

28. Ridley, *Genome,* pp. 264–265.

29. Ibid., p. 264.

30. Ann Gibbons, "From Field to Lab, New Insights on Being Human," *Science* 288 (2000): 798–800.

31. Gina Kolata, "Manic-Depression Gene Tied to Chromosome 11," *Science* 235 (1987): 1139–1140; Peter Aldhous, "The Promise and Pitfalls of Molecular Genetics," *Science* 257 (1992): 164–165.

32. "Gene Linked to Bipolar Disorder," *Champaign-Urbana (Ill.) News-Gazette,* June 16, 2003, p. A-5.

33. Avshalom Caspi et al., "Influence of Life Stress on Depression: Moderation by a Polymorphism in the 5-HTT Gene," *Science* 301 (2003): 386–389.

34. Plomin, Owen, and McGuffin, "Genetic Basis," p. 1734.

35. Robert Plomin, "The Role of Inheritance in Behavior," *Science* 248 (1990): 183–188.

36. Plomin, Owen, and McGuffin, "Genetic Basis," p. 1737.

37. Gary Taubes, "Averaged Brains Pinpoint a Site for Schizophrenia," *Science* 266 (1994): 221.

38. Constance Holden, "Sex and the Schizophrenic," *Science* 287 (2000): 2145.

39. Linda M. Brzystowicz et al., "Location of a Major Susceptibility Locus for Familial Schizophrenia on Chromosome 1 q21–22," *Science* 288 (2000): 678–682.

40. Betty Glad, "Personality, Role Strains, and Alcoholism: Key Pitman as Chairman of the Senate Foreign Relations Committee," *Politics and the Life Sciences* 7 (1988): 18–32.

41. *Powell v. Texas,* 392 U.S. 514 (1968), opinion of Justice Thurgood Marshall.

42. Plomin, Owen, and McGuffin, "Genetic Basis," p. 1734.

43. Dean Hamer and Peter Copeland, *Living with Our Genes* (New York: Anchor Books, 1998), pp. 141, 142–143.

44. Charles C. Mann, "Behavioral Genetics in Transition," *Science* 264 (1994): 1687.

45. Ernest P. Noble, "DRD2 Gene and Alcoholism," *Science* 281 (1998): 1287–1288; Howard J. Edenberg, "Genetics of Alcoholism," *Science* 282 (1998): 1269.

46. Constance Holden, "New Clues to Alcoholism Risk," *Science* 280 (1998): 1348.

47. Hamer and Copeland, *Living with Our Genes*, p. 243.

48. Jean L. Marx, "Obesity Gene Discovery May Help Solve Weighty Problem," *Science* 266 (1994): 1477–1478.

49. Anthony G. Commuzzie and David B. Allison, "The Search for Human Obesity Genes," *Science* 280 (1998): 1374–1377.

50. Jeffrey Friedman, "Genetics of Obesity" (paper presented at the American Association for the Advancement of Science Annual Meeting, Philadelphia, 1998).

51. Hamer and Copeland, *Living with Our Genes*, pp. 253–254.

52. I. Sadaf Farooqi et al., "Clinical Spectrum of Obesity and Mutations in the Melanocortin 4 Receptor Gene," *New England Journal of Medicine* 348 (2003): 1085–1095; Ruth Branson et al., "Binge Eating as a Major Phenotype of Melanocortin 4 Receptor Gene Mutations," *New England Journal of Medicine* 348 (2003): 1096–1103.

53. Douglas Madsen, "A Biochemical Property Relating to Power Seeking in Humans," *American Political Science Review* 79 (1985): 448–457.

54. Douglas Madsen, "Power Seekers Are Different: Further Biochemical Evidence," *American Political Science Review* 80 (1986): 261–269.

55. Richard E. Vatz and Lee S. Weinberg, "Biochemistry and Power-Seeking," *Politics and the Life Sciences* 10 (1991): 69–75.

56. Hamer and Copeland, *Living with Our Genes*, pp. 75–76.

57. With respect to depression, note again the New Zealand study by Caspi et al., "Influence of Life Stress," pp. 386–389, reported at note 33.

58. Hamer and Copeland, *Living with Our Genes*, pp. 66–67.

59. Klaus-Peter Lesch et al., "Association of Anxiety-Related Traits with a Polymorphism in the Serotonin Transporter Gene Regulatory System," *Science* 274 (1996): 1527–1531 at 1529.

60. Ahmad R. Hariri et al., "Serotonin Transporter Genetic Variation and the Response of the Human Amygdala," *Science* 297 (2002): 400–403.

61. Joseph LeDoux, *The Emotional Brain* (London: Weidenfeld and Nicolson, 1998), pp. 154–165.

62. David Goldman, "High Anxiety," *Science* 274 (1996): 1483; Constance Holden, "Serotonin and Bulimia," *Science* 293 (2001): 205.

63. Plomin, Owen, and McGuffin, "Genetic Basis," p. 1735.

64. Lesch et al., "Association of Anxiety-Related Traits," p. 1530.

65. Hamer and Copeland, *Living with Our Genes*, pp. 105–106; Ridley, *Genome*, pp. 149–150, 154–155.

66. Jean L. Marx, "How Stimulant Drugs May Calm Hyperactivity," *Science* 283 (1999): 306.

67. Clinton Rossiter, *Conservatism in America: The Thankless Persuasion*, 2nd ed. (New York: Vintage Books, 1962), pp. 16–17, 74, 168.

68. Auke Tellegen et al., "Personality Similarity in Twins Reared Apart and To-gether," *Journal of Personality and Social Psychology* 54 (1988): 1031–1039.

69. Hamer and Copeland, *Living with Our Genes,* pp. 30–46; Ridley, *Genome,* pp. 161–165. For recent research linking mutational recessives in the *DJ-1* gene to Parkinson's, see Vincenzo Bonifati et al., "Mutations in the *DJ-1* Gene Associated with Autosomal Recessive Early-Onset Parkinsonism," *Science* 299 (2003): 256–259.

70. See generally George Howe Colt, "Were You Born That Way?" *Life,* April 1998, pp. 38–50.

71. Constance Holden, "Happiness and DNA," *Science* 272 (1996): 1591, 1593.

72. James David Barber, *The Presidential Character* (Englewood Cliffs, N.J.: Prentice Hall, 1972).

73. D. F. Swaab and M. A. Hofman, "An Enlarged Suprachiasmatic Nucleus in Homosexual Men," *Brain Research* 537 (1990): 141; Simon LeVay, "A Difference in Hypothalamic Structure between Heterosexual and Homosexual Men," *Science* 253 (1991): 1034–1037.

74. Marcia Barinaga, "Is Homosexuality Biological?" *Science* 253 (1991): 956–957.

75. Constance Holden, "Twin Study Links Genes to Homosexuality," *Science* 255 (1992): 33; Constance Holden, "More on Genes and Homosexuality," *Science* 268 (1995): 1571.

76. Dean H. Hamer et al., "A Linkage between DNA Markers on the X Chromo-some and Male Sexual Orientation," *Science* 261 (1993): 321–327.

77. Holden, "More on Genes and Homosexuality;" Hamer and Copeland, *Living with Our Genes,* pp. 196–197.

78. Constance Holden, "A Marker for Female Homosexuality?" *Science* 279 (1998): 1639.

79. Hamer and Copeland, *Living with Our Genes,* pp. 188–192.

80. George Rice et al., "Male Homosexuality: Absence of Linkage to Microsatel-lite Markers at Xq28," *Science* 284 (1999): 665–667.

81. Hamer and Copeland, *Living with Our Genes,* p. 197.

82. Ibid., p. 199.

83. Lumsden and Wilson, *Promethean Fire,* p. 125.

84. Constance Holden, "Experts Slam Olympic Gene Test," *Science* 255 (1992): 1073.

85. Plomin, Owen, and McGuffin, "Genetic Basis," p. 1735. According to Segal, the IQ correlation between "virtual twins" (nonbiological siblings) raised together is 0.26, whereas the corresponding IQ correlation between identical twins is 0.86. Constance Holden, "Offbeat Twins," *Science* 288 (2000): 1735. When raised apart, the heritability quotient for identical twins drops to 0.76. Ridley, *Genome,* p. 83.

86. Gerald E. McClearn et al., "Substantial Genetic Influence on Cognitive Abil-ities in Twins 80 or More Years Old," *Science* 276 (1997): 1560–1563.

87. Plomin and Petrill, "Genetics and Intelligence."

88. Robert J. Sternberg, "The Holey Grail of General Intelligence," *Science* 289 (2000): 399, 401.

89. Jocelyn Kaiser, "Jigsaw Puzzle Gene," *Science* 273 (1996): 435.

90. Constance Holden, "The First Gene Marker for IQ?" *Science* 280 (1998): 681.

91. U.S. Constitution, Amendment 14.

92. Ridley, *Genome,* p. 85.

93. Richard J. Herrnstein and Charles Murray, *The Bell Curve* (New York: Free Press, 1994).

94. Throughout this chapter I have cited with approval several reports purporting to show the powerful pull of heritability between identical twins, meaning, of course, the proportion of phenotypic variance attributable to genetic effects. These findings have been challenged for many years on the ground that they fail to test adequately for environmental correlations between subjects. For example, see M. W. Feldman and S. P. Otto, "Twin Studies, Heritability, and Intelligence," *Science* 278 (1997): 1383–1384. To this literature, I reinquire, as have many others: Why are identical twins raised apart so similar in their heritability quotients, and why are "virtual twins" raised together so dissimilar in everything?

95. Richard D. Arvey et al., "Mainstream Science on Intelligence," *Wall Street Journal,* December 13, 1994, p. A18.

96. J. Philippe Rushton, *Race, Evolution, and Behavior,* special abridged ed. (New Brunswick, N.J.: Transaction, 1999), chs. 4 and 5. The quote is on p. 65.

97. J. Philippe Rushton, "*The Bell Curve:* Premises, Conclusions, and Policies" (roundtable discussion convened at the American Political Science Association Annual Meeting, Chicago, 1995).

98. Jerry Hirsch, review of *Why Race Matters: Race Differences and What They Mean,* by Michael E. Levin, *Politics and the Life Sciences* 18 (1) (1999): 159–161; Jon Beckwith, quoted in Erwin Fleissner, "Race and the Human Genome," *Hastings Center Report* 29 (4) (1999): 40–42 at 41.

99. Rushton, *Race, Evolution, and Behavior,* p. 71.

100. Natalie Angier, "Do Races Differ? Not Really, Genes Show," *New York Times,* August 22, 2000, pp. D1, D6.

101. David Berreby, "How, but Not Why, the Brain Distinguishes Race," *New York Times,* September 5, 2000, p. F3.

102. Mary-Claire King and Arno G. Motulsky, "Mapping Human History," *Science* 298 (2002): 2342–2343.

103. Pamela Sankar and Mildred K. Cho, "Toward a New Vocabulary of Human Genetic Variation," *Science* 298 (2002): 1337–1338.

104. Sharon Schmickle, "Race Controversy Arises in World of Genome Study," *Champaign-Urbana (Ill.) News-Gazette,* March 29, 2002, p. D-4.

105. Eliot Marshall, "DNA Studies Challenge the Meaning of Race," *Science* 282 (1998): 654–655.

106. Constance Holden, "Wiley Drops IQ Book after Public Furor," *Science* 272 (1996): 644; Constance Holden, "Wiley Declines to Publish Jensen Book," *Science* 273 (1996): 877.

107. Andrews and Nelkin, "Bell Curve," 13–14. Cf. the discussion in ch. 2, note 128. For other instances of Nelkin-Andrews hostility to behavioral genetics, see Ira

H. Carmen, review of *The Clash of Culture and Biology,* ed. Ronald A. Carson and Mark A. Rothstein, *Politics and the Life Sciences* 19 (1) (2000): 124–126.

108. Nicholas Wade, "For Sale: A DNA Test to Measure Racial Mix," *New York Times,* October 1, 2002, p. F4.

109. Nicholas Wade, "Race Is Seen as Real Guide to Track Roots of Disease," *New York Times,* July 30, 2002, pp. D1–D2.

110. James Q. Wilson and Richard J. Herrnstein, *Crime and Human Nature* (New York: Simon and Schuster, 1985).

111. R. C. Lewontin, Steve Rose, and Leon J. Kamin, *Not in Our Genes* (New York: Pantheon Books, 1984).

112. Quoted in Richard Stone, "HHS 'Violence Initiative' Caught in a Crossfire," *Science* 258 (1992): 212–213.

113. Quoted in Joseph Palca, "NIH Wrestles with Furor over Conference," *Science* 257 (1992): 739.

114. Eliot Marshall, "NIH Told to Reconsider Crime Meeting," *Science* 262 (1993): 23–24.

115. Wade Roush, "Conflict Marks Crime Conference," *Science* 269 (1995): 1808–1809.

116. Stone, "HHS 'Violence Initiative.'"

117. *NEA v. Finley,* 524 U.S. 569 (1998).

118. The statute was declared invalid on other grounds by the Supreme Court in *Edwards v. Aguillard,* 482 U.S. 578 (1987).

119. Colt, "Were You Born That Way?"

120. Martin Enserink, "Searching for the Mark of Cain," *Science* 289 (2000): 575–579.

121. Barbara Culliton, "XYY: Harvard Researcher under Fire Stops Newborn Screening," *Science* 188 (1975): 1284–1285.

122. H. G. Brunner et al., "Abnormal Behavior Associated with a Point Mutation in the Structural Gene for Monoamine Oxidase A," *Science* 262 (1993): 578–580; Michael Simm, "Violence Study Hits a Nerve in Germany," *Science* 264 (1994): 653.

123. Erik Stokstad, "Violent Effects of Abuse Tied to Gene," *Science* 297 (2000): 752.

124. J. Q. Wilson, "Moral Sense," 1–11. The quote is the first sentence of the article.

125. For his useful comments on sociality, I acknowledge Gary R. Johnson, "The Evolutionary Origins of Government and Politics," in *Human Nature and Politics,* ed. Albert Somit and Joseph Losco (Greenwich, Conn.: JAI Press, 1995), pp. 243–305, especially pp. 251–257.

126. J. Q. Wilson, "Moral Sense," pp. 1, 7.

127. Nicholas Wade, "First Gene for Social Behavior Identified in Whiskery Mice," *New York Times,* September 9, 1997, p. C4.

128. Larry J. Young et al., "Increased Affiliative Response to Vasopressin in Mice Expressing the V1a Receptor from a Monogamous Vole," *Nature* 400 (1999): 766–768.

129. Michael J. B. Krieger and Kenneth G. Ross, "Identification of a Major Gene Regulating Complex Social Behavior," *Science* 295 (2002): 328–332.

130. Veronica P. Volny and Deborah M. Gordon, "Genetic Basis for Queen-Worker Dimorphism in a Social Insect," *Proceedings of the National Academy of Sciences* 99 (2002): 6108–6111.

131. Y. Ben-Shahar et al., "Influence of Gene Action across Different Time Scales on Behavior," *Science* 296 (2002): 741–744; Natalie Angier, "Honeybee Shows a Little Gene Activity Goes Miles and Miles," *New York Times,* May 7, 2002, p. D3.

132. Maria B. Sokolowski, "Social Eating for Stress," *Nature* 419 (2002): 893–894.

133. Bernard Crespi and Stevan Springer, "Social Slime Molds Meet Their Match," *Science* 299 (2003): 56–57.

134. Rick L. Riolo, Michael D. Cohen, and Robert Axelrod, "Evolution of Cooperation without Reciprocity," *Nature* 414 (2001): 441–443. The coauthors of this study are political scientists and complex systems analysts, a sort of budding consilient collaboration. Cf. other aspects of the Axelrod research agenda cited in ch. 6.

135. Gretchen Vogel, "Mutations Reveal Genes in Zebrafish," *Science* 296 (2002): 1221.

136. Nicholas Wade, "Fish Genes Aid Human Discoveries," *New York Times,* July 26, 2002, p. A19.

137. Gene E. Robinson, "Sociogenomics Takes Flight," *Science* 297 (2002): 204–205. For the gene-linked nepotistic enhancements between workers and queens in the care and feeding of larvae, see Henry Fountain, "All the Queen's Relatives," *New York Times,* March 4, 2003, p. D3.

138. Neil G. Copeland et al., "Mmu 16—Comparative Genomic Highlights," *Science* 296 (2002): 1617–1618.

139. Gretchen Vogel, "Jumbled DNA Separates Chimps and Humans," *Science* 296 (2002): 719, 721.

140. Gene E. Robinson, "From Society to Genes with the Honey Bee," *American Scientist* 86 (1998): 456–462 at 457.

141. Mitch Leslie, "Tagging Honeybee Genes," *Science* 295 (2002): 1607.

142. Eric. S. Lander, "Array of Hope," *Nature Genetics Supplement* 21 (1999): 3–4.

143. This paragraph is based on Zweiger, *Transducing the Genome,* pp. 180–181.

144. Eliot Marshall, "Snipping Away at Genome Patenting," *Science* 277 (1997): 1752–1753.

145. Eliot Marshall, "'Playing Chicken' over Gene Markers," *Science* 278 (1997): 2046–2048.

146. Zweiger, *Transducing the Genome,* p. 178.

147. Eliot Marshall, "Drug Firms to Create Public Database of Genetic Mutations," *Science* 284 (1999): 406–407.

148. David G. Wang et al., "Large-Scale Identification, Mapping, and Genotyping of Single-Nucleotide Polymorphisms in the Human Genome," *Science* 280 (1998): 1077–1082.

149. Francis S. Collins, "Microarrays and Macroconsequences," *Nature Genetics Supplement* 21 (1999): 2.

150. Laura Helmuth, "Map of the Human Genome 3.0," *Science* 293 (2001): 583, 585.

151. Trisha Gura, "Can SNPs Deliver on Susceptibility Genes?" *Science* 293 (2001): 593–595.

152. Nila Patel et al., "Blocks of Limited Haploytype Diversity Revealed by High-Resolution Scanning of Human Chromosome 21," *Science* 294 (2001): 1719–1723; Pui-Yan Kwok, "Genetic Association by Whole-Genome Analysis?" *Science* 294 (2001): 1669–1670.

153. Stacey B. Gabriel et al., "The Structure of Haplotype Blocks in the Human Genome," *Science* 296 (2002): 2225–2229.

154. Jennifer Couzin, "HapMap Launched with Pledges of $100 Million," *Science* 298 (2002): 941–942.

155. Peter McGuffin, Brien Riley, and Robert Plomin, "Toward Behavioral Genomics," *Science* 291 (2001): 1232, 1249.

156. Dean Hamer, "Rethinking Behavior Genetics," *Science* 298 (2002): 71–72.

157. Elizabeth Pennisi, "Insects Rank Low among Genome Priorities," *Science* 294 (2001): 1261–1262.

158. Elizabeth Pennisi, "Chimps and Fungi Make Genome 'Top Six,'" *Science* 296 (2002): 1589, 1591.

159. John B. Spalding and R. L. Bernstein, "Genome Comparisons," *Science* 297 (2002): 1037.

6. Consilience

1. Charles E. Merriam, *New Aspects of Politics* (Chicago: University of Chicago Press, 1925, 3rd ed. 1970, p. 171. See also Glendon Schubert, "Psychobiological Politics," *Canadian Journal of Political Science* 16 (1983): 535–576.

2. Darwin, *Origin of Species*.

3. Stephen Jay Gould, "Darwinism and the Expansion of Evolutionary Theory," *Science* 216 (1982): 380–387.

4. N. A. Chagnon and W. Irons, ed., *Evolutionary Biology and Human Social Behavior* (North Scituate, Mass.: Duxbury Press, 1979), p. 251; emphasis in text.

5. G. C. Williams, *Adaptation and Natural Selection*.

6. W. D. Hamilton, "The Genetical Evolution of Social Behavior," parts 1 and 2, *Journal of Theoretical Biology* 7 (1964): 1–32.

7. R. L. Trivers, "The Evolution of Reciprocal Altruism," *Quarterly Review of Biology* 46 (1971): 35–57.

8. E. O. Wilson, *Sociobiology*, p. 4.

9. Ibid., p. 16.

10. Ibid., p. 3.

11. Ibid. This is the title of his ch. 1.

12. R. Dawkins, *The Selfish Gene* (New York: Oxford University Press, 1978).

13. E. O. Wilson, *Sociobiology*, p. 5.

14. Ibid., p. 180.

15. Ibid., p. 271.

16. Ibid., p. 272.

17. Ibid., p. 287; emphasis in text. Steven Pinker has fleshed out in greater detail than Wilson all the misperceptions and misconceptions flowing from the "brain as tabula rasa" presupposition. *The Blank Slate* (New York: Viking, 2002).

18. E. O. Wilson, *Sociobiology,* pp. 299–300.

19. For Gould's left-wing politics/science, see Caroline L. Herzenberg, "On Stephen Jay Gould," *Science* 297 (2002): 1120.

20. Nicholas Wade, "Sociobiology: Troubled Birth for New Discipline," *Science* 191 (1976): 1151–1155; Beckwith and fourteen cosigners, "Against Sociobiology."

21. E. O. Wilson, *Naturalist,* pp. 347–350.

22. Ibid., p. 332.

23. For a balanced review of the "sociobiology wars" generally in sync with my own thinking, see Ullica Segerstrale, *Defenders of the Truth* (New York: Oxford University Press, 2000). For an extended—if overdrawn—critique of sociobiology by a political scientist, see Glendon Schubert, "The Sociobiology of Political Behavior," in *Sociobiology and Human Politics,* ed. Elliott White (Lexington, Mass.: Heath, 1981), pp. 193–238.

24. Charles J. Lumsden and Edward O. Wilson, *Genes, Mind, and Culture* (Cambridge: Harvard University Press, 1981).

25. Ibid., p. 2.

26. Ibid., p. 35.

27. Ibid., p. 7.

28. Ibid., p. 27.

29. Ibid., pp. 245–246.

30. Ibid., pp. xi, 249–251.

31. Ibid., p. 253.

32. Lumsden and Wilson, *Promethean Fire,* pp. 175–180.

33. Ibid., p. 152.

34. Noam Chomsky, *Rules and Representations* (New York: Columbia University Press, 1980). The theory is fine-tuned in Marc D. Hauser et al., "The Faculty of Language: What Is It, Who Has It, and How Did It Evolve," *Science* 298 (2002): 1569–1579.

35. Martin A. Nowak et al., "Evolution of Universal Grammar," *Science* 291 (2001): 114–118. This report constructs a mathematical model to demonstrate how a universal grammar could (and probably did) arise by natural selection.

36. Michael Balter, "First Gene Linked to Speech Identified," *Science* 294 (2001): 32.

37. Michael Balter, "'Speech Gene' Tied to Modern Humans," *Science* 297 (2002): 1105.

38. Elizabeth Pennisi, "Gene Activity Clocks Brain's Fast Evolution," *Science* 296 (2002): 233, 235; Wolfgang Enard et al., "Intra- and Interspecific Variation in Primate Gene Expression Patterns," *Science* 296 (2002): 340–343.

39. Eliot Marshall, "Probing Primate Morality," *Science* 271 (1996): 904; Frans B. M. de Waal, "Primates—A Natural Heritage of Conflict Resolution," *Science* 289

(2000): 586–590; Gretchen Vogel, "Orangutans, Like Chimps, Heed the Cultural Call of the Collective," *Science* 299 (2003): 27–28.

40. "Open Peer Commentary," *Behavioral and Brain Sciences* 5 (1982): 1–37.

41. E. O. Wilson, *Consilience*. In my various references to *Consilience* cited below, I shall refer to the page numberings as they appear in the paperback ed. (New York: Vintage Books, 1999).

42. Ibid., pp. 4–5.

43. Ibid., pp. 8–9, quoting Whewell.

44. Ibid., p. 29.

45. Ibid., pp. 11, 58–60, 74.

46. Ibid., pp. 88–89.

47. Ibid., p. 93.

48. Ibid., p. 101, quoting Loomis and Sternberg.

49. Ibid., p. 138.

50. Ibid., p. 139, emphasis in text.

51. Ibid., pp. 146, 148, 171.

52. Ibid., pp. 180, 182, 183–188, 196.

53. Ibid., pp. 197–199.

54. Ibid., pp. 200–205 and the general commentary in ch. 9.

55. Ibid., pp. 220, 224–226.

56. Ibid., pp. 260–261, 263, 269, 275–276.

57. Ibid., pp. 278, 286.

58. Michael Ruse, "On E. O. Wilson and His Religious Vision," *Science* 290 (2000): 943.

59. John Dupré, "Unification Not Proved," *Science* 280 (1998): 1395.

60. Jerry Hirsch, "The Pitfalls of Heritability," *Times Literary Supplement,* February 12, 1999, p. 33.

61. Leah Ceccarelli, *Shaping Science with Rhetoric* (Chicago: University of Chicago Press, 2001).

62. Richard Rorty, "Studied Ambiguity," *Science* 293 (2001): 2399–2400.

63. Lisa Chang and L. Bryan Ray, "Whole-istic Biology," *Science* 295 (2002): 1661, quoting with approval Ludwig von Bertalanffy, *General System Theory, Foundations, Development, Applications* (New York: Braziller, 1969).

64. Michael Ruse, "The Faith of an Evolutionist," *Politics and the Life Sciences* 18 (1999): 347–349; Edward O. Wilson, "Responding to the Reviews of Elshtain, Kaye, and Ruse," *Politics and the Life Sciences* 18 (1999): 350–351.

65. Howard L. Kaye, "Consilience: E. O. Wilson's Confession of Faith," *Politics and the Life Sciences* 18 (1999): 344–346; E. O. Wilson, "Responding," p. 350.

66. Jean Bethke Elshtain, "Consilience or Creation?" *Politics and the Life Sciences* 18 (1999): 341–344.

67. Wendell Berry, *Life Is a Miracle* (Washington, D.C.: Counterpoint, 2000).

68. Ruse, "On E. O. Wilson"; Ruse, "Faith of an Evolutionist," p. 349.

69. E. O. Wilson, *Consilience*, pp. 6–7.

70. Ibid., p. 263.

71. Ibid., pp. 260–261.

72. Wilson says: "[M]uch if not all religious behavior could have arisen from evolution by natural selection." *Consilience,* p. 282. And see also David Sloan Wilson, *Darwin's Cathedral* (Chicago: University of Chicago Press, 2002). I fail to see how any of these likelihoods demonstrates the absence of a deity or deities.

73. E. O. Wilson, *Consilience,* p. 160.

74. Ibid., p. 299.

75. E. O. Wilson, "Responding," p. 351.

76. R. E. Spier, "Toward a New Human Species?" *Science* 296 (2002): 1807, 1809.

77. Elshtain, "Consilience or Creation?" p. 343.

78. E. O. Wilson, "Responding," p. 350.

79. E. O. Wilson, *Consilience,* p. 303.

80. Morley Roberts, *Bio-Politics: An Essay in the Physiology, Pathology, and Politics of the Social and Somatic Organism* (London: Dent, 1938). My account of the history and contributions of the biopolitics subfield is based on Ira H. Carmen, "Biopolitics: The Newest Synthesis?" *Genetica* 99 (1997): 173–184.

81. David Easton, *The Political System* (New York: Knopf, 1960). Curiously, the word "biology" does not appear in Easton's index. It was too early!

82. Albert Somit, "Toward a More Biologically-Oriented Political Science: Ethology and Pharmacology," *Midwest Journal of Political Science* 12 (1968): 550–567.

83. Roger D. Masters and Denis G. Sullivan, "Nonverbal Displays and Political Leadership in France and the United States," *Political Behavior* 11 (1989): 123–156, at 126, 128.

84. Herbert A. Simon, *Models of Thought* (New Haven: Yale University Press, 1979).

85. LeDoux, *Emotional Brain,* pp. 43–44, 308.

86. Paul D. MacLean, "A Triangular Brief on the Evolution of Brain and Law," *Journal of Social and Biological Structures* 5 (1982): 369–379.

87. LeDoux, *Emotional Brain,* pp. 289–292.

88. Carl E. Schwartz et al., "Inhibited and Uninhibited Infants 'Grown Up': Adult Amygdalar Response to Novelty," *Science* 300 (2003): 1952–1953; Reginald B. Adams Jr. et al., "Effects of Gaze on Amygdala Sensitivity to Anger and Fear Faces," *Science* 300 (2003): 1536.

89. Graham Wallas, *Human Nature in Politics* (Gloucester, Mass.: Smith, 1908).

90. Herbert A. Simon, "Human Nature in Politics: The Dialogue of Psychology with Political Science," *American Political Science Review* 79 (1985): 293–304 at 295.

91. Ibid., pp. 294, 301.

92. Elinor Ostrom, *Governing the Commons* (New York: Cambridge University Press, 1990).

93. Herbert A. Simon, "A Mechanism for Social Selection and Successful Altruism," *Science* 250 (1990): 1665–1668.

94. J. Q. Wilson, "Moral Sense," p. 1.

95. Ibid., p. 4.

96. Ibid., pp. 5, 8, 9.

97. Ibid., p. 7.

98. Gregory E. Kaebnick, "The Nature of the Problem," *Hastings Center Report* 32 (6) (2002): 40–42.

99. Francis Fukuyama, *Our Posthuman Future* (New York: Farrar, Straus and Giroux, 2002), pp. 7, 9, 13, 14.

100. Ibid., pp. 10, 15; emphasis in text.

101. Ibid., p. 130.

102. Ibid., p. 82.

103. Ibid., p. 10.

104. Albert Somit and Steven A. Peterson, "Rational Choice and Biopolitics: A (Darwinian) Tale of Two Theories," *PS* 32 (1999): 39 –44.

105. Nicholas Wade, "A Dim View of a 'Posthuman Future,'" *New York Times*, April 2, 2002, pp. D1, D4.

106. Karl W. Deutsch, "Mechanism, Organism, and Society: Some Models in Natural and Social Science," *Philosophy of Science* 18 (1951): 230–252 at 230.

107. E. O. Wilson, *Naturalist*, pp. 315 –320 at p. 317.

108. Robert Axelrod and William D. Hamilton, "The Evolution of Cooperation," *Science* 211 (1981): 1390–1396 at 1392.

109. Robert Axelrod and Douglas Dion, "The Further Evolution of Cooperation," *Science* 242 (1988): 1385 –1390.

110. Robert Axelrod, "The Evolution of Strategies in the Iterated Prisoner's Dilemma," in *Genetic Algorithms and Simulated Annealing*, ed. Lawrence Davis (London: Pitman, 1987), pp. 32–41. For a general discussion of this research approach, see Stephanie Forrest, "Genetic Algorithms: Principles of Natural Selection Applied to Computation," *Science* 261 (1993): 872–878.

111. Axelrod, "Evolution of Strategies," p. 41.

112. Robert Axelrod, "An Evolutionary Approach to Norms," *American Political Science Review* 80 (1986): 1095 –1111.

113. Robert Pool, "Putting Game Theory to the Test," *Science* 267 (1995): 1591–1593.

114. Axelrod, "Evolutionary Approach," p. 1104.

115. Robert Pool, "Economics: Game Theory's Winning Hands," *Science* 266 (1994): 371.

116. For a further encomium, see ch. 5 in text at note 134.

117. Martin A. Nowak and Karl Sigmund, "Tit for Tat in Heterogeneous Populations," *Nature* 355 (1992): 250–252.

118. Claus Wedekind, "Give and Ye Shall Be Recognized," *Science* 280 (1998): 2070–2071.

119. Alan G. Sanfey et al., "The Neural Basis of Economic Decision-Making in the Ultimate Game," *Science* 300 (2003): 1755 –1758.

120. Martin A. Nowak and Karl Sigmund, "Shrewd Investments," *Science* 288 (2000): 819 –820.

121. Leda Cosmides, "The Logic of Social Exchange: Has Natural Selection Shaped How Humans Reason? Studies with the Wason Selection Task," *Cognition* 31 (1989): 187 –276.

122. Henry Fountain, "Bees, Come on Down!" *New York Times,* July 29, 2003, p. D4.

123. Barry J. Richmond et al., "Predicting Future Rewards," *Science* 301 (2003): 179–180.

124. George Steiner, "Fields of Force," *New Yorker,* October 28, 1972, pp. 42–117.

125. Carmen, "Chess Algorithms," pp. 99–121.

126. E. O. Wilson, *Consilience,* p. 8.

127. Glendon Schubert, "The Evolution of Political Science: Paradigms of Physics, Biology, and Politics," *Politics and the Life Sciences* 1 (1982): 97–123 at 100.

128. John C. Wahlke, "Pre-Behavioralism in Political Science," *American Political Science Review* 73 (1978): 9–31 at 24, 27.

129. Ada W. Finifter, ed., *Political Science: The State of the Discipline* (Washington, D.C.: American Political Science Association, 1983).

130. I am indebted to Claude S. Phillips for providing detailed chronologies of the rocky relationship between political science and biopolitics. See "Consilience and the Social Sciences" (paper presented at the annual meeting of the Association for Politics and the Life Sciences, Atlanta, September 1999), and "Political Science Reaction to Sociobiology, 1975–Present" (paper presented at the annual meeting of the Association for Politics and the Life Sciences, Washington, D.C., September 2000).

131. Quoted in Robert J.-P. Hauck, "From the Editor," *PS* 24 (1991): 4.

132. Ada W. Finifter, ed., *Political Science: The State of the Discipline II* (Washington, D.C.: American Political Science Association, 1993).

133. Robert A. Dahl, "From Genes to Political Institutions" (paper presented on the occasion of the Fred Greenstein festschrift, Princeton University, April 24, 2001).

134. Ibid., p. 4; emphasis in text.

135. Catherine Serrin, "At the Frontier," *Michigan Alumnus,* Spring 2000, pp. 26–31.

136. E. O. Wilson, *Consilience,* p. 295.

137. Ibid., p. 13.

138. Cf. Robin Dunbar et al., ed., *The Evolution of Culture* (New Brunswick: Rutgers University Press, 1999).

139. Stephen R. Palumbi, "Humans as the World's Greatest Evolutionary Force," *Science* 293 (2001): 1786–1790.

140. George R. vB. Ennenga, "Artificial Evolution," *Artificial Life* 3 (1997): 51–61.

141. Edward O. Wilson, *The Future of Life* (New York: Knopf, 2002).

142. E. O. Wilson, *Consilience,* p. 168.

143. Somit and Tanenhaus, *Development of Political Science,* p. 78.

144. Heinz Eulau and Susan Zlomke, "Harold D. Lasswell's Legacy to Mainstream Political Science," *Annual Review of Political Science* 2 (1999): 75–89.

145. Larry Arnhart, *Darwinian Natural Right* (Albany: State University of New York Press, 1998), pp. 31–36.

Bibliography

Adams, Mark D., et al. "Complementary DNA Sequencing: Expressed Sequence Tags and Human Genome Project." *Science* 252 (1991): 1651–1656.

Adams, Reginald B., Jr., et al. "Effects of Gaze on Amygdala Sensitivity to Anger and Fear Faces." *Science* 300 (2003): 1536.

Aiuti, Allesandro. "Correction of ADA-SCID by Stem Cell Gene Therapy Combined with Nonmyeloablative Conditioning." *Science* 296 (2002): 2410–2413.

Aldhous, Peter. "The Promise and Pitfalls of Molecular Genetics." *Science* 257 (1992): 164–165.

Anderson, Christopher. "Genome Project Goes Commercial." *Science* 259 (1993): 300–302.

Anderson, W. French. "The Best of Times, The Worst of Times." *Science* 288 (2000): 627, 629.

———. "The First Signs of Danger." *Human Gene Therapy* 3 (1992): 359.

———. "Human Gene Therapy." *Science* 256 (1992): 812.

Andrews, Lori B. *The Clone Age.* New York: Henry Holt, 1999.

Andrews, Lori B., and Dorothy Nelkin. "The Bell Curve: A Statement." *Science* 271 (1996): 13–14.

Angier, Natalie. "Do Races Differ? Not Really, Genes Show." *New York Times*, August 22, 2000, pp. D1, D6.

———. "Gene Implant Therapy Is Backed for Children with Rare Disease." *New York Times*, March 8, 1990, pp. 1, B9.

———. "Honeybee Shows a Little Gene Activity Goes Miles and Miles." *New York Times*, May 7, 2002, p. D3.

Areen, Judith, and Patricia King. "Legal Regulation of Human Gene Therapy." *Human Gene Therapy* 1 (1990): 151–161.

Arnhart, Larry. *Darwinian Natural Right.* Albany: State University of New York Press, 1998.

Arvey, Richard D., et al. "Mainstream Science on Intelligence." *Wall Street Journal*, December 13, 1994, p. A18.

Attanasio, John B. "The Constitutionality of Regulating Human Genetic Engineering: Where Procreative Liberty and Equal Opportunity Collide." *University of Chicago Law Review* 53 (1986): 1274–1342.

Axelrod, Robert. "An Evolutionary Approach to Norms." *American Political Science Review* 80 (1986): 1095–1111.

———. "The Evolution of Strategies in the Iterated Prisoner's Dilemma." In *Genetic Algorithms and Simulated Annealing,* ed. Lawrence Davis. London: Pitman, 1987.

Axelrod, Robert, and Douglas Dion. "The Further Evolution of Cooperation." *Science* 242 (1988): 1385–1390.

Axelrod, Robert, and William D. Hamilton. "The Evolution of Cooperation." *Science* 211 (1981): 1390–1396.

Avery, O. T., MacLeod, C. M., and M. McCarty. "Studies on the Chemical Nature of the Substance Inducing Transformation of Pneumococcal Types: Induction of Transformation by a Deoxyribonucleic Acid Fraction Isolated from Pneumococcus Type III." *Journal of Experimental Medicine* 79 (1944): 137–158.

Baker, Lynne R. *Persons and Bodies: A Constitutional View.* Cambridge: Cambridge University Press, 2000.

Baker, Michael. "Korean Report Sparks Anger and Inquiry." *Science* 283 (1999): 16–17.

———. "Report Casts Doubt on Korean Experiment." *Science* 283 (1999): 617, 619.

Balkin, Jack M. "The Cloning Conundrum." *New York Times,* January 30, 2002, p. A27.

Balter, Michael. "First Gene Linked to Speech Identified." *Science* 294 (2001): 32.

———. "Generous Funding Wins a Seat at the Genome Top Table." *Science* 274 (1996): 1293.

———. "'Speech Gene' Tied to Modern Humans." *Science* 297 (2002): 1105.

Barber, James David. *The Presidential Character.* Englewood Cliffs, N.J.: Prentice Hall, 1972.

Barinaga, Marcia. "Is Homosexuality Biological?" *Science* 253 (1991): 956–957.

Beckwith, Jon. "Foreword: The Human Genome Initiative: Genetics' Lightning Rod." *American Journal of Law and Medicine* 17 (1–2) (1991): 1–13.

———. *Making Genes, Making Waves.* Cambridge: Harvard University Press, 2002.

Beckwith, Jonathan, and fourteen cosigners. "Against Sociobiology." *New York Review of Books,* November 13, 1975.

Ben-Shahar, Y., et al. "Influence of Gene Action across Different Time Scales on Behavior." *Science* 296 (2002): 741–744.

Bentley, Arthur. *The Process of Government.* Chicago: University of Chicago Press, 1908.

Berreby, David. "How, but Not Why, the Brain Distinguishes Race," *New York Times,* September 5, 2000, p. F3.

Berry, Wendell. *Life Is a Miracle.* Washington, D.C.: Counterpoint, 2000.

Bertram, Lars, et al. "Evidence for Genetic Linkage of Alzheimer's Disease to Chromosome 10q." *Science* 290 (2000): 2302–2303.

Biggin, Susan. "Italian Committee Loses Its Balance." *Science* 267 (1995): 326.

Bishop, Jerry E., and Michael Waldholz. *Genome.* New York: Simon and Schuster, 1990.

Bonifati, Vincenzo, et al. "Mutations in the *DJ-1* Gene Associated with Autosomal Recessive Early-Onset Parkinsonism." *Science* 299 (2003): 256–259.

Bonnicksen, Andrea L. *Crafting a Cloning Policy.* Washington D.C.: Georgetown University Press, 2002.

———. "Human Embryos and Genetic Testing: A Private Policy Model." *Politics and the Life Sciences* 11 (1) (1992): 53–62.

———. *In Vitro Fertilization.* New York: Columbia University Press, 1989.

———. "National and International Approaches to Human Germ-line Gene Therapy." *Politics and the Life Sciences* 13 (1) (1994): 39–49.

Bradish, Nancy. "Cloning Forces Congress to Do a Double-Take." *Bio News,* February/March, 1998, pp. 1, 3.

Branson, Ruth, et al. "Binge Eating as a Major Phenotype of Melanocortin 4 Receptor Gene Mutations." *New England Journal of Medicine* 348 (2003): 1096–1103.

Brunner, H. G., et al. "Abnormal Behavior Associated with a Point Mutation in the Structural Gene for Monoamine Oxidase A." *Science* 262 (1993): 578–580.

Brzystowicz, Linda M., et al. "Location of a Major Susceptibility Locus for Familial Schizophrenia on Chromosome 1q21–q22." *Science* 288 (2000): 678–682.

Buck v. Bell, 274 U.S. 200 (1927).

Bult, Carol J., et al. "Complete Genome Sequence of the Methanogenic Archaeon, Methanococcus jannaschii." *Science* 273 (1996): 1058–1073.

Bush, George W. "Address to the Nation." Reprinted in *The Future Is Now,* ed. William Kristol and Eric Cohen. Lanham, Md.: Rowman and Littlefield, 2002.

Byrand, S. L. "Evolutionary Wisdom or Genetic Roulette?" *Science* 278 (1997): 882.

Calabresi, Guido, and Philip Bobbitt. *Tragic Choices.* New York: Norton, 1978.

Cantor, Charles R. "Orchestrating the Human Genome Project." *Science* 248 (1990): 49–51.

Caplan, Arthur L. "His Genes, Our Genome." *New York Times,* May 3, 2002, p. A27.

Caplan, Arthur, and J. Craig Venter. "Using One's Head." *Science* 278 (1997): 1547–1548.

Carmen, Ira H. "Bioconstitutional Politics: Toward an Interdisciplinary Paradigm." *Politics and the Life Sciences* 5 (1987): 193–207.

———. "Bioethics, Public Policy, and Political Science." *Politics and the Life Sciences* 13 (1994): 79–81.

———. "Biopolitics: The Newest Synthesis?" *Genetica* 99 (1997): 173–184.

———. "Chess Algorithms of Supreme Court Decision Making: A Bioconstitutional Politics Analysis." *Political Behavior* 11 (1989): 99–121.

———. *Cloning and the Constitution.* Madison: University of Wisconsin Press, 1986.

———. "A Death in the Laboratory: The Politics of the Gelsinger Aftermath." *Molecular Therapy* 3 (2001): 425–428.

———. "Debates, Divisions, and Decisions: Recombinant DNA Advisory Committee (RAC) Authorization of the First Human Gene Transfer Experiments." *American Journal of Human Genetics* 50 (1992): 245–260.

———. "God and Man in and around the White House." Unpublished manuscript (2000).

———. "Human Gene Therapy: A Biopolitical Overview and Analysis." *Human Gene Therapy* 4 (1993): 187–193.

———. *Movies, Censorship, and the Law.* Ann Arbor: University of Michigan Press, 1966.

———. "Ownership of the Human Genome." *Science* 237 (1987): 1555.

———. *Power and Balance.* New York: Harcourt Brace Jovanovich, 1978.

———. Review of *The Clash of Culture and Biology,* ed. Ronald A. Carson and Mark A. Rothstein. *Politics and the Life Sciences* 19 (1) (2000): 124–126.

———. "Should Human Cloning Be Criminalized?" *Journal of Law and Politics* 13 (1997): 745–758.

———. "Washington Politics and Genetic Engineering Research: When Worlds Collide." *Human Gene Therapy* 7 (1996): 97–108.

Casey, Denise. "Human Genome Working Draft: First-Edition Travel Guides." *Human Genome News* 11 (3–4) (2001): 3–4.

Caspi, Avshalom, et al. "Influence of Life Stress on Depression: Moderation by a Polymorphism in the 5-HTT Gene." *Science* 301 (2003): 386–389.

Cavazzana-Calvo, Marina, et al. "Gene Therapy of Human Severe Combined Immunodeficiency (SCID)-XI Disease." *Science* 288 (2000): 669–672.

Ceccarelli, Leah. *Shaping Science with Rhetoric.* Chicago: University of Chicago Press, 2001.

Center for Biotechnology Policy and Ethics. "Baby Boom: Human Embryo Cloning Hits the Mass Media." Newsletter 3 (4) January 1, 1994.

Chadwick, Ruth. "Germ-line Gene Therapy, Autonomy, and Community." *Politics and the Life Sciences* 13 (2) (1994): 223–225.

Chagnon, N. A., and W. Irons, ed. *Evolutionary Biology and Human Social Behavior.* North Scituate, Mass.: Duxbury Press, 1979.

Chan, A. W. S., et al. "Clonal Propagation of Primate Offspring by Embryo Splitting." *Science* 287 (2000): 317–319.

Chang, Kenneth. "Company Says It Used Cloning to Create New Kidneys for Cow." *New York Times,* January 31, 2002, p. A22.

Chang, Lisa, and L. Bryan Ray. "Whole-istic Biology." *Science* 295 (2002): 1661.

Cho, Mildred K., et al. "Ethical Considerations in Synthesizing a Minimal Genome." *Science* 286 (1999): 2087–2090.

Chomsky, Noam. *Rules and Representations.* New York: Columbia University Press, 1980.

Clinton, Bill. Letter to Harold Shapiro, chair, National Bioethics Advisory Commission, February 24, 1997. Reprinted in "Cloning Human Beings." Report and Recommendations of the NBAC, June 1997.

———. "Science in the 21st Century." *Science* 276 (1997): 1951.

"Cloned Pigs Lack Gene That Hinders Transplants." *Champaign-Urbana (Ill.) News-Gazette,* August 23, 2002, p. A-1.

Collins, Francis S. "Microarrays and Macroconsequences." *Nature Genetics Supplement* 21 (1999): 2.

———. "Positional Cloning: Let's Not Call It Reverse Anymore." *Nature Genetics* 1 (1992): 3–6.

Collins, Francis S., et al. "New Goals for the U.S. Human Genome Project: 1998–2003." *Science* 282 (1998): 682–689.

Colt, George Howe. "Were You Born That Way?" *Life,* April 1998, pp. 38–50.

Commuzzie, Anthony G., and David B. Allison. "The Search for Human Obesity Genes." *Science* 280 (1998): 1374–1377.

Cook, Ann. "Cloning Humans Next? Ex-UI Student Doing It with Rabbits." *Champaign-Urbana (Ill.) News-Gazette,* June 11, 1987, p. B-9.

Cook-Deegan, Robert M. *The Gene Wars.* New York: Norton, 1994.

———. "Germ-line Gene Therapy: Keep the Window Open a Crack." *Politics and the Life Sciences* 13 (2) (1994): 217–220.

Copeland, Neil G., et al. "Mmu 16—Comparative Genomic Highlights." *Science* 296 (2002): 1617–1618.

Cosmides, Leda. "The Logic of Social Exchange: Has Natural Selection Shaped How Humans Reason? Studies with the Wason Selection Task." *Cognition* 31 (1989): 187–276.

Couzin, Jennifer. "HapMap Launched with Pledges of $100 Million." *Science* 298 (2002): 941–942.

———. "RAC Confronts in Utero Gene Therapy Proposals." *Science* 282 (1998): 27.

———. "Taking Aim at Celera's Shotgun." *Science* 295 (2002): 1817.

Crawford, Mark H. "HUGO: Genome Data Open to Scientists." *Science* 246 (1989): 1565.

Crespi, Bernard, and Stevan Springer. "Social Slime Molds Meet Their Match." *Science* 299 (2003): 56–57.

Culliton, Barbara J. "Gene Therapy Clears First Hurdle." *Science* 247 (1990): 1287.

———. "Gene Therapy Guidelines Revised." *Science* 228 (1985): 561–562.

———. "NIH Asked to Tighten Gene Therapy Rules." *Science* 233 (1986): 1378–1379.

———. "XYY: Harvard Researcher under Fire Stops Newborn Screening." *Science* 188 (1975): 1284–1285.

Culver, Charles C., and Bernard Gert. *Philosophy in Medicine.* New York: Oxford University Press, 1982.

Dahl, Robert A. "From Genes to Political Institutions." Paper presented on the occasion of the Fred Greenstein festschrift, Princeton University, April 24, 2001.

Darwin, Charles. *The Origin of Species by Means of Natural Selection.* New York: John B. Alden, 1859.

Davies, Kevin. *Cracking the Genome.* New York: Free Press, 2001.

Davis, Bernard D. "Germ-line Therapy: Evolutionary and Moral Considerations." *Human Gene Therapy* 3 (1992): 361–363.

Davis, Bernard D., and colleagues. "The Human Genome and Other Initiatives." *Science* 249 (1990): 342–343.

Dawkins, R. *The Selfish Gene.* New York: Oxford University Press, 1978.

Demaine, Linda J., and Aaron X. Fellmeth. "Natural Substances and Patentable Inventions." *Science* 300 (2003): 1375–1376.

Deutsch, Karl W. "Mechanism, Organism, and Society: Some Models in Natural and Social Science." *Philosophy of Science* 18 (1951): 230–252.

de Waal, Frans B. M. "Primates—A Natural Heritage of Conflict Resolution." *Science* 289 (2000): 586–590.

Dickson, David. "Europe Says No to Animal Patents." *Science* 245 (1989): 25.

——. "Go-Ahead for Gene Sequencing Venture." *Science* 240 (1988): 1728.

——. "A Soviet Human Genome Program?" *Science* 240 (1988): 140.

——. "Watson Floats a Plan to Carve Up the Genome." *Science* 244 (1989): 521–522.

Dulbecco, Renato. "A Turning Point in Cancer Research: Sequencing the Human Genome." *Science* 231 (1986): 1055–1056.

Dunbar, Robin, et al., ed. *The Evolution of Culture.* New Brunswick, N.J.: Rutgers University Press, 1999.

Dupré, John. "Unification Not Proved." *Science* 280 (1998): 1395.

Easton, David. *The Political System.* New York: Knopf, 1960.

Edenberg, Howard J. "Genetics of Alcoholism." *Science* 282 (1998): 1269.

Edwards v. Aguillard, 482 U.S. 578 (1987).

Eisenberg, Rebecca S. "Patenting the Human Genome." *Emory Law Journal* 39 (1990): 721–745.

Eisenstadt v. Baird, 405 U.S. 438 (1972).

Elshtain, Jean Bethke. "Consilience or Creation?" *Politics and the Life Sciences* 18 (1999): 341–344.

Enard, Wolfgang, et al. "Intra- and Interspecific Variation in Primate Gene Expression Patterns." *Science* 296 (2002): 340–343.

Ennenga, George R. vB. "Artificial Evolution." *Artificial Life* 3 (1997): 51–61.

Enserink, Martin. "Iceland OKs Private Health Data Bank." *Science* 283 (1999): 13.

——. "Opponents Criticize Iceland's Data Base." *Science* 282 (1998): 859.

——. "Physicians Wary of Scheme to Pool Icelanders' Genetic Data." *Science* 281 (1998): 890–891.

——. "Searching for the Mark of Cain." *Science* 289 (2000): 575–579.

Eramian, Dan. "The Real Message from Dolly." *Bio News,* April/May 1997, 3–5.

Ertekin-Taner, Nilufer, et al. "Linkage of Plasma Aβ42 to a Quantitative Locus on Chromosome 10 in Late-Onset Alzheimer's Disease Pedigrees." *Science* 290 (2000): 2303–2304.

"Ethical, Legal, and Social Implications Program, NCHGR Activities Report." May 1993.

Eulau, Heinz, and Susan Zlomke. "Harold D. Lasswell's Legacy to Mainstream Political Science." *Annual Review of Political Science* 2 (1999): 75–89.

European Parliament and the Council of the European Union. Common Position. April 8, 1998.

Evans, John H. *Playing God?* Chicago: University of Chicago Press, 2002.

Executive Order no. 12,975. *Federal Register* 60 (October 5, 1995): 52063–52065.

Executive Order no. 13,237, *Code of Federal Regulations,* title 3, sec. 821 (2001).

Farooqi, I. Sadaf, et al. "Clinical Spectrum of Obesity and Mutations in the Melanocortin 4 Receptor Gene." *New England Journal of Medicine* 348 (2003): 1085–1095.

Feldbaum, Carl. "Some History Should Be Repeated." *Science* 295 (2002): 975.

Feldman, M. W., and S. P. Otto. "Twin Studies, Heritability, and Intelligence." *Science* 278 (1997): 1383–1384.

Feng, Li-Xin, et al. "Generation and in Vitro Differentiation of a Spermatogonial Cell Line." *Science* 297 (2002): 392–395.

Finifter, Ada W., ed. *Political Science: The State of the Discipline.* Washington, D.C.: American Political Science Association, 1983.

———. *Political Science: The State of the Discipline II.* Washington, D.C.: American Political Science Association, 1993.

Finkel, Elizabeth, et al. "ScienceScope." *Science* 291 (2001): 807.

Fleissner, Erwin. "Race and the Human Genome." *Hastings Center Report* 29 (4) (1999): 40–42.

Fletcher, John C., and W. French Anderson. "Germ-line Gene Therapy: A New Stage of Debate." *Law, Medicine and Health Care* 20 (1992): 26–39.

Fletcher, John C., and Gerd Richter. "Human Fetal Gene Therapy: Moral and Ethical Questions." *Human Gene Therapy* 7 (1996): 1605–1614.

Fletcher, Joseph. *The Ethics of Genetic Control.* Garden City, N.Y.: Doubleday Anchor, 1974.

Forrest, Stephanie. "Genetic Algorithms: Principles of Natural Selection Applied to Computation." *Science* 261 (1993): 872–878.

Fountain, Henry. "All the Queen's Relatives." *New York Times,* March 4, 2003, p. D3.

———. "Bees, Come on Down!" *New York Times,* July 29, 2003, p. D4.

Frank, Lone. "Estonia Prepares for National DNA Database." *Science* 290 (2002): 31.

Frankel, Mark S., and Audrey R. Chapman. "Facing Inheritable Genetic Modifications." *Science* 292 (2001): 1303.

Friedman, Jeffrey. "Genetics of Obesity." Paper presented at the American Association for the Advancement of Science Annual Meeting, Philadelphia, 1998.

Friedman, Theodore, and Johann Olav Koss. "Gene Transfer and Athletics—An Impending Problem." *Molecular Therapy* 3 (2001): 819–820.

Fukuyama, Francis. "A Milestone in the Conquest of Nature." Reprinted in *The Future Is Now,* ed. William Kristol and Eric Cohen. Lanham, Md.: Rowman and Littlefield, 2002.

———. *Our Posthuman Future.* New York: Farrar, Straus and Giroux, 2002.

———. "Separating Good Biotech from Bad." Reprinted in *The Future Is Now,* ed. William Kristol and Eric Cohen. Lanham, Md.: Rowman and Littlefield, 2002.

Gabriel, Stacey B., et al. "The Structure of Haplotype Blocks in the Human Genome." *Science* 296 (2002): 2225–2229.

"Gene Linked to Bipolar Disorder." *Champaign-Urbana (Ill.) News-Gazette,* June 16, 2003, p. A-5.

George, Robert P. "Stem-Cell Research: Don't Destroy Human Life." Reprinted in *The Future Is Now*, ed. William Kristol and Eric Cohen. Lanham, Md.: Rowman and Littlefield, 2002.

Gibbons, Ann. "From Field to Lab, New Insights on Being Human." *Science* 288 (2000): 798–800.

Gilbert, Robert E. *Mortal Presidency*. New York: Fordham University Press, 1998.

Glad, Betty. "Personality, Role Strains, and Alcoholism: Key Pitman as Chairman of the Senate Foreign Relations Committee." *Politics and the Life Sciences* 7 (1988): 18–32.

Goffeau, A., et al. "Life with 6000 Genes." *Science* 274 (1996): 546–567.

Goldman, David. "High Anxiety." *Science* 274 (1996): 1483.

Gordon, John W. "Genetic Enhancement in Humans." *Science* 283 (1999): 2023–2024.

Gould, Stephen Jay. "Darwinism and the Expansion of Evolutionary Theory." *Science* 216 (1982): 380–387.

Greely, Henry T. "Genomics Research and Human Subjects." *Science* 282 (1998): 625.

———. "Legal, Ethical, and Social Issues in Human Genome Research." *Annual Review of Anthropology* 27 (1998): 473–502.

Green, Ronald M., et al. "Overseeing Research on Therapeutic Cloning." *Hastings Center Report* 32 (2002): 27–32.

Griswold v. Connecticut, 381 U.S. 479 (1965).

Grompe, Marcus. "Adult versus Embryonic Stem Cells: It's Still a Tie." *Molecular Therapy* 6 (2002): 303–305.

Group of Advisors on the Ethical Implications of Biotechnology. "Ethical Aspects of Prenatal Diagnosis." *Politics and the Life Sciences* 15 (2) (1996): 329–334.

Guenin, Louis M. "Morals and Primordials." *Science* 292 (2001): 1659–1660.

Gura, Trisha. "Can SNPs Deliver on Susceptibility Genes?" *Science* 293 (2001): 593–595.

———. "Gene Therapy and the Germ Line." *Molecular Therapy* 6 (2002): 2–4.

Hall, Stephen S. "President's Bioethics Council Delivers." *Science* 297 (2002): 322–324.

Hamer, Dean. "Rethinking Behavior Genetics." *Science* 298 (2002): 71–72.

Hamer, Dean H., and Peter Copeland. *Living with Our Genes*. New York: Anchor Books, 1998.

Hamer, Dean H., et al. "A Linkage between DNA Markers on the X Chromosome and Male Sexual Orientation." *Science* 261 (1993): 321–327.

Hamilton, W. D. "The Genetical Evolution of Social Behavior," parts 1 and 2. *Journal of Theoretical Biology* 7 (1964): 1–32.

Hammond, Thomas H., and Gary J. Miller. "The Core of the Constitution." *American Political Science Review* 81 (1987): 1155–1174.

Handberg, R. B. "Talking about the Unspeakable in a Secretive Institution: Health and Disability among Supreme Court Justices." *Politics and the Life Sciences* 8 (1989): 70–73.

Hanna, Kathi E., Robert M. Cook-Deegan, and Robyn Y. Nishimi. "Finding a Forum for Bioethics in U.S. Public Policy." *Politics and the Life Sciences* 12 (1993): 205–219.

Hariri, Ahmad R., et al. "Serotonin Transporter Genetic Variation and the Response of the Human Amygdala." *Science* 297 (2002): 400–403.

Harris, William F., II. "Bonding Word and Polity: The Logic of American Constitutionalism." *American Political Science Review* 76 (1982): 34–45.

Harris v. McRae, 448 U.S. 297 (1980).

Hauck, Robert J-P. "From the Editor." *PS* 24 (1991): 4.

Hauser, Marc D., et al. "The Faculty of Language: What Is It, Who Has It, and How Did It Evolve." *Science* 298 (2002): 1569–1579.

Hedges, S. Blair, and Sudhir Kumar. "Vertebrate Genomes Compared." *Science* 297 (2002): 1283–1285.

Helmuth, Laura. "Map of the Human Genome 3.0." *Science* 293 (2001): 583, 585.

Herrnstein, Richard J., and Charles Murray. *The Bell Curve*. New York: Free Press, 1994.

Herzenberg, Caroline L. "On Stephen Jay Gould." *Science* 297 (2002): 1120.

Hirsch, Jerry. "The Pitfalls of Heritability." *Times Literary Supplement*, February 12, 1999, p. 33.

———. Review of *Why Race Matters: Race Differences and What They Mean*, by Michael E. Levin. *Politics and the Life Sciences* 18 (1) (1999): 159–161.

Holden, Constance. "All Quiet on the Bioethics Front." *Science* 281 (1998): 169.

———. "Battle Heats Up over Cloning." *Science* 295 (2002): 2009.

———. "Boxer Genes." *Science* 277 (1997): 321.

———. "Calf Cloned from Bovine Cell Line." *Science* 277 (1997): 903.

———. "Carbon-Copy Clone Is the Real Thing." *Science* 295 (2002): 1443–1444.

———. "Clone Rangers." *Science* 284 (1999): 2083.

———. "Company Gets Funds to Clone Baby." *Science* 289 (2000): 2271.

———. "Didactics of *Gattaca*." *Science* 278 (1997): 1019.

———. "Experts Slam Olympic Gene Test." *Science* 255 (1992): 1073.

———. "The First Gene Marker for IQ?" *Science* 280 (1998): 681.

———. "Happiness and DNA." *Science* 272 (1996): 1591, 1593.

———. "HHS Inks Cell Deal; NAS Calls for More Lines." *Science* 293 (2001): 1966–1967.

———. "A Marker for Female Homosexuality?" *Science* 279 (1998): 1639.

———. "More on Genes and Homosexuality." *Science* 268 (1995): 1571.

———. "New Clues to Alcoholism Risk." *Science* 280 (1998): 1348.

———. "NIH's List of 64 Leaves Questions." *Science* 293 (2001): 1567.

———. "Offbeat Twins." *Science* 288 (2000): 1735.

———. "Primate Parthenotes Yield Stem Cells." *Science* 295 (2002): 779–780.

———. "Scholars' Group Defends Cloning." *Science* 276 (1997): 1341.

———. "Serotonin and Bulimia." *Science* 293 (2001): 205.

———. "Sex and the Schizophrenic." *Science* 287 (2000): 2145.

———. "Smoking Genes." *Science* 289 (2000): 2271.

———. "Sperm Factories." *Science* 266 (1994): 1482.

———. "Twin Study Links Genes to Homosexuality." *Science* 255 (1992): 33.

———. "UN Weighs In on Cloning." *Science* 278 (1997): 1407.

———. "Where HUGOing?" *Science* 255 (1992): 27.

———. "Wiley Declines to Publish Jensen Book." *Science* 273 (1996): 877.

———. "Wiley Drops IQ Book after Public Furor." *Science* 272 (1996): 644.

———. "Would Cloning Ban Affect Stem Cells?" *Science* 293 (2001): 1025.

Holden, Constance, and Jocelyn Kaiser. "Report Backs Ban; Ethics Panel Debuts." *Science* 295 (2002): 601–602.

Holden, Constance, and Gretchen Vogel. "'Show Us the Cells,' U.S. Researchers Say." *Science* 297 (2002): 923–925.

Home Building and Loan Association v. Blaisdell, 290 U.S. 398 (1934).

Hotchkiss, R. D. "Portents for a Genetic Engineering." *Journal of Heredity* 56 (1965): 197–202.

Hudson, Kathy L., et al. "Genetic Discrimination and Health Insurance: An Urgent Need for Reform." *Science* 270 (1995): 391–393.

HUGO Ethics Committee—Statement on Cloning. *Eubios Journal of Asian and International Bioethics* 9 (1999): 70.

"Human Gene Marker/Therapy Clinical Protocols." *Human Gene Therapy* 9 (1998): 2805–2852.

Human Genome Organization. "General Assembly of HUGO Members." *Genome Digest* 4 (2) (April 1997): 4–5.

Humpherys, David, et al. "Epigenetic Instability in ES Cells and Cloned Mice." *Science* 293 (2001): 95–97.

Jaenisch, Rudolph, and Ian Wilmut. "Don't Clone Humans!" *Science* 291 (2001): 2552.

Johnson, Gary R. "The Evolutionary Origins of Government and Politics." In *Human Nature and Politics,* ed. Albert Somit and Joseph Losco. Greenwich, Conn.: JAI Press, 1995.

Juengst, Eric T. "Germ-line Gene Therapy: Back to Basics." *Journal of Medical Philosophy* 16 (1991): 587–592.

———. "The NIH 'Points to Consider' and the Limits of Human Gene Therapy." *Human Gene Therapy* 1 (1990): 425–433.

Kaebnick, Gregory E. "The Nature of the Problem." *Hastings Center Report* 32 (6) (2002): 40–42.

Kahn, Patricia. "Sequencers Split over Data Release." *Science* 271 (1996): 1798–1799.

Kaiser, Jocelyn. "British Genome Boost." *Science* 270 (1995): 903.

———. "Cloned Pigs May Help Overcome Rejection." *Science* 295 (2002): 25, 27.

———. "Cloning Ban Waits in the Wings." *Science* 277 (1997): 1191

———. "Commercial Gene Kingdom Splits Up." *Science* 276 (1997): 1959.

———. "Congress Bans Embryo Study Funds." *Science* 271 (1996): 585.

———. "Jigsaw Puzzle Gene." *Science* 273 (1996): 435.

———. "Population Databases Boom, From Iceland to the U.S." *Science* 298 (2002): 1158–1161.

———. "Seeking the Cause of Induced Leukemias in X-SCID Trial." *Science* 299 (2003): 495.

Kaiser, Jocelyn, et al. "Celera, NIH Make a Deal." *Science* 293 (2001): 191.

Kansas v. Colorado, 206 U.S. 46 (1907).

Kass, Leon R. "Genetic Tampering." Reprinted in *The Future Is Now,* ed. William Kristol and Eric Cohen. Lanham, Md.: Rowman and Littlefield, 2002.

———. "Making Babies—The New Biology and the 'Old' Morality." *Public Interest* 26 (1972): 18–56.

———. "Making Babies—The New Biology and the 'Old' Morality." Reprinted in *The Future Is Now,* ed. William Kristol and Eric Cohen. Lanham, Md.: Rowman and Littlefield, 2002.

———. "Making Babies' Revisited." Reprinted in *The Future Is Now,* ed. William Kristol and Eric Cohen. Lanham, Md.: Rowman and Littlefield, 2002.

———. "New Beginnings in Life." In *The New Genetics and the Future of Man,* ed. Michael P. Hamilton, 15–63. Grand Rapids, Mich.: Eerdmans, 1972.

———. "Preventing a Brave New World." Reprinted in *The Future Is Now,* ed. William Kristol and Eric Cohen. Lanham, Md.: Rowman and Littlefield, 2002.

———. "Why Not Immortality?" Reprinted in *The Future Is Now,* ed. William Kristol and Eric Cohen. Lanham, Md.: Rowman and Littlefield, 2002.

———. "The Wisdom of Repugnance." Reprinted in *The Future Is Now,* ed. William Kristol and Eric Cohen. Lanham, Md.: Rowman and Littlefield, 2002.

Kaye, Howard L. "Consilience: E. O. Wilson's Confession of Faith." *Politics and the Life Sciences* 18 (1999): 344–346.

Keller, Evelyn Fox. "Nature, Nurture, and the Human Genome Project." In *The Code of Codes,* ed. Daniel J. Kevles and Leroy Hood. Cambridge: Harvard University Press, 1993.

Kestenbaum, David. "Cloning Plan Spawns Ethics Debate." *Science* 279 (1998): 315.

Kevles, Daniel J. *In the Name of Eugenics.* New York: Knopf, 1985.

———. "Out of Eugenics: The Historical Politics of the Human Genome." In *The Code of Codes,* ed. Daniel J. Kevles and Leroy Hood. Cambridge: Harvard University Press, 1993.

Kevles, Daniel J., and Leroy Hood. "Reflections." In *The Code of Codes,* ed. Daniel J. Kevles and Leroy Hood. Cambridge: Harvard University Press, 1993.

Kiernan, Vincent. "Cosmetic Uses of Genetic Engineering May Soon Be a Reality." *Chronicle of Higher Education,* October 3, 1997, pp. A17–A18.

King, Mary-Claire, and Arno G. Motulsky. "Mapping Human History." *Science* 298 (2002): 2342–2343.

Kinkead, Gwen. "To Study Disease, Britain Plans a Genetic Census." *New York Times,* December 31, 2002, pp. D5, D8.

Kitcher, Philip. "1953 and All That: A Tale of Two Sciences." *Philosophical Review* 93 (1984): 335–373.

Knoppers, Bartha Maria. "Genetic Benefit Sharing." *Science* 290 (2000): 49.

Knoppers, Bartha Maria, and Ruth Chadwick. "The Human Genome Project: Under an International Ethical Microscope." *Science* 265 (1994): 2035–2036.

Knoppers, Bartha Maria, and Sonia LeBris. "Genetic Choices: A Paradigm for Prospective International Ethics?" *Politics and the Life Sciences* 13 (2) (1994): 228–230.

———. "Recent Advances in Medically Assisted Conception: Legal, Ethical, and Social Issues." *American Journal of Law and Medicine* 27 (4) (1991): 329–361.

Kolata, Gina. "Manic-Depression Gene Tied to Chromosome 11." *Science* 235 (1987): 1139–1140.

———. "Using Genetic Tests, Ashkenazi Jews Vanquish a Disease." *New York Times,* February 18, 2003, pp. D1, D6.

Kolberg, Rebecca. "Human Embryo Cloning Reported." *Science* 262 (1993): 652–653.

Koshland, Daniel E., Jr. "Ahead of Schedule and on Budget." *Science* 266 (1994): 199.

———. "Sequences and Consequences of the Human Genome." *Science* 246 (1989): 189.

Krieger, Michael J. B., and Kenneth G. Ross. "Identification of a Major Gene Regulating Complex Social Behavior." *Science* 295 (2002): 328–332.

Krimsky, Sheldon. "Human Gene Therapy: Must We Know Where to Stop Before We Start?" *Human Gene Therapy* 1 (1990): 171–173.

Kristof, Nicholas D. "Interview with a Humanoid." *New York Times,* July 23, 2002, p. A19.

Kuhn, Thomas S. *The Structure of Scientific Revolutions.* 2nd ed. Chicago: University of Chicago Press, 1970.

Kwok, Pui-Yan. "Genetic Association by Whole-Genome Analysis?" *Science* 294 (2001): 1669–1670.

Lander, Eric S. "Array of Hope." *Nature Genetics Supplement* 21 (1999): 3–4.

———. "In Wake of Genetic Revolution, Questions about Its Meaning." *New York Times,* September 12, 2000, p. D5.

Lanza, Mark P., et al. "Science over Politics." *Science* 283 (1999): 1849–1850.

Lanza, Robert P., et al. "Cloned Cattle Can Be Healthy and Normal." *Science* 294 (2001): 1893–1894.

Lasswell, H. D., and Abraham Kaplan. *Power and Society.* New Haven: Yale University Press, 1950.

LeDoux, Joseph. *The Emotional Brain.* London: Weidenfeld and Nicolson, 1998.

Lesch, Klaus-Peter, et al. "Association of Anxiety-Related Traits with a Polymorphism in the Serotonin Transporter Gene Regulatory System." *Science* 274 (1996): 1527–1531.

Leslie, Mitch. "Tagging Honeybee Genes." *Science* 295 (2002): 1607.

LeVay, Simon. "A Difference in Hypothalamic Structure between Heterosexual and Homosexual Men." *Science* 253 (1991): 1034–1037.

Lewin, Roger. "Genome Planners Fear Avalanche of Red Tape." *Science* 244 (1989): 1543.

———. "Genome Projects Ready to Go." *Science* 240 (1988): 602–604.

———. "Politics of the Genome." *Science* 235 (1987): 1453.

———. "Proposal to Sequence the Human Genome Stirs Debate." *Science* 232 (1986): 1598–1600.

Lewontin, R. C. "The Dream of the Human Genome." *New York Review of Books,* May 28, 1992, pp. 31–40.

Lewontin, R. C., Steve Rose, and Leon J. Kamin. *Not in Our Genes.* New York: Pantheon Books, 1984.

Llewellyn, Karl N. "The Constitution as an Institution." *Columbia Law Review* 34 (1934): 1–40.

Lowi, Theodore J. *The End of Liberalism.* Chicago: Norton, 1969.

———. *The Politics of Disorder.* New York: Norton, 1971.

Lumsden, Charles J., and Edward O. Wilson. *Genes, Mind, and Culture.* Cambridge: Harvard University Press, 1981.

———. *Promethean Fire.* Cambridge: Harvard University Press, 1983.

Lyon, Jeff, and Peter Gorner. *Altered Fates.* New York: Norton, 1995.

MacLean, Paul D. "A Triangular Brief on the Evolution of Brain and Law." *Journal of Social and Biological Structures* 5 (1982): 369–379.

Madsen, Douglas. "A Biochemical Property Relating to Power Seeking in Humans." *American Political Science Review* 79 (1985): 448–457.

———. "Power Seekers Are Different: Further Biochemical Evidence." *American Political Science Review* 80 (1986): 261–269.

Makalowski, Wojciech. "Not Junk After All." *Science* 300 (2003): 1246–1247.

Mann, Charles C. "Behavioral Genetics in Transition." *Science* 264 (1994): 1687.

March, James G., and Johan P. Olsen. *Rediscovering Institutions.* New York: Free Press, 1989.

Marshall, Eliot. "Bioethics Panel Urges Broader Oversight." *Science* 292 (2001): 1466–1467.

———. "Biomedical Groups Derail Fast-Track Anticloning Bill." *Science* 279 (1998): 1123–1124.

———. "Celera and *Science* Spell Out Data Access Provisions." *Science* 291 (2001): 1191.

———. "Claim of Human-Cow Embryo Greeted with Skepticism." *Science* 282 (1998): 1390–1391.

———. "A Clash over Testing for Alzheimer's Disease." *Science* 280 (1998): 1004.

———. "Clinton and Blair Back Rapid Release of Data." *Science* 287 (2000): 1903.

———. "Clinton Urges Outlawing Human Cloning." *Science* 276 (1997): 1640.

———. "Commercial Firms Win U.S. Sequencing Funds." *Science* 285 (1999): 310.

———. "The Company That Genome Researchers Love to Hate." *Science* 266 (1994): 1800–1802.

———. "DNA Studies Challenge the Meaning of Race." *Science* 282 (1998): 654–655.

———. "Drug Firms to Create Public Database of Genetic Mutations." *Science* 284 (1999): 406–407.

———. "Ethicists Back Stem Cell Research, White House Treads Cautiously." *Science* 285 (1999): 502.

———. "Fraud Strikes Top Genome Lab." *Science* 274 (1996): 908–910.

———. "Gene Therapy a Suspect in Leukemia-Like Disease." *Science* 298 (2002): 34–35.

———. "Gene Therapy on Trial." *Science* 288 (2000): 951–957.

———. "Genome Researchers Take the Pledge." *Science* 272 (1996): 477–478.

———. "Genome Teams Adjust to Shotgun Marriage." *Science* 292 (2001): 1982.

———. "Genomic's Odd Couple." *Science* 275 (1997): 778.

———. "How a Bland Statement Sent Stocks Sprawling." *Science* 287 (2000): 2127.

———. "Mammalian Cloning Debate Heats Up." *Science* 275 (1997): 1733.

———. "Moratorium Urged on Germ Line Gene Therapy." *Science* 289 (2000): 2023.

———. "New Chair of Bioethics Panel Wants National Debate on Issues." *Science* 293 (2001): 1243.

———. "NIH Told to Reconsider Crime Meeting." *Science* 262 (1993): 23–24.

———. "NIH to Produce a 'Working Draft' of the Genome by 2001." *Science* 281 (1998): 1774–1775.

———. "Panel Approves Gene Trial for 'Normals.'" *Science* 275 (1997): 1561. .

———. "Panel Reviews Risks of Germ Line Changes." *Science* 294 (2001): 2268–2269.

———. "Panel Urges Cloning Ethics Boards." *Science* 275 (1997): 22.

———. "Panel Weighs a Law against Cloning." *Science* 276 (1997): 1185.

———. "'Playing Chicken' over Gene Markers." *Science* 278 (1997): 2046–2048.

———. "Probing Primate Morality." *Science* 271 (1996): 904.

———. "Public Group Completes Draft of the Mouse." *Science* 296 (2002): 1005.

———. "Ruling May Free NIH to Fund Stem Cell Studies." *Science* 283 (1999): 465–466.

———. "Second Child in French Trial Is Found to Have Leukemia." *Science* 299 (2003): 320.

———. "Snipping Away at Genome Patenting." *Science* 277 (1997): 1752–1753.

———. "A Strategy for Sequencing the Genome 5 Years Early." *Science* 267 (1995): 783–784.

———. "Talks of Public-Private Deal End in Acrimony." *Science* 287 (2000): 1723–1724.

———. "Varmus Grilled over Breach of Embryo Research Ban." *Science* 276 (1997): 1963.

———. "Varmus Puts His Stamp on NIH." *Science* 270 (1995): 1290.

———. "A Versatile Cell Line Raises Scientific Hopes, Legal Questions." *Science* 282 (1998): 1014–1015.

———. "What to Do When Clear Success Comes with an Unclear Risk?" *Science* 298 (2002): 510–511.

———. "Whose Genome Is It, Anyway?" *Science* 273 (1996): 1788–1789.

Marshall, Eliot, and Elizabeth Pennisi. "Hubris and the Human Genome." *Science* 280 (1998): 994–995.

———. "NIH Launches the Final Push to Sequence the Genome." *Science* 272 (1996): 188–189.

Marshall, Eliot, and Gretchen Vogel. "Cloning Announcement Sparks Debate and Scientific Skepticism." *Science* 294 (2001): 1802–1803.

Marx, Jean L. "Cloning Sheep and Cattle Embryos." *Science* 239 (1988): 463–464.

———. "Genome Project Plans Described." *Science* 260 (1993): 152–153.

———. "How Stimulant Drugs May Calm Hyperactivity." *Science* 283 (1999): 306.

———. "New Gene Tied to Common Form of Alzheimer's." *Science* 281 (1998): 507, 509.

———. "Obesity Gene Discovery May Help Solve Weighty Problem." *Science* 266 (1994): 1477–1478.

Masterman, Margaret. "The Nature of a Paradigm." In *Criticism and the Growth of Knowledge,* ed. Imray Lakatos and Alan Musgrave. Cambridge: Harvard University Press, 1970.

Masters, Roger D., and Denis G. Sullivan. "Nonverbal Displays and Political Leadership in France and the United States." *Political Behavior* 11 (1989): 123–156.

McClearn, Gerald E., et al. "Substantial Genetic Influence on Cognitive Abilities in Twins 80 or More Years Old." *Science* 276 (1997): 1560–1563.

McGrath, James, and Davor Solter. "Inability of Mouse Blastomere Nuclei Transferred to Enucleated Zygotes to Support Development in Vitro." *Science* 226 (1984): 1317–1319.

McGuffin, Peter, Brien Riley, and Robert Plomin. "Toward Behavioral Genomics." *Science* 291 (2001): 1232, 1249.

McKibben, Bill. "Unlikely Allies against Cloning." *New York Times,* March 27, 2002, p. A23.

McLaren, Anne. "Cloning: Pathways to a Pluripotent Future." *Science* 288 (2000): 1775–1780.

McNeill, Donald G., Jr. "Religious Sect Says It Will Announce the First Cloned Baby." *New York Times,* December 27, 2002, p. A13.

Mehlman, Maxwell, and Jeffrey Botkin. *Access to the Genome: The Challenge to Equality.* Washington, D.C.: Georgetown University Press, 1997.

"Memorandum on the Prohibition on Federal Funding for Cloning of Human Beings." *Weekly Compilation of Presidential Documents* 33 (March 4, 1997): 281.

Mendel, Gregor. "Versuche über Pflanzen-Hybriden." *Verhandlungen Des Naturforschenden Vereines, Abnahndlungen, Brunn,* 4 (1866): 3–47.

Merriam, Charles E. *New Aspects of Politics.* Chicago: University of Chicago Press, 1925. 3rd ed., 1970.

Merton, Robert F. *Social Theory and Social Structure.* New York: Free Press, 1968.

Milwaukee Pub. Co. v. Burleson, 255 U.S. 407 (1921).

Morell, Virginia. "Huntington's Gene Finally Found." *Science* 260 (1993): 28–30.

Morton, Oliver. "First Dolly, Now Headless Tadpoles." *Science* 278 (1997): 798.

Motulsky, Arno G. "Impact of Genetic Manipulation on Society and Medicine." *Science* 219 (1983): 135–136.

Munson, Ronald. *Intervention and Reflection,* 3rd ed. Belmont, Calif.: Wadsworth, 1988.

Murphy, Walter F., James Fleming, and Sotirios A. Barber. *American Constitutional Interpretation.* 2nd ed. Westbury, N.Y.: Foundation Press, 1995.

Myers, Amanda, et al. "Susceptibility Locus for Alzheimer's Disease on Chromosome 10." *Science* 290 (2000): 2304–2305.

Nance, Walter E., and John Fletcher. "Human Reproductive Cloning." *Science* 297 (2002): 1477.

National Institutes of Health. Meeting of the Human Gene Therapy Subcommittee of the Recombinant DNA Advisory Committee (RAC), March 30, 1990. Transcript of Proceedings. Gaithersburg, Md.: StenoTech, Inc., 1990.

NEA v. Finley, 524 U.S. 569 (1998).

NIH Recombinant DNA Advisory Committee. "The Revised 'Points to Consider' Document." Reprinted in *Human Gene Therapy* 1 (1990): 93–103.

Noble, Ernest P. "DRD2 Gene and Alcoholism." *Science* 281 (1998): 1287–1288.

Normile, Dennis. "Bid for Better Beef Gives Japan a Leg Up on Cattle." *Science* 282 (1998): 1975–1976.

———. "Human Cloning Ban Allows Some Research." *Science* 290 (2000): 1872.

Normile, Dennis, and Elizabeth Pennisi. "Team Wrapping Up Sequence of First Human Chromosome." *Science* 285 (1999): 2038–2039.

Nowak, Martin A., and Karl Sigmund. "Shrewd Investments." *Science* 288 (2000): 819–820.

———. "Tit for Tat in Heterogeneous Populations." *Nature* 355 (1992): 250–252.

Nowak, Martin A., et al. "Evolution of Universal Grammar." *Science* 291 (2001): 114–118.

Nowak, Rachel. "Bacterial Genome Sequence Bagged." *Science* 269 (1995): 468–470.

———. "Breast Cancer Gene Offers Surprises." *Science* 265 (1994): 1796–1799.

———. "Genetic Testing Set for Takeoff." *Science* 265 (1994): 464–467.

"Open Peer Commentary." *Behavioral and Brain Sciences* 5 (1982): 1–37.

Ostrom, Elinor. *Governing the Commons.* New York: Cambridge University Press, 1990.

Palca, Joseph. "Genome Projects Are Growing Like Weeds." *Science* 245 (1989): 131.

———. "NIH Wrestles with Furor over Conference." *Science* 257 (1992): 739.

Palumbi, Stephen R. "Humans as the World's Greatest Evolutionary Force." *Science* 293 (2001): 1786–1790.

Parens, Erik. "Is Better Always Good? The Enhancement Project." In *Enhancing Human Traits,* ed. Erik Parens. Washington, D.C.: Georgetown University Press, 1998.

Parens, Erik, and Eric Juengst. "Inadvertently Crossing the Germ Line." *Science* 292 (2001): 397.

Patel, Nila, et al. "Blocks of Limited Haplotype Diversity Revealed by High-Resolution Scanning of Human Chromosome 21." *Science* 294 (2001): 1719–1723.

Pennisi, Elizabeth. "Academic Sequencers Challenge Celera in a Sprint to the Finish." *Science* 283 (1999): 1822–1823.

———. "Chimps and Fungi Make Genome 'Top Six.'" *Science* 296 (2002): 1589, 1591.

———. "Chromosome 21 Done, Phase Two Begun." *Science* 288 (2000): 939.

———. "Cloned Mice Provide Company for Dolly." *Science* 281 (1998): 495, 497.

———. "Finally, the Book of Life and Instructions for Navigating It." *Science* 288 (2000): 2304–2307.

———. "Fruit Fly Genome Yields Data and a Validation." *Science* 287 (2000): 1374.

———. "Fruit Fly Researchers Sign Pact with Celera." *Science* 283 (1999): 767.

———. "Funders Reassure Genome Sequencers." *Science* 280 (1998): 1185.

———. "Gene Activity Clocks Brain's Fast Evolution." *Science* 296 (2002): 233, 235.

———. "Insects Rank Low among Genome Priorities." *Science* 294 (2001): 1261–1262.

———. "Mouse Sequencers Take Up the Shotgun." *Science* 287 (2000): 1179–1181.

———. "Transgenic Lambs from Cloning Lab." *Science* 277 (1997): 631.

———. "Worming Secrets from the *C. elegans* Genome." *Science* 282 (1998): 1972–1974.

Pennisi, Elizabeth, and Dennis Normile. "Perseverance Leads to Cloned Pig in Japan." *Science* 289 (2000): 1118–1119.

Pennisi, Elizabeth, and Gretchen Vogel. "Clones: A Hard Act to Follow." *Science* 288 (2000): 1722–1727.

Pennisi, Elizabeth, and Nigel Williams. "Will Dolly Send in the Clones?" *Science* 275 (1997): 1415–1416.

Peterson, Steven A. "The Health of the Polity: Results from a National Sample." *Politics and the Life Sciences* 10 (1991): 45–56.

Phillips, Claude S. "Consilience and the Social Sciences." Paper presented at the Association for Politics and the Life Sciences Annual Meeting, Atlanta, 1999.

———. "Political Science Reaction to Sociobiology 1975—Present." Paper presented at the Association for Politics and the Life Sciences Annual Meeting, Washington, D.C., 2000.

"Physicist Says He'll Clone Himself with Wife's Help." *Champaign-Urbana (Ill.) News-Gazette,* September 7, 1998, p. A-5.

Pickrell, John. "Experts Assail Plan to Help Childless Couples." *Science* 291 (2001): 2061, 2063.

Pinker, Steven. *The Blank Slate.* New York: Viking, 2002.

Planned Parenthood v. Casey, 505 U.S. 833 (1992).

Plomin, Robert. "The Role of Inheritance in Behavior." *Science* 248 (1990): 183–188.

Plomin, Robert, Michael J. Owen, and Peter McGuffin. "The Genetic Basis of Complex Human Behaviors." *Science* 264 (1994): 1733–1739.

Plomin, Robert, and Stephen A. Petrill. "Genetics and Intelligence: What's New?" *Intelligence* 24 (1997): 53–77.

Pool, Robert. "Economics: Game Theory's Winning Hands." *Science* 266 (1994): 371.

———. "Putting Game Theory to the Test." *Science* 267 (1995): 1591–1593.

Powell v. Texas, 392 U.S. 514 (1968).

President's Commission for the Study of Ethical Problems in Medicine and Biomedical and Behavioral Research. *Splicing Life: A Report on the Social and Ethical Issues of Genetic Engineering with Human Beings.* Washington, D.C.: Government Printing Office, 1982.

President's Council on Bioethics. "Human Cloning and Human Dignity: An Ethical Inquiry." Washington, D.C., July 2002. http://www.bioethics.gov/topics/cloning_index.html (accessed May 4, 2004).

Preston, Richard. "The Genome Warrior." *New Yorker,* June 12, 2000, pp. 66–83.

Quigg, Donald J., assistant secretary of the Department of Commerce and Commissioner of Patents and Trademarks. "Release," April 21, 1987.

Ramsey, Paul. "Shall We Reproduce?" *Journal of the American Medical Association* 220 (1972): 1346–1350, 1480–1485.

Randall, Richard S. *Freedom and Taboo.* Berkeley: University of California Press, 1989.

"Researchers Make Strain of Pigs That Carry Human Genes." *Champaign-Urbana (Ill.) News-Gazette,* October 22, 2002, p. A-2.

Rice, George, et al. "Male Homosexuality: Absence of Linkage to Microsatellite Markers at Xq28." *Science* 284 (1999): 665–667.

Richmond, Barry J., et al. "Predicting Future Rewards." *Science* 301 (2003): 179–180.

Ridley, Matt. *Genome.* New York: Perennial, 2000.

Riolo, Rick L., Michael D. Cohen, and Robert Axelrod. "Evolution of Cooperation without Reciprocity." *Nature* 414 (2001): 441–443.

Roberts, Leslie. "Academy Backs Genome Project." *Science* 239 (1988): 725–726.

———. "Carving Up the Human Genome." *Science* 242 (1988): 1244–1246.

———. "Controversial from the Start." *Science* 291 (2001): 1182–1188.

———. "Ethical Questions Haunt New Genetic Technologies." *Science* 243 (1989): 1134–1136.

———. "The Genetic Map Is Back on Track after Delays." *Science* 248 (1990): 805.

———. "Genome Center Grants Chosen." *Science* 249 (1990): 1497.

———. "Genome Patent Fight Erupts." *Science* 254 (1991): 184–186.

———. "Genome Project: An Experiment in Sharing." *Science* 248 (1990): 953.

———. "Genome Project Under Way, at Last." *Science* 243 (1989): 167–168.

———. "HUGO Takes on Role as Marriage Broker." *Science* 254 (1991): 932.

———. "New Game Plan for Genome Mapping." *Science* 245 (1989): 1438–1440.

———. "NIH Gene Patents, Round Two." *Science* 255 (1992): 912–913.

———. "NIH Takes New Tack on Gene Mapping." *Science* 258 (1992): 1573.

———. "Plan for Genome Centers Sparks a Controversy." *Science* 246 (1989): 204–205.

———. "Report Card on the Genome Project." *Science* 253 (1991): 376.

———. "Rumors Fly over Rejection of NIH Claim." *Science* 257 (1992): 1855.

———. "Scientists Voice Their Opposition." *Science* 256 (1992):1273–1274.

———. "Taking Stock of the Genome Project." *Science* 262 (1993): 20–22.

———. "Watson versus Japan." *Science* 246 (1989): 576, 578.

———. "Who Owns the Human Genome?" *Science* 237 (1987): 358–361.

———. "Why Watson Quit as Project Head." *Science* 256 (1992): 301–302.

Roberts, Morley. *Bio-Politics: An Essay in the Physiology, Pathology, and Politics of the Social and Somatic Organism.* London: Dent, 1938.

Robertson, John A. "Liberty, Identity, and Human Cloning." *Texas Law Review* 76 (1998): 1371–1456.

Robins, Robert S., and Henry Rothschild. "Ethical Dilemmas of the President's Physician." *Politics and the Life Sciences* 7 (1988): 3–11.

Robinson, Gene E. "From Society to Genes with the Honey Bee." *American Scientist* 86 (1998): 456–462.

———. "Sociogenomics Takes Flight." *Science* 297 (2002): 204–205.

Roe v. Wade, 410 U.S. 113 (1973).

Rorty, Richard. "Studied Ambiguity." *Science* 293 (2001): 2399–2400.

Rossiter, Clinton. *Conservatism in America: The Thankless Persuasion.* 2nd ed. New York: Vintage Books, 1962.

Rothenberg, Karen, et al. "Genetic Information and the Workplace: Legislative Approaches and Policy Challenges." *Science* 275 (1997): 1755–1757.

Roush, Wade. "Conflict Marks Crime Conference." *Science* 269 (1995): 1808–1809.

Ruse, Michael. "The Faith of an Evolutionist." *Politics and the Life Sciences* 18 (1999): 347–349.

———. "On E. O. Wilson and His Religious Vision." *Science* 290 (2000): 943.

Rushton, J. Philippe. "*The Bell Curve:* Premises, Conclusions, and Policies." Round-table discussion convened at the American Political Science Association Annual Meeting, Chicago, 1995.

———. *Race, Evolution, and Behavior.* Special abridged ed. New Brunswick, N.J.: Transaction, 1999.

Safire, William. "The Crimson Birthmark." *New York Times,* January 21, 2002, p. A19.

Saltus, Richard. "Crash Effort to Map Human Genes Urged." *Boston Globe,* June 9, 1986, p. 53.

Sandel, Michael J. *Liberalism and the Limits of Justice.* 2nd ed. Cambridge: Cambridge University Press, 1998.

Sanfey, Alan G., et al. "The Neural Basis of Economic Decision-Making in the Ultimate Game." *Science* 300 (2003): 1755–1758.

Sankar, Pamela, and Mildred K. Cho. "Toward a New Vocabulary of Human Genetic Variation." *Science* 298 (2002): 1337–1338.

Schechtman, Marya. *The Constitution of Selves.* Ithaca: Cornell University Press, 1996.

Schmeck, Harold M., Jr. "U.S. Sets Guidelines in Using Gene Transplants in Humans." *New York Times,* September 24, 1985, pp. 1, 18.

Schmickle, Sharon. "Race Controversy Arises in World of Genome Study." *Champaign-Urbana (Ill.) News-Gazette,* March 29, 2002, p. D-4.

Schnieke, Angelika E., et al. "Human Factor IX Transgenic Sheep Produced by Transfer of Nuclei from Transfected Fetal Fibroblasts." *Science* 278 (1997): 2130–2133.

Schubert, Glendon. "The Evolution of Political Science: Paradigms of Physics, Biology, and Politics." *Politics and the Life Sciences* 1 (1982): 97–123.

———. "Psychobiological Politics." *Canadian Journal of Political Science* 16 (1983): 535–576.

———. "The Rhetoric of Constitutional Change." *Journal of Public Law* 16 (1967): 16–50.

———. "The Sociobiology of Political Behavior." In *Sociobiology and Human Politics,* ed. Elliott White. Lexington, Mass.: Heath, 1981.

Schwartz, Carl E., et al. "Inhibited and Uninhibited Infants 'Grown Up': Adult Amygdalar Response to Novelty." *Science* 300 (2003): 1952–1953.

ScienceScope. "De-Celeration." *Science* 296 (2002): 635.

Seed, Richard G. "Inevitable Human Cloning . . . On the Road to Immortality." Unpublished outline in the author's possession.

Segerstrale, Ullica. *Defenders of the Truth.* New York: Oxford University Press, 2000.

Serrin, Catherine. "At the Frontier." *Michigan Alumnus,* Spring 2000, 26–31.

Shapiro, David. "Germ-line Gene Therapy: Do We Need an International Approach?" *Politics and the Life Sciences* 13 (2) (1994): 233–234.

Shapiro, Harold T. "Ethical and Policy Issues of Human Cloning." *Science* 277 (1997): 195.

Shattuck, Roger. *Forbidden Knowledge: From Prometheus to Pornography.* New York: St. Martin's, 1996.

Shouse, Ben. "Less Can Be More, U.K. Study Finds." *Science* 293 (2001): 1238–1239.

———. "Revisiting the Numbers: Human Genes and Whales." *Science* 295 (2002): 1457.

Simerly, Calvin, et al. "Molecular Correlates of Primate Nuclear Transfer Failures." *Science* (2003): 297.

Simm, Michael. "Violence Study Hits a Nerve in Germany." *Science* 264 (1994): 653.

Simon, Herbert A. "Human Nature in Politics: The Dialogue of Psychology with Political Science." *American Political Science Review* 79 (1985): 293–304.

———. "A Mechanism for Social Selection and Successful Altruism." *Science* 250 (1990): 1665–1668.

———. *Models of Thought.* New Haven: Yale University Press, 1979.

Skinner v. Oklahoma, 316 U.S. 535 (1942).

Sokolowski, Maria B. "Social Eating for Stress." *Nature* 419 (2002): 893–894.

Somit, Albert. "Toward a More Biologically-Oriented Political Science: Ethology and Pharmacology." *Midwest Journal of Political Science* 12 (1968): 550–567.

Somit, Albert, and Joseph Tanenhaus. *The Development of Political Science.* Boston: Allyn and Bacon, 1967.

Somit, Albert, and Steven A. Peterson. "Rational Choice and Biopolitics: A (Darwinian) Tale of Two Theories." *PS* 32 (1999): 39–44.

Spalding, John B., and R. L. Bernstein. "Genome Comparisons." *Science* 297 (2002): 1037.

Spier, R. E. "Toward a New Human Species?" *Science* 296 (2002): 1807, 1809.

Steiner, George. "Fields of Force." *New Yorker,* October 28, 1972, pp. 42–117.

Sternberg, Robert J. "The Holey Grail of General Intelligence." *Science* 289 (2000): 399, 401.

Stokstad, Erik. "Violent Effects of Abuse Tied to Gene." *Science* 297 (2002): 752.

Stolberg, Sheryl Gay. "Bush Makes Fervent Bid to Get Senate to Ban Cloning Research." *New York Times,* April 11, 2002, p. A23.

———. "Bush Rallies Opponents of Cloning." *New York Times,* April 10, 2002, p. A19.

———. "Bush's Advisers on Ethics Discuss Human Cloning." *New York Times,* January 18, 2002, p. A19.

———. "Key Republican Backs Cloning in Research." *New York Times,* May 1, 2002, p. A20.

———. "Some for Abortion Rights Lean Right in Cloning Fight." *New York Times,* January 24, 2002, p. A23.

———. "Total Ban on Cloning Research Appears Dead." *New York Times,* June 14, 2002, p. A31.

Stone, Richard. "Brits and EC at Odds over Gene Patenting." *Science* 256 (1992): 727.

———. "Germ Cell Gene Panel." *Science* 255 (1991): 841.

———. "HHS 'Violence Initiative' Caught in a Crossfire." *Science* 258 (1992): 212–213.

———. "NIH to Size Up Growth Hormone Trials." *Science* 257 (1992): 739.

———. "NIH to Study 'Germ-line' Therapy." *Science* 266 (1994): 1631.

———. "Nonprofit to Launch Gene-Mapping Effort." *Science* 267 (1995): 443.

———. "Taking Honest Abe's DNA Fingerprint." *Science* 256 (1992): 446.

Sun, Marjorie. "NIH and DOE Draft Genome Pact." *Science* 241 (1988): 1596.

Swaab, D. F., and M. A. Hofman. "An Enlarged Suprachiasmatic Nucleus in Homosexual Men." *Brain Research* 537 (1990): 141.

Szathmáry, Eörs, et al. "Can Genes Explain Biological Complexity?" *Science* 292 (2001): 1315.

Taubes, Gary. "Averaged Brains Pinpoint a Site for Schizophrenia." *Science* 266 (1994): 221.

Tellegen, Auke, et al. "Personality Similarity in Twins Reared Apart and Together." *Journal of Personality and Social Psychology* 54 (1988): 1031–1039.

"Testicle Transplant Successful." *Champaign-Urbana (Ill.) News-Gazette,* August 15, 2002, p. A-2.

Thompson, Larry. "Healy Approves an Unproven Treatment." *Science* 259 (1993): 172.

Thomson, J. A., et al. "Embryonic Stem Cell Lines Derived from Human Blastocysts." *Science* 282 (1998): 1145–1147.

Travis, John. "New Piece in Alzheimer's Puzzle." *Science* 261 (1993): 828–829.

Trivers, R. L. "The Evolution of Reciprocal Altruism." *Quarterly Review of Biology* 46 (1971): 35–57.

Truman, David B. *The Governmental Process.* New York: Knopf, 1951.

U.S. Department of HHS, PHS, NIH. *Recombinant DNA Research* 16 (January 1994): 239–243.

United States v. Morrison, 529 U.S. 598 (2000).

Vatz, Richard E., and Lee S. Weinberg. "Biochemistry and Power-Seeking." *Politics and the Life Sciences* 10 (1991): 69–75.

Venter, J. Craig. "Human Genome Agreements." *Science* 275 (1997): 601–602.

———. "A Part of the Human Genome Sequence." *Science* 299 (2003): 1183–1184.

Vogel, Gretchen. "Bush Grapples with Stem Cells, Cloning." *Science* 292 (2001): 2409, 2411.

———. "Bush Squeezes between the Lines on Stem Cells." *Science* 293 (2001): 1242–1245.

———. "Cloning: Could Humans Be Next?" *Science* 291 (2001): 808–809.

———. "Cloning Bills Proliferate in U.S. Congress." *Science* 292 (2001): 1037.

———. "Human Cloning Plans Spark Talk of U.S. Ban." *Science* 292 (2001): 31.

———. "In Contrast to Dolly, Cloning Resets Telomere Clock in Cattle." *Science* 288 (2000): 586–587.

———. "Jumbled DNA Separates Chimps and Humans." *Science* 296 (2002): 719, 721.

———. "Mutations Reveal Genes in Zebrafish." *Science* 296 (2002): 1221.

———. "NIH Pulls Plug on Ethics Review." *Science* 292 (2001): 415–416.

———. "Orangutans, Like Chimps, Heed the Cultural Call of the Collective." *Science* 299 (2003): 27–28.

———. "Researchers Get Green Light for Work on Stem Cells." *Science* 289 (2000): 1442–1443.

———. "Rumors and Trial Balloons Precede Bush's Funding Decision." *Science* 293 (2001): 186–187.

———. "Stem Cells Not So Stealthy After All." *Science* 297 (2002): 175–177.

———. "U.N. Split over Full or Partial Cloning Ban." *Science* 298 (2002): 1316–1317.

Vogelstein, Bert, et al. "Please Don't Call It Cloning!" *Science* 295 (2002): 1237.

Volny, Veronica P., and Deborah M. Gordon. "Genetic Basis for Queen-Worker Dimorphism in a Social Insect." *Proceedings of the National Academy of Sciences* 99 (2002): 6108–6111.

von Bertalanffy, Ludwig. *General System Theory, Foundations, Development, Applications.* New York: Braziller, 1969.

Wade, Nicholas. "After Scare, a Gene Therapy Trial Proceeds." *New York Times,* January 8, 2002, p. F6.

———. "Clinton Asks Study of Bid to Form Part-Human, Part-Cow Cells." *New York Times,* November 15, 1998, p. 25.

———. "A Dim View of a 'Posthuman Future.'" *New York Times,* April 2, 2002, pp. D1, D4.

———. "First Gene for Social Behavior Identified in Whiskery Mice." *New York Times,* September 9, 1997, p. C4.

———. "Fish Genes Aid Human Discoveries." *New York Times,* July 26, 2002, A19.

———. "For Sale: A DNA Test to Measure Racial Mix." *New York Times,* October 1, 2002, p. F4.

———. "A Genomic Treasure Hunt May Be Striking Gold." *New York Times,* June 18, 2002, pp. D1, D4.

———. "Race Is Seen as Real Guide to Track Roots of Disease." *New York Times,* July 30, 2002, pp. D1-D2.

———. "Scientist Reveals Genome Secret: It's Him." *New York Times,* April 27, 2002, p. 1.

———. "Sociobiology: Troubled Birth for New Discipline." *Science* 191 (1976): 1151–1155.

Wahlke, John C. "Pre-Behavioralism in Political Science." *American Political Science Review* 73 (1978): 9 –31.

Wallas, Graham. *Human Nature in Politics.* Gloucester, Mass.: Smith, 1908.

Walters, LeRoy. "Human Gene Therapy: Ethics and Public Policy." *Human Gene Therapy* 2 (1991): 115 –122.

Wang, David G., et al. "Large-Scale Identification, Mapping, and Genotyping of Single-Nucleotide Polymorphisms in the Human Genome." *Science* 280 (1998): 1077–1082.

Watson, James D. "The Human Genome Project: Past, Present, and Future." *Science* 248 (1990): 44 –49.

———. "In Defense of DNA." *New Republic,* June 25, 1977, pp. 13 –14.

———. "Moving toward the Clonal Man—Is This What We Want?" U.S. Senate, 92nd Cong., 1st sess. *Congressional Record* 117, no. 61 (April 29, 1971): 12,751–12,752.

———. "A Personal View of the Project." In *The Code of Codes,* ed. Daniel J. Kevles and Leroy Hood. Cambridge: Harvard University Press, 1993.

Watson, J. D., and F. H. C. Crick. "A Structure of Deoxyribose Nucleic Acid." *Nature* 171 (1953): 737–738.

Wedekind, Claus. "Give and Ye Shall Be Recognized." *Science* 280 (1998): 2070–2071.

Wexler, Nancy. "Clairvoyance and Caution: Repercussions from the Human Genome Project." In *The Code of Codes,* ed. Daniel J. Kevles and Leroy Hood. Cambridge: Harvard University Press, 1993.

Williams, G. C. *Adaptation and Natural Selection.* Princeton: Princeton University Press, 1966.

Wilmut, I. M., et al. "Viable Offspring Derived from Fetal and Adult Mammalian Cells." *Nature* 385 (1997): 810–813.

Wilson, David Sloan. *Darwin's Cathedral.* Chicago: University of Chicago Press, 2002.

Wilson, Edward O. *Consilience.* New York: Knopf, 1998. Reprint, New York: Vintage Books, 1999.

———. *The Future of Life.* New York: Knopf, 2002.

———. *Naturalist.* New York: Warner Books, 1995.

———. "Responding to the Reviews of Elshtain, Kaye, and Ruse." *Politics and the Life Sciences* 18 (1999): 350–351.

———. *Sociobiology: The New Synthesis.* Cambridge: Harvard University Press, 1975.

Wilson, James Q. "The Moral Sense." *American Political Science Review* 87 (1993): 1–11.

———. "The Paradox of Cloning." Reprinted in *The Ethics of Human Cloning,* ed. Leon R. Kass and James Q. Wilson. Washington, D.C.: AEI Press, 1998.

———. "Sex and Family." In *The Ethics of Human Cloning,* ed. Leon R. Kass and James Q. Wilson. Washington, D.C.: AEI Press, 1998.

Wilson, James Q. and Richard J. Herrnstein. *Crime and Human Nature.* New York: Simon and Schuster, 1985.

Winston, Robert M. L. "Germ-line Gene Therapy: An Exaggerated Threat." *Politics and the Life Sciences* 13 (2) (1994): 237–238.

Winston, Robert M. L., and Alan H. Handyside. "New Challenges in Human in Vitro Fertilization." *Science* 260 (1993): 932–936.

Wivel, Nelson A., and LeRoy Walters. "Germ-line Gene Modification and Disease Prevention: Some Medical and Ethical Perspectives." *Science* 262 (1993): 533–538.

Wolf, Susan M. "Ban Cloning? Why NBAC Is Wrong." *Hastings Center Report* 27 (1997): 12–15.

Young, Larry J., et al. "Increased Affiliative Response to Vasopressin in Mice Expressing the V1a Receptor from a Monogamous Vole." *Nature* 400 (1999): 766–768.

Zanjani, Esmail I., and W. French Anderson. "Prospects for in Utero Human Gene Therapy." *Science* 285 (1999): 2084–2088.

Zimmer, Carl. "Tinker, Tailor: Can Venter Stitch Together a Genome from Scratch?" *Science* 299 (2003): 1006–1007.

Zweiger, Gary. *Transducing the Genome.* New York: McGraw Hill, 2001.

Index

AAAS, 212

Abbott Laboratories, 204

Abortion: congressional disputation over, 84, 108, 159; European views, 260n14; fetal research and, 88, 152–153; genetic implications, xiii, 72–73, 81, 126–127, 162, 169; Italian politics and, 111; as political issue, 47–48, 80–81, 119, 152; pre-embryos and, 82–84, 114, 159; as religious issue, 152; Supreme Court policy on, 79–81, 98, 152, 230

ABS Global, 103

Ackerman, Bruce, 271n90

ADA enzyme deficiency, 141, 145–149, 151

Adams, Mark D., 256nn47 52 56

Adams Jr., Reginald B., 282n88

Adaptation theory: abortion and, 48, 127; APSA founders' view of, 4, 14; Aristotle on, 247–248; author's constitutional elite and, 164; bans on germline enhancement, 163; in the context of evolutionary theory, 75, 208, 210; culture and, 14, 211, 218; Fukuyama on, 233; game theory, 235–236, 241; gender specialization, 195; genetics of eating, 176; the "green beard" phenomenon, 200; group selection and, 169; H. Simon on, 229–230; homosexuality and, 184; human cloning and, 123–124; the human genome and, 75, 127; J. Q. Wilson on, 118; in political science study of, 224; in proposed graduate course on consilience, 245–246; public policy

and, 4, 14; rational choice and, 230; sociality and, 197; in the Supreme Court and, 214

Adeno-associated viral vectors, 148, 157

Adenoviral vectors, 148–149, 155

Adkins v. Children's Hospital, 77

Adler, Reid, 45–46

Advanced Cell Technology, Inc. (ACT), 89, 112, 120–122, 125, 128

African Americans. *See* Race

Africans, 190, 205

Aiuti, Allesandro, 268n30

Alcoholism, 175–176, 191

Aldhous, Peter, 273n31

Alekhine, Alexander, 241

Allele(s): in alcoholism studies, 175; in the APOE susceptibility gene, 172–174; in the context of SNPs, 74, 191, 204–205; in fire ants, 198; among geographic clusters, 191; in honeybees, 199; in intelligence studies, 186; orchestrating epigenetic rules, 214; in serotonin research, 179; in slime molds, 200

Alpha thalassemia, 153

Altruism, 200, 235–236. *See also* Reciprocal altruism

Alzheimer's disease, 26, 33, 52, 67, 172–174, 177

America (Americans): Asilomar moratorium, driving force behind, 132; Barber's research on presidents of, 182; *The Bell Curve*'s recommendations for, 188; breast cancer among Jewish